Conventional and Contemporary Practices of Plant Breeding

Patrick O. Ayiecho · Julius O. Nyabundi

Conventional and Contemporary Practices of Plant Breeding

Springer

Patrick O. Ayiecho
Genetics and Plant Breeding
Maseno University
Maseno, Kenya

Julius O. Nyabundi
Agronomy and Crop Ecology
and Physiology
Maseno University
Maseno, Kenya

ISBN 978-3-031-74997-1 ISBN 978-3-031-74998-8 (eBook)
https://doi.org/10.1007/978-3-031-74998-8

This Springer imprint is published by the registered company Springer Nature Switzerland AG
The registered company address is: Gewerbestrasse 11, 6330 Cham, Switzerland

To Our Parents and Those Who Taught Us

Preface

The desire to write this book came as a result of the challenges experienced while teaching crop breeding for resistance to biotic and abiotic stress factors to plant breeding and plant pathology students in the Faculty of Agriculture of the University of Nairobi. The intention was to bring into one text the principles and methods of crop breeding for resistance to stress conditions which can be complemented with relevant information from other sources to teach the course.

Plant breeding, which is considered as an art, science and plant evolution directed by human beings, started as early as the Neolithic times by early plant domesticators and cultivators and was later integrated with the science of genetics after the rediscovery of Mendelian work at the beginning of twentieth century. The perfection of plant breeding methods and introduction of biometrics, field plot techniques, cytogenetics, mutagenesis, molecular genetics, recombinant DNA technology, genetic engineering, bioinformatics, genomics and other omics technologies for use in plant selection and genetic alteration of plant characteristics have made the science and practice of plant breeding more sophisticated and rewarding in terms of improved crop performance and food security and a lucrative commercial activity.

This text is an overview of various plant breeding topics, bringing information from olden times of crop evolution and primitive unconscious breeding efforts by early cultivators through the ages to the present-day molecular approach of genetic engineering, transgenesis and use of genomic methods, including historically important old work. The book which covers various principles and methods of plant breeding, targets students in agricultural and plant sciences, especially those taking plant breeding as a key unit/course in universities in developing countries. The text will also be of relevance to plant breeding lecturers and professional plant breeders. Since the text has extensive and detailed coverage of resistance to stress breeding it will be of great use to breeders in developing countries where crop production and food security have serious challenges due to various stress factors to which crops are often exposed, particularly taking into consideration problems associated with climate change. Breeding for Striga resistance has been given special attention since it is one of the factors causing cereal crop failure in the developing countries within the tropics and subtropics. The tables and diagrams in the text have been designed to give additional information or clarify what is covered in the text.

The book assumes that the target beneficiaries have knowledge of agronomy and basic genetics with exposure to biometry, field plot techniques, cytogenetics, cell biology, plant biochemistry, molecular genetics, plant pathology, plant entomology and plant physiology. However, chapters on basic population genetics and basic quantitative genetics have been included because, in some cases, basic genetics courses to undergraduate students do not necessarily cover such basic topics. The two chapters have been included to take care of such deficiencies.

We have made references to several literary information and various scientific works by other scientists and authors. We are thankful to the authors whose works are cited here and the respective publishers for allowing us to use their materials in this text. They are individually and duly recognised at the end of the respective chapters where the information is used.

The book was written while we were employees of Maseno University. We are grateful for the office space and library facilities availed to us at the university. We also made visits to the University of Nairobi library to obtain some necessary literary materials. We are thankful to the librarian of the Faculties of Agriculture and Veterinary Sciences Library of the University of Nairobi for the support. While the pictures in the text were done by the senior author the diagram illustrations were drawn by Denis Ngeno to whom we are grateful. Cecilia Gaku and Caroline Ambuyo in the Vice-Chancellor's office, Maseno University, helped with printing of drafts of this work whenever there was a need for such services. We are indebted to Prof. Paul Oleche of Maseno University for converting the text to LaTeX format and sorting out associated challenges, adding hyperlink feature, proofreading the manuscript, formatting and validating citations and cross-references. We also thank our families for their patience and bearing with us while spending a lot of time writing and organising the text.

Maseno, Kenya
July 2024

Patrick O. Ayiecho, Ph.D.
poayiecho@maseno.ac.ke

Julius O. Nyabundi, Ph.D.
nyabundijo@maseno.ac.ke

Contents

About the Authors

Patrick O. Ayiecho MGH (born 1953), also known as Patrick Ayiecho Olweny is a graduate of University of Nairobi (B.Sc. Agric. and M.Sc. Plant Breeding) and University of California (Ph.D. Genetics). Dr. Ayiecho is a currently professor of genetics and plant breeding in Maseno University and also served in a similar capacity at the University of Nairobi for several years. In addition to teaching genetics and plant breeding in the two universities, he has undertaken plant breeding research in a number crops like barley, grain amaranths, field beans, sesame and sorghum, focusing on crop improvement for resistance to abiotic and biotic stress factors, especially drought resistance and sorghum resistance to *Striga*. He has successfully guided over 40 M.Sc. and Ph.D. students to accomplish their graduate studies in both universities. During the 1990s, he obtained some German Academic Exchange Service (DAAD) fellowships to visit a number of German research and academic institutions for the purpose of advising and guiding DAAD-sponsored Kenyan Ph.D. students in German institutions. In 1999, he was awarded the International Atomic Energy Agency fellowship for scientific visits to two Indian institutions, the Bhabha Atomic Research Centre and the Indian Agricultural Research Centre and two Pakistani institutions, the Nuclear Institute for Agriculture and the Nuclear Institute of Agriculture and Biology in connection with his induced mutagenesis research for crop improvement research activities. Dr. Ayiecho served as an acting Head of the Department of Crop Science at the University of Nairobi and the Coordinator for International Relations and Interlink in

Maseno University. He is the founding secretary general of Plant Breeders Association of Kenya and has also served as a Member of the Parliament and an Assistant Minister for Education in the Government of Kenya for 10 years during which he was a Member of the African Parliamentary Network Against Corruption and Global Organisation of Parliamentarians Against Corruption.

Julius O. Nyabundi OGW (born 1954), is a graduate of the University of Nairobi (B.Sc. Agric. and M.Sc. Agronomy) and the University of California (Ph.D. Ecology). Dr. Nyabundi who is currently a Professor of Horticulture in Maseno University served as a senior lecturer at the University of Nairobi for several years. In addition to teaching agronomy, crop ecology and physiology in the two universities, he has undertaken agronomy, crop physiology and plant water relations research for a number crops like sorghum, finger millet, tomatoes, field beans, sesame and sorghum performance under Striga infestation. He has successfully guided over 30 M.Sc. and Ph.D. students to accomplish graduate studies in both universities. Dr. Nyabundi is currently serving as the Vice-Chancellor of Maseno University and previously held various management and administrative positions in a number of institutions in Kenya. In Maseno University, he served as the Head of Horticulture Department, the Dean of Faculty of Science, the Dean of Faculty of Agriculture and Food Security, the Director of Botanic Garden, the Director of StudentWelfare Services and the Acting Deputy Vice-Chancellor in Charge of Administration, Finance and Development. He served as the Managing Director of Chemelil Sugar Company for 5 years and later as the Principal of Muranga University College.

Abbreviations and Acronyms

2,4-D	2,4-Dichlorophenoxyacetic acid
2-AP	2-Aminopurine
5-BU	5-Bromouracil
A	Adenine
ABA	Abscisic acid
AFLPs	Amplified fragment length polymorphisms
AfricaRice	African Rice Centre
ALDH	Aldehyde dehydrogenase
ATP	Adenine triphosphate
Avr gene	Avirulence gene
BAP	Benzyl aminopurine
BCE	Before Christian Era
BMV	Brome mosaic virus
Bt	*Bacillus thuringiensis*
C	Cytosine
CaMV	Cauliflower mosaic virus
CE	Christian Era
CGIAR	Consortium of International Agricultural Research Centres
CGMS	Cytoplasmic-genic sterility
CHA	Chemical hybridising agents
CIAT	International Centre for Tropical Agriculture
CIMMYT	International Maize and Wheat Improvement Centre
CIP	International Potato Centre
CMS	Cytoplasmic male sterility
CRISPR	Clustered Regularly Interspaced Short Palindromic Repeats
CRISPR-Cas9	Clustered Regularly Interspaced Short Palindromic Repeats—Associated Protein 9
CRRS	Combined reciprocal S2 and Full sib recurrent selection
DArTseq	Diversity array technology sequencing
DEGs	Differentially expressed genes
DEPs	Differentially expressed proteins

DIMBOA	2-4-dihydroxy-7-methoxy-1, 4-benzoxazin-3-one
DNA	Deoxyribonucleic acid
DSM	Diallel selective mating
dsRNA	Double-stranded RNA
EAGA	Extended agar gel assay
EES	Ethyl ethane sulphonate
EGMS	Environment-sensitive genic male sterility
EI	Ethylene imine
EMS	Ethyl-methane sulphonate
ENU	N-ethyl-N-nitroso urea
EO	Ethyl oxide
ETI	Effector triggered immunity
ETS	Effector triggered susceptibility
G	Guanine
G × E	Genotype-environment interaction
GBS	Genotype-by-sequencing
GCA	General combining ability
GEBVs	Genomic estimated breeding values
GIS	Geographic information system
GLAM	Green leaf area at maturity
GMO's	Genetically modified organisms
GMS	Genic male sterility
GWAS	Genome-wide association studies
HF	Hydrogen fluoride
HGMS	Humidity-sensitive genic male sterility
HIPV	Herbivore-induced plant volatile
HIV	Human immunodeficiency virus
HR	Homologous recombination
HT	Herbicide-tolerant
IAA	Indole-3-acetic acid or Indole acetic acid
IBA	Indole-3-butyric acid
IBPGR	International Board for Plant Genetic Resources
ICARDA	International Centre for Agricultural Research in Dryland Areas
ICP	Insecticidal crystalline proteins
ICRISAT	International Crops Research Institute for Semi-Arid Tropics
IITA	International Institute for Tropical Agriculture
IPGRI	International Plant Genetic Resource Institute
IRRI	International Rice Research Institute
ISSRs	Inter-simple sequence repeats
KEPHIS	Kenya Plant Health Inspectorate Services
LEA	Late embryogenesis abundant proteins
LED	Light-emitting diode
LRR	Leucine-rich repeat
MABC	Marker-assisted backcrossing
MARS	Marker-assisted recurrent selection

MAS	Marker assisted selection
MBOA	6-methoxybenzoxaline
MCMoV	Maize chlorotic mottle virus
MiMe	Mitosis instead of meiosis
miRNAs	MicroRNAs
MLND	Maize lethal necrosis disease
MN	Meganucleases
MNU	Methyl-N-nitroso urea
MSAP	Methylation sensitive amplified polymorphism
mtDNA	Mitochondrial DNA
MYB	Myeloblastosis
NAA	α-naphthalene acetic acid
NHEJ	Non-homologous end-joining
NLRs	Nucleotide-binding site leucine-rich repeat proteins
Omics	Cellular/molecular level assays involving genomics, transcriptomics, proteomics, metabolomics, phenomics and epigenomics
OPVs	Open-pollinated varieties
orf	Open reading frames
PAMPS	Pathogen-associated molecular patterns
PCR	Polymerase chain reaction
PEG	Polyethylene glycol
PHD	Homeodomain
PRR	Pattern recognition receptors
PTGMS	Photoperiod thermo-sensitive genic male sterility
PTI	PAMP-triggered immunity
QTLs	Quantitative trait loci
R gene	Resistance gene
RAPDs	Random amplified polymorphic DNAs
rDNA	Recombinant DNA
RFLPs	Restriction fragment length polymorphisms
RISC	RNA-induced silencing complex
RLK	Receptor-like kinase
RLP	Receptor-like proteins
RNA	Ribonucleic Acid
RNAi	RNA interference
RNase	Ribonuclease
ROS	Reactive oxygen species
SA	Sodium azide
SAAT	Sonication-assisted Agrobacterium-mediated transformation
SCA	Specific combining ability
SCMT	Silicon carbide-mediated transformation
siRNA	Short virus-derived small interfering RNAs
SNPs	Single nucleotide polymorphisms
SPD	Single pod descent

SSD	Single seed descent
SSRs	Simple sequence repeats
STMS	Sequence-tagged microsatellites
STRs	Short tandem repeats
T	Thymine
T-DNA	Tumour inducing (Ti) plasmid DNA
TALENS	Transcription activator-like effector nucleases
TGMS	Thermo-sensitive or temperature-sensitive genic male sterility
TMV	Tobacco mosaic virus
USA	United States of America
vir gene	Virulence gene
VNTRs	Variable number of tandem repeats
vsiRNAs	Short virus-derived small interfering RNAs
WAKs	Wall-associated protein kinases
WARDA	West African Rice Development Association
ZFNs	Zinc finger nucleases

List of Figures

List of Tables

Chapter 1
Introduction and Objectives of Plant Breeding

1.1 Introduction

Plant breeding is genetic manipulation and evolutionary process of plants directed by human beings for the purpose of economic exploitation of plants and prevention of undernourishment. Breeding is key to addressing issues of food security and alleviating the problems of undernourishment through development of high yielding, good quality, stable and well adapted crops and crop varieties. This is particularly important taking into consideration the challenges associated with climate change, increasing world population and insufficient crop yields accompanied by poor nutritive quality of the produced food, especially in the developing countries. These constraints have made food intake of large populations in developing countries, especially children below the age of five, too low for normal health. While some problems leading to undernourishment can be addressed through socio-economic approaches and removing cultural barriers, the agricultural scientist, especially the breeder, is the most important as regards the reducing cost of agricultural production and improving the profits to the farmer. The value of crop plant breeding has been known since the start of the science of agriculture. In the early 18th century, an Anglo-Irish satirist and political pamphleteer poet, Jonathan Swift, recognised and appreciated the role of the agriculturalist in regard to food production. Among cost reducing technologies in agricultural production there is none which is more effective than plant breeding as it leads to more profits to the farmer through development of high yielding, good quality, disease and pest resistant and well adapted climate smart crop varieties.

Plant breeding originated with agriculture when the early cultivators went looking for good plants from the fields for their consumption. Scientific developments later led to perfection of plant breeding as it is practised today. In 700BC the Babylonians pollinated date palm artificially. Thomas Fairchild, an English gardener produced the plant hybrids in 1717 and between 1760 and 1766 Joseph Kolreuter, a German botanist, made crosses in tobacco and detected incompatibility in some plants he worked with. In late 18th century and early 19th century Thomas Andrew Knight, a

P. O. Ayiecho and J. O. Nyabundi, *Conventional and Contemporary Practices of Plant Breeding*, https://doi.org/10.1007/978-3-031-74998-8_1

British botanist and horticulturalist, used artificial hybridisation to develop several fruit varieties while John Le Couteur and Patrick Shireff conducted selection and progeny testing to improve some cereal crops in 1840's. In the mid-nineteenth century Louis de Vilmorin applied progeny testing for improvement of sugar beets using a method later referred to as Vilmorin method or Vilmorin isolation principle and at the beginning of the 20th century Nilsson Ehle and Johansen developed and refined individual plant selection method and pure line selection theory for self-pollinated crops respectively.

Plant breeding as it practised today is a science based on the genetic principles developed by Gregory Mendel in 19th century and other geneticists that followed him in the 20th century. Cytogenetic principles and biometrics were later used in refining plant breeding methods. The discovery of heterosis phenomenon and inbreeding depression by G. H. Shull, E. M. East and other maize breeders in early 20th century led to perfection of hybrid varieties development methods in maize and other cross-pollinated crops. Importance of induced mutations was discovered in early 20th century when the scientists of that time observed that X-rays could create genetic variation in Drosophila and plants. Mutation breeding and biotechnology methods like tissue culture, cell culture, callus culture, protoplast culture and genetic engineering are later additions to plant breeding, complementing the conventional breeding methods and substantially improving the expertise of the agricultural scientist to manipulate the plant genotype.

The modern plant breeder is faced with a more dynamic situation than the previous traditional breeder/crop domesticator of Mesolithic and Neolithic eras and subsequent periods who was basically a hunter and a gatherer. The present-day plant breeder has to contend with the challenge of developing well adapted and climate smart high yielding crop varieties to prevent undernourishment and help economic well-being of farmers. The breeder has to develop high quality varieties of food crops, industrial crops and medicinal plants with resistance to abiotic and ever-changing biotic stresses, particularly taking into consideration the challenges associated with climate change. To execute his/her duties successfully the breeder needs the understanding of botany, agronomy, plant physiology, entomology, nematology, genetics and cytogenetics, biotechnology and molecular biology of the gene, plant biochemistry, plant pathology, bacteriology, virology, biometrics, market trends, farmers needs and production problems in target areas. Since a plant breeder will not be a specialist in all these areas the breeder needs to work hand in hand with specialists in these various disciplines of plant science for accomplishment of plant breeding objectives.

Among the plant species on earth only few thousands can be categorised as having some economic value and a few hundreds as being of major economic importance in the world trade. Furthermore, a much a smaller number of the plant species that were domesticated by the Stone Age people are important as regards food supplies among which wheat, maize, rice, sorghum, millets, barley, potatoes, cassava, taro, yams, sweet potatoes, tomatoes, cabbage, oil palm, coconut, rapeseed, apples, grapes, oranges, beans, soybeans, groundnuts, bananas, plantains, sugarcane, sugar beet and peas currently stand between human beings and starvation. Of these crops only three cereals, maize, rice and wheat, contribute more than half of food to humans world-wide.

1.2 Objectives of Plant Breeding

The prime objective of plant breeding is to manipulate the genotype of the plants identified and available for human use to develop cultivars with desired features to prevent undernourishment, promote food security and for utilisation in economically desired purposes. The activities of plant breeding are focused on improving agronomic and economic worth of the plants such as high yields, improved quality, disease and pest resistance, resistance to parasitic weeds, adaptation to different climatic conditions and climate change resilience, adaptation to other forms of abiotic stress, elimination of dormancy, uniform maturity and shortening of maturity period, non-shattering of seeds at maturity, non-sprouting of seeds in the field before harvesting, plant type and consumer acceptability. Yield takes various forms like seed yield, fruit yield or vegetative yield as dictated by the plant part that is of economic importance. Improved yield is a consequence of improved adaptability of the crop to environmental conditions, resistance and tolerance to various stress factors. Resistance breeding activities concentrate on making crop production economical to the farmers through development of varieties with resistance or tolerance to abiotic stress factors and resistance to diseases, pests and parasitic weeds and resilient to climate change. Without resistant varieties management of crop diseases, pests and parasitic weeds is costly to the farmers since in absence of resistance the farmers resort to the expensive and environment unfriendly pesticides. For certain production situations the breeder has to develop varieties to address undesirable temperature conditions, mineral toxicity or deficiency, salinity, alkalinity, wetland conditions, submergence and water-logging, dry-land conditions and drought. Quality attributes also influence the economic worth, suitability and the kind of use to be given to the crop produce. Quality is determined by the part of the crop to be used and can be in form of grain or seed size, colour, milling and baking characteristics, cooking quality, nutritional content in regard to sugar, oil, vitamin and protein contents, essential amino acid mineral contents, absence of toxic substances, keeping quality and animal feed quality attributes like palatability, leafiness, texture and nutritive value. Uniformity of maturity is also an important objective for crop quality control as well. Breeders also address plant type in relation to the plant structure. For example, short varieties may be developed to address lodging problems and make mechanical harvesting easy. Short varieties generally have high harvest indices and in dicotyledon crops short varieties tend to have determinate growth or semi-determinate growth with uniform maturity. In maize varieties with low ear placement tend to be lodging resistant. Breeders aim at developing varieties with improved photosynthetic efficiency, fast growth rate, desired shoot and root characteristics, desirable leaf canopy, slow leaf senescence, high harvest index and varieties responsive to better soil fertility and moisture status without undesired excessive growth and lodging. In cereals additional features that the breeder focuses on are dwarf or semi-dwarf varieties with erect synchronous tillers, thick leaves having low senescence, erect or steeply inclined leaves for good solar radiation penetration and short leaves with high leaf area index. In other cereals where tillering is not desired like maize the plant should have a single erect stem with a large ear or head.

For successful development of improved varieties, the breeder has to define the objectives well, have adequate human and capital resources and the necessary infrastructure, develop and use appropriate methodology and be able to assess and review the work as it progresses. The breeder needs to assemble diverse germplasm materials and have working relations with the relevant gene banks since effectiveness of plant breeding is conditioned on genetic variability at the breeder's disposal. Maintenance of genetic diversity through collection and evaluation of germplasm is both an objective and an activity of plant breeding.

Problems

1.1 Which plant breeding objectives address challenges of climate change as regards food production?

1.2 Give any four aspects of crop plant quality addressed by plant breeding.

Chapter 2
Origin of Agriculture, Crop Domestication and Centers of Diversity

2.1 Origin of Agriculture and Crop Centres of Diversity

According to space scientists, palaeontologists and other scientists dealing with the evolution of the universe one of the theories about the origin of the universe suggests that the universe was created through the Big Bang in about 13.7 billion years ago followed by formation of the solar system about 4.65 billion years ago (Earle, 2019). That is, the earth was formed about 9 billion years after the Big Bang, preceding formation of the oceans on it in 4.4–4.3 billion years ago. Abiogenesis on earth probably occurred around 4.1 billion years ago when bacteria were created followed later by appearances of large aquatic organisms like sponges, flatworms, jelly fish and various arthropods, other aquatic organisms, terrestrial animals and plants between 600 and 220 million years ago (Payne et al., 2020). About 3.2 million years ago the earliest hominids (*Australopithecus*) arose from their ancestral hominins, followed by *Homo erectus* round 1,800,000 years ago and modern humans (*Homo sapiens)* in 300,000 years ago, all of them appearing first in Africa (Godfrain & Vercauteren Drubbel, 2019; Scerri et al., 2018). Information about origin and evolution of various forms of life on earth is based on fossilised remains, carbon isotope radiometric data and biogeochemistry. The early humans lived as gatherers of seeds of annual plants and edible organs of other plants like tubers, rhizomes, fruits and nuts and hunters for animal meat until they invented agriculture.

Agriculture is believed to have originated during the Neolithic period 10,000–13,000 years ago (Godfrain & Vercauteren Drubbel, 2019; Purugganan, 2019) in different parts of the old world (Africa, Asia, and Europe) and new world (Americas and Australia) along the river valleys like Indus, Euphrates, Ganges, Yangtze, Tigris, Niger, Nile, Yellow (Huang Ho) and Amazon, either simultaneously or at different times and independently when various human communities turned to agriculture and got engaged in growing and harvesting plants as crops and keeping animals to supplement the food they obtained from hunting and gathering. There were global changes in climatic conditions due to transition from the last ice age (Pleistocene) to the

P. O. Ayiecho and J. O. Nyabundi, *Conventional and Contemporary Practices of Plant Breeding*, https://doi.org/10.1007/978-3-031-74998-8_2

post-glacial age (Holocene) which created warm and wet conditions that favoured the domestication and cultivation of annual crops like cereal grasses followed by legumes in different areas around the globe (Atkins et al., 1998; Bar-Yosef, 2017). This was the beginning of agriculture which transformed the human being from hunter/gatherer to the sedentary husbandman. There are also archaeological evidences suggesting pre-domestication cultivation of certain wild plants much earlier than the actual domestication period (Gross & Olsen, 2010). Agriculture became the sole food production activity of human beings that is today the primary source of food for direct use, feeds and raw materials for industrial food products processing and medicinal plants. Agriculture might have originated in isolated discrete defined places and/or in vast areas without sharply definable centres around the globe. After about 6000 years from the time of its invention the culture of cultivation of various crops and rearing of livestock established itself replacing hunting and gathering or being practised together with hunting and gathering in many regions across the globe. Considering the period in which the human being has occupied the earth agriculture is a recent prehistoric activity which has been practised for a short time, taking up only a small proportion of the time the human has inhabited the earth.

Information on ancient agricultural activities is based on archaeological or historical data which give probable dates when and where specific crops were domesticated. The domestication of a crop species in a given area is ascertained by the presence of its wild and weedy relatives and other ancestors, archaeobotanical remains, palynology, palaeobotany, history, art and archaeology and radiocarbon dating of past remains; genetic and cytogenetic analyses, chemotaxonomy, morphology, ecology and distribution of plant species, linguistics, oral traditions, cultivation methods, religion, magic and witchcraft and other cultural practices of human communities and other information like geology and hydrology (Harlan, 1971). Useful information is also obtained through allozyme and deoxyribonucleic acid (DNA) molecular analyses, including ancient DNA.

In the old-world ancient agriculture is believed to have originated in regions like Western and Central Asia, South East Asia, parts of Europe, Abyssinia, western Africa and parts of the Sahel while in the new-world the regions involved in origin of agriculture were South America, Mesoamerica (Central America), North America and Australia. Most of the early agricultural activity regions are within the tropical and subtropical areas and are also confined along the river valleys where the fertile soils in alluvial plains and availability of water favoured sedentary human life, early domestication and cultivation of a diversity of crops.

Western and Central Asia or Near East includes Mediterranean and Asia Minor, parts the Mesopotamia and the Fertile Crescent/the Levant (covers Iraq, Syria, Lebanon, Jordan, Palestine, Israel, Egypt and Turkey) and includes the present Iran, Kuwait, Saudi Arabia, Azerbaijan, Afghanistan, north-west India, Tajikistan, Uzbekistan, Turkmenistan, and Tibet. The ancient Mesopotamia, considered to be the original birthplace of agriculture, also comprised of Sumerian, Babylonian, Assyrian and Chaldean civilizations. Agricultural activities were initiated in Western and Central Asian region around 11000 BCE with cultivation of a number of cereals (einkorn and emmer wheat, barley), legumes (pea, chickpea, vetch, lentils, broad bean), flax and vegetables as early as 10000 BCE to 6000 BCE (Purugganan & Fuller, 2011). Other crops domesticated in this region include cabbages, mustard, rape, kales, carrot, garlic, pear, pomegranate, cotton, safflower, linseed, pear, rye, oat, cauliflower, sesame, date, apple, cherry, turnip, pistachio, almond, onion, grapes, figs and olive.

South East Asia centre encompasses Malaysia, South China, Indonesia, Philippines, Java, Siam, New Guinea and Indo-China (Thailand, Cambodia, Vietnam, Myanmar/Burma) and Laos. Agriculture probably originated in various parts of this region between 11500 BCE and 6200 BCE with ancient cultivation of yams, sago, palm and banana, mango, cocoyam (taro), peach and banana (Atkins et al., 1998). Rice is believed to have originated in Indo-China area and was also cultivated in China along the Yangtze, Yellow and Wei rivers in 11500BCE to 6200BCE (Purugganan, 2019) and later spread across the rest of south eastern Asia. Diploid cotton types originated in this region about 3000BC while tea is believed to have been brought under cultivation in the Tibetan mountains, central Asia and south east China more than 2000 years ago. *Gossypium arboretum, G. herbaceum,* cucumbers and pigeon pea are believed to be native to India where they were under cultivation 3000 years ago. Sesame *(Sesamum indicum)* was under cultivation in some Asiatic countries (India, Palestine and Syria) around 3600BCE–3000BCE (Nayar & Mehra, 1970). Coconut, sugarcane and breadfruit were domesticated in New Guinea where agriculture was already practised by 7000 BCE. Sugarcane was under cultivation in New Guinea in about 7000 BCE and in south eastern Asian islands by 3500 BCE while coconut was cultivated in the New Guinea around 3000 BCE (Atkins et al., 1998). Banana and some root crops were also cultivated in New Guinea in 5000 BCE (Doughty, 2010). Moth bean, lime, tangerine, nutmeg, horse gram, asparagus bean (yard-long bean), pepper, arrow root, breadfruit, clove, soybean, citrus and jute were domesticated in South East Asia as well.

West Africa, the Sahel and Abyssinia region which cover parts of Nigeria, Benin, Togo, Ghana, Niger, Mali, Burkina Faso, Chad, Cameroon and Ethiopia had ancient agricultural activities by 6000 BCE (Meyer & Purugganan, 2013). Sorghum was domesticated and cultivated in western to eastern African stretch before 2000 BCE (Doughty, 2010) while millets (finger millet, bulrush millet and pearl millet), cowpea, yams and African rice (*Oryza glaberrima*) originated and were cultivated in this region between 3000 BCE and 1000 BCE. Sesame, which has over twenty of its wild relatives like *Sesamum capense (alatum), S. radiatum, S. angolense, S. calycinum, S. anguistifolium and S. latifolium, etc.,* in tropical African countries, was cultivated in Africa for several hundreds of years during the pre-Christian era (Nayar & Mehra, 1970). *Coffea arabica* was domesticated around 1700 BCE in Ethiopia while *C. canephora* arose in west and central Africa about the same time. Other crops believed to have been domesticated in the African region include safflower, oil palm, onion, linseed, watermelon, kola, castor, lettuce, teff, niger, okra, tamarind, ensete, black eyed pea, and some forms of wheat and barley (Singh, 1990). Egyptian civilisation belongs to the Near East/Fertile Crescent which falls under Western and Central Asia and it is in Egypt where an irrigation system using the shaduf developed as early as 3500 BCE along the Nile river with cultivation of papyrus for paper, date palm, sesame, castor bean, pines (resins), spices, perfumes, cosmetics, lemon, pomegranate, cucumber, artichoke, lentils, garlic, leek, lettuce, mint, endive, chicory, radish, melons, wheat, barley, grapes, figs, olives and collections of drugs and medicinal plants.

In South America ancient agriculture originated in Amazonia (Peru, Bolivia, Ecuador, Colombia, Venezuela, Chile, Brazil and Paraguay) around 9000 BCE with the earliest known crop cultivation being undertaken in Colombia. There was early cultivation of root crops namely, cassava, potato, arrow root and cocoyam and other crops like squash and bottle gourd in this region between 8000 BCE and 7000BCE while cocoa, field beans, squash pepper and maize were domesticated between 8000 BCE and 5000BCE (Atkins et al., 1998). Field bean cultivation was initiated in the Huaca Prieta of Peru and southern slopes of the Andes and northern Argentina. Cultivation of ancient maize, which was already going on in Mexico, was undertaken in this region by 3000 BCE. Other crops domesticated in this region are pineapples, amaranth, cotton, quinoa, groundnuts, rubber, quinine tree (*Cinchona calisaya*), pumpkin, guava, tobacco, papaya, sweet potato, tomato and avocado (Singh, 1990).

Mesoamerica (Central America) centre covers present day Mexico to Honduras, where agriculture had been initiated by 8000 BCE with cultivation of crops like *Dioscorea* yams, field bean, squashes, avocado and cassava between 8000 BCE and 5000 BCE (Atkins et al., 1998; Doughty, 2010). The cucurbits (squashes and pumpkins) were also domesticated around the same time while maize, a crop of world-wide importance, was brought under cultivation in the Tehuacan in about 5200 BCE when its wild relative teosinte (*Euchlaena mexicana*) was transformed into maize through domestication (Atkins et al., 1998). Primitive maize was grown in Tehuacan valley until modern maize arose through possible natural hybridisation between primitive maize and teosinte. Chilli peppers had been domesticated here around 4000 BC while *Amaranthus* and yellow sapote were cultivated first time in the Tehuacan valley between 5000 BCE and 2000 BCE. Lima bean (*Phaseolus lunatus*), Upland cotton (*Gossypium hirsutum*), tomato, cocoa, sweet potato, groundnut, chenopods, vanilla, sisal, guava and papaya are also natives of Central America, particularly southern Mexico (Singh, 1990).

Other areas in which agricultural activities took place in ancient times leading to domestication of some crops are North America, Australia and European-Siberian regions. There are records of ancient agricultural activities by 3000 BCE in eastern North America where sunflower, chenopods, blueberry, cranberry, Jerusalem artichoke, plums, cherries, and some cucurbits were domesticated more than 3000 years ago (Doughty, 2010). Crops like oat, sugar beet, strawberry, cherry, raspberry and plum are also believed to have undergone some form of domestication in parts of Europe (Atkins et al., 1998). Though North American region is the home of the progenitor species of strawberry (*Fragaria x ananassa*), the presently cultivated octoploid strawberry originated through hybridisation of two progenitor octoploid species, namely, *F. virginiana* and *F. chiloensis*, in Europe about 300 years ago (Edger et al., 2019). While crops that were domesticated and cultivated in European-Siberian region in ancient times are gooseberry, parsnip, plum, pear, black mustard, chicory, peach, apple, turnip and cabbage. Australia is considered as the domestication centre for macadamia (Esquinas-Alcazar, 1993).

In 1886 Alphonse de Candolle suggested that every crop had a region of origin where it arose and was also domesticated and in early twentieth century Nikolai Vavilov of the Institute of Plant Industry at Leningrad proposed that each crop had a region believed to be its centre of origin. Between 1916 and 1940 Nikolai Vavilov conducted exploratory studies across the world and collected various crop plant species which helped him identify areas where cultivated plant species and their relatives had great diversity, making him conclude that each crop species had a primary centre of diversity which could be equated to the primary centre of origin or centre of domestication. Vavilov observed that the regions with great diversity for a given crop coincided with regions of diversity for its wild and weedy relatives and were, therefore, the areas of origin or centres of domestication for the crop. Vavilov originally enlisted eight centres of origin which fell within various regions where agriculture was believed to have originated, namely, China Centre (central and western China), Hindustan Centre (Burma, Assam, Malaya Archipelago, Java, Borneo, Sumatra, Philippines), Central Asia Centre (north western India, Afghanistan, Tajikistan, Uzbekistan), Asia Minor Centre (parts of Asia Minor, Iran, highlands of Turkmenistan, Transcaucasia), Mediterranean Centre, Abyssinian Centre (Ethiopia and Eritrea), Central American Centre (Mexico) and South American Centre comprising of Peru, Bolivia, Ecuador, Colombia, Chile, Brazil and Paraguay (Maxted & Vincent, 2021; Singh, 1990). Vavilov and his colleagues later expanded, subdivided and re-clustered the centres of origin and added other centres, giving a total of twelve vast centres of origin for the various crops around the world, namely, Near Eastern Centre, Mediterranean Centre, Africa—Abyssinia Centre, North America Centre, Central Asia Centre, Central America-Mexican Centre, European-Siberian Centre, Chinese-Japanese Centre, Hindustan Centre, South America Centre, Indochinese-Indonesia Centre and Australia Centre (Esquinas-Alcazar, 1993; Maxted & Vincent, 2021). These regions (Table 2.1) have evidences of ancient agriculture and old civilisations and are important genetic resource centres due to their wealth in wide genetic diversity for various crops and their relatives. There are further suggestions that there exist more than twenty separate regions in which various crops underwent domestication during ancient times independently or simultaneously. Apart from centres of origin which are considered as primary centres, there are secondary centres of variability which are also centres of diversity but are not necessarily centres of origin since diversity expanded from primary centres to such centres through migration (Singh, 1990). These are areas to which ancient cultivation expanded as domestication was spreading from the original centres because of movement of the ancient cultivators. As the early products of domestication expanded from the primary centres of origin, they encountered varied environments and cultural practices, developed new genotypic and phenotypic variations due to rare mutations and hybridisation with relatives. The secondary centres of diversity arose as further domestication and diversification took place, leading to rich diversity for the cultivated forms beyond centres of origin implied by Vavilov.

Table 2.1 Centres of diversity for some crops (based on Esquinas-Alcazar, 1993; Singh, 1990)

Centre	Areas included	Primary centre for	Secondary centre for
Near Eastern and Asia Minor Centre	Asia Minor which includes Saudi Arabia, Turkey Iraq, most of Syria, Iran	*Crocus*, carrot (*Daucus carota),* various species of *Triticum,* peppermint (*Mentha* spp.), *Trifolium resupinatum,* common vetch (*Vicia sativa),* alfalfa *(Medicago sativa*), barley (*Hordeum vulgare*), artichoke (*Cyanara cardunculus*), oat (*Avena sativa*), chestnuts (*Castanea sativa),* lettuce (*Lactuca sativa*), lupine (*Lupinus* spp.), Faba bean (*Vicia faba*), lentil (*Lens esculentum*), pea (*Pisum sativum*), melon, *Vitis* spp., pear (*Pyrus communis*), apricot (*Prunus armeniaca*), other *Prunus* spp., safflower (*Carthamus tinctorius*), sesame (*Sesamum indicum*), flax (*Linum usitatissimum*), cabbage (*Brassica oleracea* var. *capitata*), kale (*Brassica* spp.), *Allium* spp., fig (*Ficus* spp.), sugar beet (*Beta vulgaris*), pomegranate (*Punica granatum*), apple (*Malus domestica*), almond (*Prunus amygdalus*),	Rape (*Brassica campestris*), black mustard (*Brassica nigra*), Indian mustard (*Brassica juncea*), turnip (*Brassica rapa* subsp. *rapa*)., apricot (P*runus armeniaca*)
Mediterranean	Western parts of Syria, Spain, Italy, Greece, Israel, Lebanon, Jordan, northern parts of Egypt, Libya, Algeria, Tunisia, Morocco	Lentil (*Lens esculentum*), Faba bean (*Vicia faba*), radish (*Raphanus sativus*), cereals like barley (*Hordeum vulgare*), oat (*Avena sativa*), durum wheat (*Triticum durum/turgidum*), emmer wheat (*Triticum dicoccum*), *Lathyrus* spp., pea (*Pisum sativum*), clovers (*Trifolium* spp.), chicken pea (*Cicer arietinum*), vetch (*Vicia sativa*), rape (*Brassica campestris*), mustard, onion (*Allium cepa*), olive (*Olea europaea*), cumin (Cuminum cyminum), celery (*Apium graveolens*), grape vines (*Vitis* spp.), fennel (*Foeniculum vulgare*), linseed (*Linum usitatissimum)*, lettuce (*Lactuca sativa*), artichoke (*Cyanara cardunculus*), asparagus (*Asparagus officinalis*), lavender (*Lavandula angustifolia*), peppermint (*Mentha* spp.), lupine (*Lupinus* spp.), cabbage (*Brassica oleracea* var. *capitata*), turnip (*Brassica rapa* subsp. *rapa*), beet root (*Beta vulgaris*), red raspberry (*Rubus ideaus*)	Sweet pepper (*Capsicum annuum*), garlic (*Allium sativum*), okra (A*belmoschus esculentus*)

(continued)

Table 2.1 (continued)

Centre	Areas included	Primary centre for	Secondary centre for
Africa - Abyssinia Centre	Most of eastern, western, central and southern African countries and parts of the Sahel	Coffee (*Coffea Arabic* and *C. canephora*), onion (*Allium* spp.), barley (*Hordeum vulgare*), *Triticum durum* (*turgidum*), emmer wheat (*Triticum dicoccum*), Bambara groundnut (*Vigna subterranean*), cowpea (*Vigna unguiculata*), finger millet (*Eleusine coracana*), African rice (*Oryza glaberrima*), sorghum(*Sorghum bicolor*), pearl millet (*Pennisetum americanum*), kola (*Cola acuminata*), bottle gourd (*Lagenaria siceraria*), gram, lentil (*Lens esculentum*), hyacinth bean (*Dolichos lablab/Lablab purpureum*), Enset (*Ensete ventricosum*), grass pea (*Lathyrus sativus*), linseed (*Linum usitatissimum*), yam (*Dioscorea* spp.), safflower (*Carthamus tinctorius*), sesame (*Sesamum indicum*), African cucumber or Kiwano melon (*Cucumis metuliferus*), castor (*Ricinus communis*), thistle seed/niger seed (*Guizotia abyssinica*), oil palm (*Elaeis guineensis*), *Gossypium* spp., kenaf (*Hibiscus cannabinus*), water melon (*Citrullus lanatus*), okra (*Abelmoschus esculentus*), teff (*Eragrostis tef*), jute mallow/Nalita jute (*Corchorus olitorius*)	Faba bean(*Vicia faba*), mustard (*Brassica* spp.), *Allium* spp.
North America Centre	USA	Jerusalem artichoke (*Helianthus tuberosus*), blueberry (*Rubus leucodermis*), strawberry (*Fragaria x ananassa*), sunflower (*Helianthus annuus*), American cranberry (*Vaccinium macrocarpon*), red raspberry (*Rubus ideaus*), buffalo gourd (*Cucurbita foestidissima*), green amaranth (*Amaranthus viridis*)	

(continued)

Table 2.1 (continued)

Centre	Areas included	Primary centre for	Secondary centre for
Central Asia Centre	N. W. India, Afghanistan, Tajikistan, Uzbekistan, Kazakhstan, Kyrgyzstan, Turkmenistan, parts of Iran	Faba bean (*Vicia faba*), wheat (*Triticum aestivum*), club wheat (*Triticum compactum*), safflower (*Carthamus tinctorius*), almond (*Prunus amygdalus*), hemp (*Cannabis* spp.), Levant cotton (*Gossypium herbaceum*), radish (*Raphanus sativus*), spinach (*Spinacia oleracea*), musk melon (*Cucumis melo*), sesame (*Sesamum indicum*), apricot (*Prunus armeniaca*), grape (*Vitis* spp.), apple (*Malus domestica*), beetroot (*Beta vulgaris*), plum (*Prunus domestica*), walnut (*Juglans* spp.), carrot (*Daucus carota*), linseed (*Linum usitatissimum*), chickpea (Cicer arietinum), mung bean/green gram (*Vigna radiata*), pear (*Pyrus communis*), onion (*Allium cepa*), garlic (*Allium sativum*), pea(*Pisum sativum*), lentil (*Lens culinaris*)	Oat (*Secale cereale*), eggplant (*Solanum melongena*), cauliflower (*Brassica oleracea* var. *botrytis*)
Central America Centre	Mexico, Honduras, Panama, Guatemala, Nicaragua, El Salvador, Costa Rica	Cotton (*Gossypium hirsutum and Gossypium purpurascens*), field bean (*Phaseolus vulgaris*), Lima bean (*Phaseolus lunatus*), pumpkin (*Cucurbita spp.*), pepper/chillies (*Capsicum annuum*), *Amaranthus* spp., *Chenopodium spp.*, sisal (*Agave sisalana*), hemp (*Cannabis* spp.), potato (*Solanum tuberosum*), tobacco (*Nicotiana tabacum*), sweet potato (*Ipomea batatas*), guava (*Psidium guajava*), avocado (*Persea americana*), tomato (*Lycopersicon esculentum*),maize (*Zea mays*), arrow root (*Canna edulis*), papaya (*Carica papaya*), yams, cassava	
European Centre	Most European countries, parts of Russia (mainly Siberia	Celery (*Apium graveolens*), hemp (*Cannabis* spp.), chicory (*Cichorium intybus*), pistachio (*Pistacia vera*), lettuce (*Lactuca sativus*), black mustard (*Brassica nigra*), hop (*Humulus lupulus*), plum (*Prunus domestica*), apple (*Pyrus/Malus* spp.), pear (*Pyrus communis*), European gooseberry (*Ribes uva-crispa*), parsnip (*Pastinaca sativa*), cabbage (*Brassica oleracea* var. *capitata*)	*Allium* spp., cauliflower (*Brassica oleracea botrytis*), spinach (*Spinacia oleracea*), carrot (*Daucus carota*)

(continued)

Table 2.1 (continued)

Centre	Areas included	Primary centre for	Secondary centre for
Chinese-Japanese Centre	Eastern parts of China, Japan	Soybean (*Glycine max*), lychee (*Litchi chinensis*), proso millet (*Panicum miliaceum*), foxtail millet (*Setaria italica*), naked oat (*Avena nuda*), peach (*Prunus persica*), Adzuki bean (*Vigna angularis*), mustard greens/Indian mustard (*Brassica juncea*), mandarin orange (*Citrus nobilis/Citrus reticulata*), pear (*Pyrus communis*), *Brassica spp., Allium spp.*, brinjal (*Solanum melongena*), bamboo (*Bambusa vulgaris*), ramie (*Boehmeria nivea*), tea (*Camellia sinensis*), plum (*Prunus divaricata*), apricot (*Prunus armeniaca) ash gourd* (*Benincasa hispida*), peach (*Prunus persica*)	Sesame (*Sesamum indicum*), watermelon (*Citrullus lunatus*), turnip (*Brassica rapa*), cow pea (*Vigna unguiculata*)
Hindustan Centre	Most of India, Bangladesh, parts of Pakistan, northern parts of Myanmar, Assam.	Mango (*Mangifera indica*), hemp (*Cannabis spp.*), mung bean (*Vigna radiata*), pigeon pea (*Cajanus cajan*), chickpea (*Cicer arietinum*), rice (*Oryza sativa*), little millet/Indian millet (*Panicum sumatrense*), black gram (*Vigna mungo*), rice bean (*Vigna umbellate*), moth bean (*Vigna aconitifolia*), hyacinth bean (*Dolichos lablab/Lablab purpureum*), horse gram (*Macrotyloma uniflorum*), brinjal (*Solanum melongena*), jute mallow/Nalita jute (*Corchorus* spp.), cucumber (*Cucumis sativus*), mustard greens/Indian mustard (*Brassica juncea*), rats tail radish (*Raphanus caudatus*), yam (*Dioscorea* spp.), *Citrus spp.*, banana (*Musa* spp.), indigo (*Indigofera* spp.), sesame (*Sesamum indicum*), Indian lettuce (*Lactuca indica*), Indian radish (*Raphanus indicus*), cotton (*Gossypium arboretum*), turmeric (*Curcuma* spp.), ginger (*Zingiber officinale*), cardamom (*Elettaria cardamomum*), betelnut/Areca nut (*Areca catechu*), sugarcane (*Saccharum officinarum*), black pepper (*Piper nigrum*)	Watermelon (*Citrullus lanatus*), melon, *Amaranthus* spp., bottle gourd (*Lagenaria siceraria*) okra (*Abelmoschus esculentus*)

(continued)

Table 2.1 (continued)

Centre	Areas included	Primary centre for	Secondary centre for
South America Centre: Peru-Chile; Brazil-Paraguay	Peru, Ecuador, Argentina, Bolivia, Colombia, Ecuador, Chile, Brazil, Paraguay	Maize (*Zea mays*), pineapple (*Ananas comosus*), tomato (*Lycopersicon esculentum*), sweet potato (*Ipomea batatas*), tobacco (*Nicotiana tabacum*), Sea Island/Egyptian cotton (*Gossypium barbadense*), cassava (*Manihot esculenta*), groundnut (*Arachis hypogea*), rubber tree (*Havea* brasiliensis), tannia/cocoyam (*Xanthosoma sagittifolium*), field bean (*Phaseolus vulgaris*), Lima bean (*Phaseolus lunatus*), *Amaranthus spp.*, *Chenopodium spp.*, pumpkin (*Cucurbita spp.*), lupins (*Lupinus spp.*), papaya (*Carica papaya*), cacao tree (*Theobroma cacao*), quinine tree (*Cinchona* spp.), guava (*Psidium guajava*), passion fruit (*Passiflora edulis*), potato	
Indochinese-Indonesia Centre	Vietnam, Laos, Cambodia, Thailand, Malaysia, Singapore, New Guinea, Borneo, Sumatra, Philippines, Myanmar, Indonesia	Mango (*Mangifera indica*), ginger (*Zingiber officinale*), yam (*Dioscorea* spp.), banana (*Musa* spp.), taro (*Colocasia esculenta*), rice (*Oryza sativa*), rice bean (*Vigna umbellata*), breadfruit, (*Artocarpus altilis*), bamboo (*Bambusa vulgaris*), winged bean (*Psophocarpus tetragonolobus*), Rambutan (*Nephelium lappaceum*), nutmeg (*Myristica fragrans*), lime (*Citrus aurantifolia*), clove (*Syzygium aromaticum*), coconut(*Cocos nucifera*), *ash gourd* (*Benincasa hispida*), Durian (*Durio zibethinus*), betelnut/Areca nut (*Areca catechu*)	Cucumber (*Cucumis sativus*), cabbage (*Brassica oleracea* var. *capitata*), *Amaranthus* spp.
Australia Centre	Australia	Macadamia (*Macadamia tetraphylla*), Australian millet (*Panicum decompositum*)	

2.2 Crop Domestication Process

2.2.1 Domestication Patterns

Crop domestication is the alteration of a plant from wild state through artificial selection leading to its inability to compete with its wild relatives and other plants and making it suitable to agricultural cultivation to fulfil human needs. Crop domestication was previously viewed as involving localised and sharply defined crop centres of origin. However, modern views of crop domestication advocate for evolutionary patterns which spread out over large areas and along migratory routes as the ancient cultivator communities interacted and moved with their crops which they grew along their routes of migration, making the sharply defined Vavilovian centres of origin diffuse into ill-defined expansive regions. Domestication process took place across large culturally connected areas of long distances and was gradual and protracted. There is advocacy for multiple regions or a geographical continuum for evolution and large ecological regions for domestication of some crops like millets and sorghum in the African domestication region (Harlan, 1971). Some centres of domestication and diversity may not necessarily coincide with centres of origin for some crops as is known for a number of crops like certain wheats. Wheat is known to have diverse landraces in Ethiopia but no wild relatives in Ethiopia, implying that wheat might not have originated in Ethiopia but in the Near East where its wild relatives are native. The concept of centres and non-centres has also been proposed where a centre, as implied by Vavilov, refers to a distinct area in which a crop is believed to have originated while non-centre refers to a situation where domestication is believed to have occurred through a geographically diffused region, in form of multiple events or multiple centres spanning large areas spreading over 5000–10000 km (Abbo et al., 2010; Harlan, 1971). That is, a crop may have single domestication event through domestication from its ancestor in a region followed by dispersion or a crop species may have multiple domestication events. A crop having multiple domestication events arose from its predecessor species through independent domestication process in which selection for it was done in different regions and by different communities simultaneously or at different times so that over time different strains and divergent gene pools arose in the different domestication centres. It is evident from Table 2.1 that several crops have more than one domestication centres, e.g., maize (Central America and South America), cabbage and other Brassicas (Near East, China and Europe), wheat and barley (Near East, Mediterranean and Abyssinia), cassava (Central America and South America), oat (Near East and Mediterranean), cotton (Central America, South America and India), yam (western Africa and South America), onion (China, India, Near East and Africa, Indochinese-Indonesia, Hindustan), sugarcane (India, south east Asia and New Guinea), apricot (China and Near East), field bean (Central America and South America), sesame (India, Near East, western and eastern Africa and Egypt) and rice (China and India), cassava (South America and Central America), tomato (South America and Central America) and potato (South America and Central America).

Selection and Domestication Traits

Early domestication was achieved by ancient cultivators who consciously selected genetically variant plants and retained plants with desired traits like preferred fruit and grain size, non-shattering feature, adaptability to cultivation conditions, palatability, colour, vigour, increased fertility, early maturity, shoot architecture, desired branching and tillering patterns, reduced height, vigorous fast growth, ease of harvesting, suitability for storage and yield. The plants were also consciously selected for absence of toxic substances, desirable taste, rapid germination, reduced or loss of seed dormancy, uniformity and accompanied by unconscious selection for some desirable traits in the domesticates. As the ancient cultivators selected the preferred plants, they observed that some plants tasted better, more appealing to the eye, had no harmful effects because of lack of toxic substances and better adapted than their relatives and that among some cereals and legumes, there were some with problems of shattering while others had tough husks and pods making them difficult to thresh. The cultivators retained the non-shattering types from which it was easy to obtain grain or seeds thereby eliminating shattering of fruiting bodies through loss or reduction of abscission zones within such organs. The ancient cultivators selected for lowered or elimination of toxic substances and undesirable taste, reduced crop plant height and indeterminate nature for crops in which the fruit or the seed was the desired product. The focus on these domestication traits varied according species under domestication and the part/organ of the plant to be used. The desired domestication traits that were subjects of selection for domestication, also referred to as domestication syndrome (Harlan, 1971), help distinguish the crop from its wild relatives.

The first plants to be selected for cultivation were probably those which grew fast to produce a crop within a season when conditions were favourable while long maturity plants like fruit plants which were conveniently collected from the wild were brought under cultivation much later. The ancient cultivators operated as traditional farmers and untrained breeders when they selected plants visually on trial-and-error basis with the gradual net effect of evolution and improvement of the crops they selected. Their skill, common sense and proper judgement helped them identify and retain plants that were preferred for their nourishment. Each season they would harvest from the plants they selected for food and keep some seed of the selected plants for planting in next season. As the early cultivators domesticated plants through selection, they also introduced plant materials they acquired from other communities into their traditional agriculture. Seed exchange among communities took place and as farmers migrated, they moved with their seeds into new environments giving opportunities to new hybridisation and recombination which expanded genetic variability for further artificial and natural selection. Natural hybridisation of wild relatives of crops with cultivated forms lead to introgression of novel desirable traits into cultivated populations resulting into new heterogeneous populations and further evolution of adapted landraces. Such hybridisation also led to the creation of some polyploid species and modification of reproductive system as well. The plants they domesticated for their subsistence and other benefits were handed to the present-day farmers and agricultural scientists through thousands of generations of similar activities. Because crop evolution took place under different environments and varied cultural practices in different regions it created several highly variable and adaptable strains referred to as landraces for various crops across the globe.

Crop domestication took a protracted time and has been a continuous process dating from the origin of agriculture hundreds of centuries back to the present when the modern farmers, the trained plant breeders and other plant scientists are also involved in further plant domestication for direct or indirect agricultural use. Two factors are necessary for crop evolutionary process: genetic diversity and selection which comprises of artificial and natural forces. Selection undertaken by cultivators and plant breeders is always accompanied and more often countered by natural selection as the artificially selected plants are subjected to the rigour of natural selection forces through biotic factors like diseases, pests, browsers and grazers and competition and abiotic stresses like varied weather conditions, altitude, soil nutrient status and pH, mineral toxicity and dryland conditions, drought, temperature variations, water-logging and submergence, wetland conditions and many other natural selection forces. While nature favoured the plants suited for survival through natural selection, the cultivators consciously or unconsciously selected and retained those plants with desired traits but not necessarily favoured by natural forces. Genetic diversity inherent in plant populations arises through hybridisation creating recombinants and occasional natural point genetic mutations and chromosomal structural and numerical aberrations. Though domestication of many crops like *Fragaria x ananassa* (strawberry), *Triticum aestivum* (wheat), *Coffea arabica* (Arabica coffee), *Saccharum officinarum* (sugarcane), *Gossypium barbadense* (Sea Island cotton), *Gossypium hirsutum* (Upland cotton), *Nicotiana tabacum* (tobacco), *Brassica carinata (*Ethiopian mustard*), B. juncea* (leaf mustard or brown mustard) and *B. napus* (rape), is associated with natural hybridisation between related species followed by polyploidisation, several crops are known to have arisen from single wild species without interspecific hybridisation like *Brassica oleracea* (broccoli), *Phaseolus vulgaris* (field bean), *Daucus carota subsp. sativus* (carrot), *Sorghum bicolor* (sorghum), *Vigna unguiculata* (cowpea), *Camellia sinensis* (tea), *Oryza sativa* (rice) and many others.Occasional hybridisation between crops and their weedy and wild relatives lead to introgression, polyploidy and other forms of new variants to select from by the ancient cultivators. Diversity within and between crop populations is important to farmers because of heterogeneous soils, production conditions, risk factors, consumer preferences and uses of different products from various crop species.

2.2.2 Morphological and Molecular Changes in Crop Plant Species Due to Domestication

The original plants the ancient cultivators brought under cultivation originated from their wild relatives, some of which are extinct today.The cultivators differentiated the undesirable plants and the desirable plant species and ignored the species which had resentful features. For example, as cassava was being domesticated for its roots the early cultivators might have realised through experience that the poisonous hydrocyanic acid was present in the tubers of some plants and the problem could be solved by totally avoiding plants suspected to have this poisonous secondary metabolite. Because of natural selection, artificial selection and management by the cultivators, occasional hybridisation with wild and weedy relatives, rare spontaneous point mutations and chromosome structural and numerical changes, the crops underwent several morphological, physiological and molecular evolutionary changes making the present-day crop plants very divergent from their original and wild progenitors.

Changes associated with domestication took place in the traits that were targeted by early cultivators for selection leading to reduced shattering, reduced or loss of dormancy, more vigorous growth, improved fertility, more synchronised flowering and uniform maturity, determinacy or semi-determinacy, elimination of seed dormancy, increased apical dominance with robust growth of the main shoot, altered secondary metabolite contents for improved palatability and elimination of toxic compounds and bitterness in edible parts. There were modifications in plant architecture in regard to reduced branching and tillering characteristics, changes in maturity duration, photoperiod requirements and reproductive features. Many crops ended up with enlarged features like plant size, fruiting/seeding organs and grain/seed size in crops like tomato, maize, field beans, sunflower and squash as compared to their wild relatives. In some crop species there were morphological and structural modifications which lead to fruit anatomy changes as in tomato and changes in spikelet and inflorescence structure in cereals. In some cereals there was reduced enclosure of grains by glumes, lemmas, and palea for easy threshing. Selection for simple morphological architecture led to reduced tillering in some cereals and less branching in non-cereals accompanied with higher yields as compared to their wild relatives. Crop evolution resulted in increased height or vegetativeness and altered shoot and root characteristics in crops in which vegetative parts are of economic importance as dictated by the targeted vegetative part. In some crops polyploidy and asexual reproduction was promoted at the expense of sexual production as for sugarcane, potato, pineapple, strawberries, sweet potato and banana. Domestication altered maturity duration and reproduction making some seed propagated staple crop plant species, e.g., wheat, rice, barley, finger millet, African rice, tomato, soybean, groundnut, field beans, cowpea, chickpea, garden pea and several others become annual and self-pollinated with elimination of possible self-incompatibility. Through domestication crop plants underwent changes that perfected their features with alteration from wild state such that they cannot survive in the wild and are unable to compete with their wild relatives and other plants and give the desired economic yields under appropriate management. The changes due to domestication are more pronounced and evident in annual crops than in perennial crops.

Domestication caused gradual genetic diversity reduction due to selection and breeding of similar genotypes with the target traits. This led to only a small fraction of the genes and alleles in the wild progenitor gene pools being retained in the domesticates which became more genetically uniform as compared to their progenitors. Reduced genetic diversity in the domesticates is because the domesticators selected only a restricted proportion of genotypes for the target traits from the diverse crop progenitor populations and sampled only small amounts of the progenitor genes controlling such traits (Abbo et al., 2003). Genetic investigations to understand the process and genetics of domestication is obtained through DNA (including ancient DNA), candidate gene and quantitative trait loci (QTLs) analyses. Some genes and QTLs concerned with domestication, phenotypic changes and major contrasting differences between the crops and their relatives are known to exist for various crops and their predecessors. A number of domestication genes and QTLs associated with various domestication traits have been identified in a number of crops like beans, maize, sorghum, rice and their relatives through genetic, QTL, ancient DNA, genome-wide association (GWAS) and whole genome analyses making it easy to understand the process and genetic basis of domestication, know the extent of DNA diversity reduction and identify genes and QTLs that were preferred by domestication (Meyer & Purugganan, 2013). For quantitative traits, the QTLs in a few segments of the genome carry the genes distinguishing the crop and its relatives and it is also known that genomic changes associated with domestication are reflected by DNA nucleotide diversity changes whereby various domesticates lost high proportions of their DNA nucleotide diversity as compared to their progenitors. This is evident in random amplified polymorphic DNAs (RAPDs), inter-simple sequence repeats (ISSRs) and sequence-tagged microsatellites (STMS) analyses for chickpea (*Cicer arietinum*) which has less DNA diversity than its progenitor *Cicer reticulatum* (Abbo et al., 2003). Similarly, several other crops, like wheat, maize, barley, rice, etc., are known to have less DNA nucleotide diversity as compared to their respective progenitors. In addition to genome-wide changes in the domesticates, there are domestication-related specific genomic changes associated with modified traits due to DNA nucleotide mutations which affect gene expression levels and protein modifications. In general, domestication led to higher reduction of diversity in annual and autogamous species than other species and down-regulation of expression for several genes in domesticates as compared to their wild progenitors.

2.3 Migration and Spread of Crops

The different communities of original cultivators in various regions domesticated plants independently though it is possible they would also receive other ideas regarding domestication from one another through migration and other forms of movement. The early products of domestication and founder species underwent widespread dispersal and were not restricted to their centres of domestication and centres of diversity only. The spread resulted in development of new diversity and landraces suited to the prevailing local conditions in different areas. Independently domesticated crops spread to different parts of the world slowly as a result of diffusion through migration as the migrating people took their farming technology with them or passing on the ideas through cultural contacts or trade, taking the crops beyond their natural original homes. As ancient cultivators migrated, they carried along the seeds of their crops for planting to produce food wherever they established new homes. The introduced plants were subjected to new and varied environmental and cultivation conditions and spatial isolation. Environmental variations which were due to climatic conditions, soil differences, altitude, fauna and flora populations, isolation, manipulation by the cultivators, occasional interspecific and interpopulation hybridisation and inherent variability in the selected plant populations lead to evolution of new forms of the various plant species brought under cultivation, hence the evolution of variable landraces, subspecies and species with divergent adaptation across the globe. The early cultivators and their neighbouring communities exchanged food-stuff and planting materials through trade and other forms of interaction. In case of invasion by hostile communities, the invaders would bring with them their food-stuff and planting materials or would take away similar materials back home. The spread of agricultural technology from the Near East to Europe started as early as 9500 BCE and reached north western Europe about 8500–7000 years ago (Pinhasi et al., 2005). The spread of the technology also took wheat and barley cultivation to the Nile valley in Africa by 6000 BCE while barley surfaced along the Indus around 3500 BCE and by 2000 BCE some of crops that were already domesticated had gone beyond their various centres of origin and domestication through migration, invasion, cultural and trade interactions (Atkins et al., 1998).

Seafarers took crops of Asian and New Guinea origin to Madagascar and East African coast and they in turn took back crops domesticated in African to Asia and Oceania. Crops like banana, coconut, rice, sugarcane, mango and others reached East African coast and spread hinterland. Crops of South America were taken to Mexico and the Caribbean and later, along with those domesticated in Mesoamerica, moved further northwards and to eastern USA and to Europe so that by sixth century CE maize and potato were under cultivation in parts of Europe. Extensive spread across the continents came with the explorers like Christopher Columbus 500 years ago when crops of the New World and Old World spread further across the globe. There were massive exchanges of crop plants between Africa, Europe, Americas, Asia and Australia. Maize, tomato, potato, sweet potato, beans and other crops of the Americas moved to Europe, Africa and Asia while wheat, barley, rice and other crops of the east and Africa moved further westwards. Crop spread initiated by the early agricultural communities continued into modern (historic) times and the eras of the Islamic empire trade across the continents, explorations initiated in the fifteenth century CE, Christian missionaries, colonialism and the colonial farmers who settled in the newly established colonies. With migration, invasion and exploration, transnational trade across the globe by the Spanish, French, Portuguese, Germans, British, Dutch and Arabs and research over centuries many crops have spread and are cultivated in other regions as major cash or staple crops (Table 2.2). This has been accompanied by establishment of botanic gardens, exchange of genetic materials by breeders, globalisation of trade and changes in farming systems and crops cultivated across the world sustaining world wide spread of crop materials. Maize, beans, sweet potatoes, potatoes, cassava which are natives of South America and Central America are now staple food crops in Africa. Peach, known to be a native of China since 2000 BCE, moved westwards to the Mediterranean where it is an important crop. Sugarcane, a

Table 2.2 Some major world crops: centres of domestication and major producers

Crop	Centre of domestiction	Major cultivation centres
Wheat	Western and Central Asia, Near East	USA, India, Russia, China
Maize	South and Central America	USA, China, Brazil, India
Rice	Indo-China	India, China, Indonesia, Bangladesh
Sorghum	Western and Eastern Africa	USA, India, Mexico, Nigeria, China, Sudan
Beans	South and Central America	Brazil, Myanmar, India, China, Africa
Barley	Western and Central Asia, Near East	Russia, Germany, France, Ukraine
Potatoes	South and Central America	China, India, Russia
Tomato	South and Central America	USA, Turkey, Egypt, China, India
Cotton	India, Pakistan, Central and South America, Africa	India, China, USA, Brazil, Pakistan, West African countries, Egypt
Tea	Eastern Asia (China)	China, Kenya, India, Sri Lanka, Uganda, Tanzania
Sugarcane	South East Asia (New Guinea)	Brazil, India, China, Thailand, Mauritius
Cassava	South America	Nigeria, Thailand, Indonesia, Brazil
Sweet potato	South and Central America	Ethiopia, Angola, USA, Rwanda
Coffee	Abyssinia and central Africa	Brazil, Vietnam, Colombia, Indonesia, Kenya
Pineapple	South America	Hawaii, Costa Rica, Brazil, Mexico, Kenya, China, India, Chile, Thailand
Cacao	South America (Andes)	Ecuador, Brazil, Malaysia, Indonesia, West Africa
Soybean	Eastern Asia (China)	USA, Brazil, Argentina, China
Sunflower	North America	Ukraine, Russia, European Union

native to New Guinea was spread by Arabs to Palestine, Sicily, Spain and the Greek islands and is now an important crop in Brazil and Mauritius and other parts of tropical Africa. Soybean similarly spread from China to become an important global crop, especially in America. Coffee which originated in tropical Africa is a major cash crop in Brazil.

Problems

2.1 What features do crop centres of diversity have in common?

2.2 Why is agriculture considered to have occupied only a short duration of human history on earth?

2.3 Distinguish centres and non-centres in regard to crop domestication

2.4 Which major morphological features are associated with domestication of crop plants?

2.5 What kind of information is used to make deductions about crop domestication in ancient times?

References

Abbo, S., Berger, J., & Turner, N. C. (2003). Evolution of cultivated chickpea: Four bottlenecks limit diversity and constraint adaptation. *Functional Plant Biology, 30*, 1081–1087. https://doi.org/10.1071/FPO3084

Abbo, S., Lev-Yadun, S., & Gopher, A. (2010). Agricultural origins: Centres and noncentres, a Near Eastern reappraisal. *Critical Reviews in Plant Sciences, 29*, 317–328. https://doi.org/10.1080/07352689.2010.502823

Atkins, P. J., Simmons, I. G., & Roberts, B. K. (1998). The origins and spread of agriculture. In *Land, People and Time*. Hodder Arnold. London. HODDER Education, Disclaimer.

Bar-Yosef, O. (2017). Multiple origins of agriculture in Eurasia and Africa. In *On Human Nature: Biology, Psychology, Ethics, Politics and Religion* (pp. 297–331). Elsevier. https://doi.org/10.1016/B978-012-420190-3-00019-3

Doughty, C. E. (2010). The development of agriculture in the Americas: An ecological perspective. *Ecosphere, 1* (6), cart 21. https://doi.org/10.1890/ES10-00098.1

Earle, S. (2019). Origin of earth and the solar system. In S. Earle (Ed.), *Physical Geology*, 2nd edn. ISBN 978-1-77420-028-5 and ISBN 978-1-77420-027-8. BCcampus. http://open.bccampus.ca

Edger, P. P., Poorten, T. J., & VanBuren, R., et al. (2019). Origin and evolution of the octoploid strawberry genome. *Nature Genetics, 51*, 541–547 (2019). https://doi.org/10.1038/s41588-019-0356-4

Esquinas-Alcazar, J. T. (1993). Plant genetic resources. In M. D. Hayward, N. O. Bosemark & I. Ramagosa (Eds.), *Plant Breeding: Principles and Prospects*. Springer Dordrecht. https://doi.org/10.1007/978-94-011-1524-7_4

Godfrain, T., & Vercauteren Drubbel, R. (2019). A brief account of human evolution for young minds. *Frontiers Young Minds, 7*, 22. https://doi.org/10.3389/frym.2019.00022

Gross, B. L., & Olsen, K. M. (2010). Genetic perspectives on crop domestication. *Trends in Plant Science, 15*(9), 529–537. https://doi.org/10.1016/j.tplants.2010.05.008

Harlan, J. R. (1971). Agricultural origins: Centres and noncentres. *Science, 174*, 468–474. https://doi.org/10.1126/science.174.4008.468

Maxted, N., & Vincent, H. (2021). Review of congruence between global crop wild relative hotspots and centres of crop origin/diversity. *Genetic Resources and Crop Evolution, 68*, 1283–1297. https://doi.org/10.1007/s10722-021-01114-7

Meyer, R. S., DuVal, A. E., & Jensen, H. R. (2012). Patterns and processes in crop domestication: An historical review and quantitative analysis of 203 global food crops. *New Phytologist, 196*, 29–48. https://doi.org/10.1111/j.1469-8137.2012.04253.x

Meyer, R. S., & Purugganan, M. D. (2013). Evolution of crop species: Genetics of domestication and diversification. *Nature Review Genetics, 14*(12), 840–852. https://doi.org/10.1038/nrg3605

Nayar N. M., & Mehra, K. L. (1970). Sesame: Its uses, botany, cytogenetics and origin. *Economic Botany, 24*, 20–31. https://doi.org/10.1007/BF02860629

Payne, J. L., Bachan, A., Heim, N. A., Hull, P. M., & Knope, M. L. (2020). The evolution of complex life and the stabilisation of earth system. *Interface Focus, 10*, 20190106. https://doi.org/10.1098/rsfs.2019.0106

Pinhasi, R., Fort, J., & Ammerman, A. J. (2005). Tracing the origin and spread of agriculture in Europe. *PLoS Biology, 3*(12), ed410. https://doi.org/10.1371/journal.pbio.0030410

Purugganan, M. D. (2019). Evolutionary insights into the nature of plant domestication. *Current Biology, 29*, R705–R714. https://doi.org/10.1016/j.cub.2019.05.053

Purugganan, M. D., & Fuller, D. (2011). Archaeological data reveal slow rates of evolution during plant domestication. *Evolution, 65*, 171–183. https://doi.org/10.1111/j.1558-5646.2010.01093x

Scerri, E. M. L., Thomas, M. G., Manica, A., Gunz, P., Stock, J. T., Stringer, C., Grove, M., Groucut, H. S., Timmermann, A., Rightmire, G. P., d'Errico, F., Tryo, C. A., Drake, N. A., Brooks, A. S., Dennel, R. W., Durbin, R., Henn, B. M., Lee-Thorp, J., de Menocal, P., Chikhi, L., et al. (2018). Did our species evolve in subdivided populations across Africa, and why does it matter? *Trends in Ecology and Evolution, 33*, 582–594. https://doi.org/10.1016/j.tree.2018.05.005

Singh, B. D. (1990). *Plant Breeding Principles and Methods*. Kalyani Publishers.

References

Chapter 3
Gerplasm Resources: Exploration and Conservation

3.1 Germplasm Resources

Plant germplasm resources comprise of invaluable plant materials belonging to plant species and its relatives that serve as genetic base for crop improvement and are often obtained from centres of diversity, gene banks, breeding programmes and farmers. The availability of diverse germplasm resources, which include landraces and other forms of farmers' varieties, elite breeding materials, genetically enhanced materials, obsolete and modern varieties and wild and weedy crop relatives, is essential for success in plant breeding programs and maintenance of high production levels. Genetic variation within and among plant populations and species is attributable to point and chromosomal mutations, genetic recombination, gene flow between different populations and species and random genetic changes arising from drift. These genetic changes are responsible for variability in DNA sequences, protein and other biochemical features, morphological and physiological characteristics and other important features in plant populations and the germplasm resources obtained from them. The germplasm can be used directly by enforcing selection on introduced or locally available materials or they can be used in the transfer of specific traits to otherwise well adapted cultivars. Germplasm collections serve as sources of desired genes and gene combinations for breeding materials and can be used to address production problems when calamities arise. They are used as security against occurrence of calamities that destroy crops by transferring appropriate characters from them to the breeding materials or superior cultivars deficient of certain specific attributes like resistance or tolerance to diseases, insect pests, parasitic weeds, drought, heat, low temperatures, salinity, mineral stress and resilience to climate change. Such characters may be simply inherited single gene or multigene qualitative or polygenically inherited quantitative traits.

Every plant breeding institution or plant breeder needs to assemble and maintain extensive diverse germplasm collections with working germplasm stocks which are properly documented, held in storage and can be made available for use as need

P. O. Ayiecho and J. O. Nyabundi, *Conventional and Contemporary Practices of Plant Breeding*, https://doi.org/10.1007/978-3-031-74998-8_3

arises. The various germplasm resources that can be assembled, conserved and made available to breeders and breeding institutions as sources of genes are categorised as follows.

1. **Landraces (primitive cultivars or folk varieties)**. These are products of crop domestication under various traditional farming methods of different civilisations and natural environments in which they evolved. They are adapted to the regions in which the respective crop species were domesticated and often contain gene combinations that evolved over several decades in the regions where the crop has been cultivated for centuries. They have wide genetic diversity making them rich germplasm resources and are part of the indigenous knowledge of the cultivators that developed them.
2. **Advanced cultivars**. These are modern, cultivated and obsolete varieties, usually high yielding and well adapted, developed by plant breeders and breeding institutions. Within a given country or region or regions having similar climatic conditions, advanced cultivars tend to be related because of exchange of breeding materials between breeders, leading to limited genetic diversity among them. Some of such cultivars have been developed using the genetically diverse landraces and occasionally with genes from weedy and wild relatives, induced mutations, biotechnology and genetic engineering.
3. **Weedy and wild relatives**. Some wild species and crop progenitors are closely related to the cultivated species and it is possible to transfer genes from them through interspecific hybridisation. The weedy and wild species related to cultivated crops carry important genes which can be used to improve crop species and address catastrophic situations.
4. **Induced mutations**. These are products of artificial mutagenesis undertaken by the plant breeder with objectives of creating and increasing variability in breeding collections to develop genotypes with specific traits like resistance to biotic and abiotic stress factors, quality parameters, plant architecture and even yield attributes. Artificial mutagenesis usually leads to wide variability which may be used directly after selection or used in hybridisation followed by selection and backcrossing.
5. **Products of modern biotechnology and genetic engineering**. Biotechnology methods create variability, especially in form of somaclonal and gametoclonal variability which puts several selectable variants at the disposal of the breeder for use to improve the economically important plant species.

3.2 Genetic Erosion and Genetic Vulnerability

During the early part of the twentieth century when Nikolai Vavilov explored the crop centres of origin around the world, he observed extensive and diverse genetic resources for most crops and their weedy and wild relatives. Such diversity existed and was maintained in farmland using traditional agriculture and other human activities, occasional spontaneous chromosomal and point mutations, intraspecific and interspecific hybridisations with sympatric wild and weedy relatives in natural grasslands, wetlands, rangeland and forests. Some decades after Vavilov's explorations it was realised that the genetic resources were in great danger and rapidly being destroyed and eroded as populations in Vavilovian centres reflected a dwindling trend for genetic diversity originally described by Nikolai Vavilov and many landraces, wild and weedy relatives of a number of crops showed trends of decline in their natural habitats. This process of genetic erosion takes place both within and outside the Vavilovian centres of diversity and has been observed for several major crops like cotton, wheats, barley, rice, maize, potato, sugar beet, tomato, soybean, sorghum, millets, a number of grain legumes, oil crops, vegetable and fruit crops around the world.

Genetic erosion, the gradual and permanent decline in the diversity of cultivated crops and their relatives, has continued over the years and is mainly attributable to modern human activities touching on mismanagement of centres of diversity and genetic resources therein. Change of land use pattern is one of the significant causes of genetic erosion. Increasing human populations, field clearance and deforestation to create land for food production and pressure on land for varied human activities destroys grassland, rangeland, forest, wetland and riverine ecosystems leading to loss or endangering of invaluable germplasm. Irreplaceable germplasm heritage is destroyed due to clearance of vegetation on large tracks of land to create room for housing and urbanisation projects, wood fuel, timber harvesting, mineral mining, dams and irrigation projects, highways, industrial construction and various other infrastructure projects. In Kenya a number of local sesame (*Sesamum indicum*) wild relatives populations like *Sesamum angolense*, *S. anguistifolium*, *S. latifolium* and *S. alatum* have been destroyed due to infrastructure developments and urban expansion of major cities like Mombasa and Nairobi around which such species existed in abundance up to late 1980's. Biodiversity, including genetic diversity, is threatened by modern agricultural technology like agricultural mechanisation, development of genetically uniform high yielding and profitable varieties, genetically modified crop varieties, possibilities of gene flow between modern varieties and the landraces, wild and weedy crop relatives, land clearing for modern agriculture, use of pesticides and herbicides with associated pollution interfering with natural ecosystems and the changes of economic activities and agricultural systems in the traditional farming communities. Political instability and civil strife make farmers flee areas of conflict leaving their fields with traditional varieties unattended and they may also consume the stored seeds of their traditional varieties and later turning to cultivation of modern varieties resulting in loss of useful resources. In some cases, the fighting forces invade villages destroying and helping themselves to the crops in the field, produce in storage and reserved seeds as well. Climate change phenomenon is highly destructive to plant genetic diversity due to extremely high temperatures, prolonged droughts, wildfires, accelerated desertification, heavy rainstorms and floods. Prolonged droughts cause loss of genetic diversity when immature crops and their relatives dry up in the fields. While wild fires, usually associated with prolonged droughts and high temperatures, decimate grasslands, rangeland, forests, wetlands and farmers' fields having important plant species. Floods submerge farms and wash down the growing crop, destroying stored seed as well. Overgrazing and over-browsing by livestock cause total destruction of ecosystems in grasslands, rangelands, islands and arid lands eliminating important plant species. In some situations, newly introduced plant species may affect plant diversity through competitive advantage over local flora and invasiveness in their new environment threatening the local biodiversity, ecosystems, habitats and even animal and human wellbeing. An example is *Prosopis juliflora* which was introduced in semiarid parts of Kenya in early 1980's for environmental stabilisation but turned out to be undesirably fast growing, highly invasive, suppressing and eliminating some native plants, thereby destroying plant diversity, habitat, ecosystem and biodiversity.

Genetic erosion may also take place in the process of germplasm conservation. Genetic erosion takes place during germplasm collections in the field through inadequate sampling from large populations leading to random loss of alleles as a result of genetic drift, improper handling of collected materials in storage by physical mixing and undesired gene flow between different germplasm materials when planted out for multiplication, evaluation and rejuvenation. Contamination and losses may arise in assembled germplasm causing changes in genetic structure of the collections unless extreme care is taken in the management of the stored resources. The breeder who needs the germplasm most as her/his raw material has also contributed significantly towards genetic erosion by developing genetically narrow high yielding varieties as the farmers are enticed to drop genetically diverse traditional landraces and replacing them with the high yielding modern pure line or hybrid varieties across the globe. For example, green revolution in wheat, barley and rice in 1970's led to superior modern varieties replacing diverse landraces and related wild and weedy species having broad genetic diversity and gene combinations necessary for adaptation to varied environmental and farming conditions, resistance and tolerance to various forms of stress and resilience to climate change. Similar genetic erosion has been observed for several other major crops like cotton, maize, potato, sugar beet, tomato, soybean, sorghum, millets, a number of grain legumes, vegetable and fruit crops around the world. Hence, broad genetic diversity contained in landraces, weedy and wild crop relatives may be lost and replaced by the narrow genetic base in modern superior cultivars developed from them, accompanied by modification or destruction of wild and weedy relatives' habitats. Though the widely cultivated high yielding genetically uniform modern varieties are highly responsive to modern agricultural technology and inputs, they are often vulnerable to pest and disease outbreaks, non-resilient to climate change and are not adaptable to resource poor small scale farming environments.

Genetic engineering has led to development of genetically modified plants with transgenic varieties being grown widely in many countries around the world. There are fears that genetically modified plants will affect plant genetic resource diversity, especially landraces, wild and weedy crop relatives through gene flow and introgression which may result in more invasive weedy plants that are resistant to both biological and artificial control methods. The introgressed genes may also make the weedy relatives more aggressive and competitive, especially if the introgressed transgenes are for better fitness and tolerance or resistance against biotic and abiotic stress factors. Gene flow between transgenic varieties and other germplasm depends on the reproductive system, sexual compatibility, hybrid fertility, pollen dispersal mechanisms and spatial separation between the crop germplasm materials and the transgenic crop field. Gene flow is highest for cases of allogamous species and lowest for autogamous species. There are possibilities that transgenic varieties also have the risk of replacing conventionally developed varieties and landraces the same way the products of green revolution displaced the landraces in centres of diversity for wheat, barley and rice.

The dwindling of genetic diversity has led to serious genetic vulnerability of many crops because of demands for high yields, uniformity and appropriate quality by growers, processors, consumers and agricultural engineers, especially for the extensively bred staple crops. Such demands have led to the general trend of developing genetically uniform crops with narrow genetic bases in the major crops. Narrowing the genetic base of crop plants humans depend on for survival has exposed the crops to possible dangers of catastrophic biotic and abiotic vagaries. Some known historical examples of catastrophic situations due to widespread cultivation of susceptible crop cultivars with narrow genetic base, reflecting genetic vulnerability of the crops, include the Irish potato famine of 1840's when the potato crop was wiped out by *Phytophthora infestans,* destruction of coffee in Sri Lanka by coffee leaf rust (*Hemileia vastatrix*) in late 1800's making the country shift to tea cultivation, serious wheat shortage in USA in 1917 because of stem rust (*Puccinia graminis*) outbreak on wheat, 1943 famine in India when rice was wiped out by brown spot disease (*Cochliobolus miyabeanus*), destruction of maize hybrids having T-cytoplasm by southern corn leaf blight (*Cochliobolus heterostrophus/Bipolaris maydis)* in 1970–71 in USA and elimination of taro by leaf blight in western Samoa in 1993. In 2019 to 2021 there was serious desert locust (*Schistocerca gregaria*) outbreak in East Africa which invaded Yemen, Ethiopia, Somali and Kenya, ravaging food crops on farmers' fields and other vegetation, depriving livestock of pasture feed and pushing some parts of the region to near-famine situation. Genetic vulnerability of the crops that are the major staple foods of the world due to the varietal uniformity makes diverse germplasm a necessity in crop improvement since the germplasm, particularly landraces and introduced cultivars, weedy and wild crop relatives, comprise the reservoirs for important traits like disease and pest resistance, resistance to parasitic weeds like *Striga, Orobanche* and *Cuscuta*, adaptability, resistance to abiotic stress conditions, climate change resilience, crop quality attributes, yield and yield related traits.

3.3 Need for Genetic Conservation

Even though a lot of germplasm resource materials have been lost substantial genetic diversity is still existent in regions of diversity for various plant species and have great potentials for sustenance of production and should be conserved for economic well-being of humankind and protection of environment. Among the plant species that were domesticated by the Stone Age people only a small number are important in the world trade and food supplies, namely, wheat, barley, maize, rice, sorghum, millets, potatoes, cassava, taro, yams, sweet potatoes, tomatoes, cabbage, oil palm, coconut, rapeseed, apples, grapes, oranges, soybeans, field beans, groundnuts, bananas, oil palm, plantains, Brassicas, sugarcane, sugar beet, cotton and peas. Furthermore, only three cereal crops, maize, wheat and rice, comprise more than 50% of the world food requirements. The few plant species that comprise the major food crops are under threat due to genetic vulnerability arising from narrow genetic base and are liable to destruction by any unforeseen pest biotype and pathogen race outbreaks and other environmental hazards. The resources are not only facing the declining genetic diversity for the various crops and their relatives but there is a major concern among the conservationists, gene banks, plant breeders, various scientists, organisations allied to Consortium of International Agricultural Research Centres (CGIAR), formerly Consultative Group on International Agricultural Research, and other research bodies and policy makers that some plant species are being pushed the point of extinction.

The dwindling of germplasm resources necessitates the conservation of the available germplasm through reinstatement and maintenance of genetic diversity as a remedy against genetic erosion. The catastrophic epidemics reflecting genetic vulnerability of crops humans rely on justify conservation of plant genetic resources as a security against unpredictable catastrophic incidences since diverse germplasm have the genes needed for developing secure and better plant varieties with resistance to diseases and pests, productivity and local adaptation to abiotic changes, especially vagaries associated with climate change. Such genes are found in the landraces and local cultivars that evolved in the traditional agricultural systems over thousands of years and also in the wild and weedy crop relatives. The remedy to genetic erosion does not lie on stopping the breeder from doing crop improvement or preventing the farmers from cultivating the modern varieties but lies on conservation and utilisation of germplasm and indigenous knowledge associated with it. Sustained development of superior crop varieties for food security, prevention of catastrophic pest and disease outbreaks and abiotic conditions depend on assemblage and conservation of broad genetic base germplasm, evaluation, documentation and utilisation of such genetic resources.

Various stakeholders including farmers, people holding indigenous knowledge, ethnobotanists, plant breeders, government supported national research organisations, non-governmental organisations, international organisations and gene banks are engaged in plant germplasm resource conservation and utilisation activities. The main player in international coordination and networking of plant genetic resources collection, documentation, characterisation, evaluation, conservation and utilisation is International Plant Genetic Resource Institute (IPGRI) which also operates as Bioversity International (and in alliance with CIAT as Alliance of Bioversity International-CIAT). It works in close collaboration with national government supported gene banks in various countries, other CGIAR organisations like International Crops Research Institute for Semi-Arid Tropics (ICRISAT), International Potato

Centre (CIP), International Maize and Wheat Improvement Centre (CIMMYT), International Rice Research Institute (IRRI), International Institute for Tropical Agriculture (IITA) and International Centre for Agricultural Research in Dryland Areas (ICARDA) African Rice Centre (AfricaRice), formerly West African Rice Development Association (WARDA) which conduct research for conservation and utilisation of resources and improvement of specific crops that fall within their mandates for sustained agricultural production and food security. Svalbard Global Seed Vault in Norway was established 2008 to serve as safe long term storage facility for duplicate seeds of genetic resources conserved in gene banks around the world as a security against catastrophic loss of such resources elsewhere.

3.4 Exploration and Collection of Genetic Resources

Germplasm exploration and collection refers to eco-geographic exploratory surveys and collection of diverse genetic resources from fields, centres of diversity and non-centres and farmers' fields for conservation storage and utilisation. Eco-geographical exploratory surveys for genetic resources are important in germplasm conservation since they help to determine and understand the amount, pattern, relationships and geographic and ecogeographic distribution of genetic diversity available for different crop species in various regions and ecosystems, identify the endangered resources due to modern developments in agricultural production system, industrialisation, urbanisation, climate change and socioeconomic trends and point out the gaps with respect to what is known to have been collected and conserved in gene banks (Williams, 2023). Remote sensing and geographic information system (GIS) data enhance exploration surveys and collecting, and can be useful for data analysis, identify areas in which the resources have certain specific traits and point out specific areas for conducting collection. The data obtained from the exploratory surveys help in locating promising sites having genetically diverse materials for field research and collecting, devise germplasm collection methods, determine agricultural activities, natural vegetation and its response to climatic conditions, and may also be used to design in-situ and on-farm conservation. The surveys take into consideration other ecological factors like altitude and slope, geology and soil types, presence of toxic minerals and saline conditions, latitude and longitude since crops and their relatives show ecotypic differentiation related to their habitats. The presence of other biotic factors like associated pests and other insects, diseases, parasitic plants, information provided by the farmers and other local communities and various forms of indigenous knowledge within the communities are also important considerations when gathering data on genetic diversity. Since information obtained during surveys and collection is useful in assessing the genetic and breeding potential of the collected materials it is important to collect data on plant architecture, presence of various biotic and abiotic stresses at the collection areas, nutritive and physical quality and any other observable traits.

Once germplasm resources are located and identified in the centres of diversity, non-centres and even in peripheral regions of various species distribution and farmers' fields, efforts are initiated to sample the available landraces and other cultivars, weedy and wild relatives for ex-situ conservation in gene banks and botanical gardens as different types of collections. Germplasm collection is done at the time when the seeds or vegetative reproductive organs or tissues are in the right stage for such exercise like the time when the farmers are harvesting the landraces and during the season when the wild and weedy relatives have mature seeds for collection. Collection is done while having in mind the patterns of ecological differentiation since such differentiation is related to population genetic structure. Areas with much ecological and genetic variability receive more frequent and closer sampling as it is known that distances between sampling points, frequency of sampling and sample size are determined by distribution of genetic structure within and between populations (Williams, 2023). While undertaking the collecting exercise sampling sites should have significant distances between them since adjacent areas have plant populations and cultivars which are genetically close, hence materials from adjacent communities are likely to have genetic similarities. Collection locations should be well-spaced (at least 10 km apart) unless they have clearly distinct cultivars and the locations have different altitudes or have physical barriers between them like hills, mountains, significant water bodies, large forests and that the communities are ethnically different (Witcombe & Erskine, 1984). If there are no such barriers and there are low genetic and ecological variations sampling sites can have significant distances between them to minimise duplication because plant populations from adjacent areas and communities under such situations are likely to have genetic similarities.

Since a lot of plant genetic resource materials are already collected for most of the of major crops like maize, rice, wheat, potato, cassava, sorghum, millet, tomato, cotton, cowpea, pigeon pea, groundnut, common bean, barley, soybean, etc., for which several diverse samples have been collected since 1980's and conserved, new sampling should bear mind what is already in conserved stocks, materials inadequately represented in the stocks that need gap filling and endangered materials that need to be rescued due to land use and agricultural changes. At least 7.5 million accessions of plant genetic resources including crop plant wild and weedy relatives, landraces, obsolete cultivars, modern cultivars and other breeding materials belonging to over 7,000 genera, are held in various forms of gene banks around the world (Priyanka et al., 2021). Some of these gene banks located in at least 30 countries have facilities for safe long term germplasm storage. Priorities for any new collection need to be given to endangered species and those suffering from habitat loss, any previously collected germplasm that has been lost and needs to be replaced, changing crop improvement needs and socio-economic changes in farming communities. It is important that collection is made in areas where the species are under various forms of environmental stress and where genotypes could have been selected for adaptability to the stress conditions. Plant materials can be collected from the farmers' fields and adjacent fields or from weedy and wild crop relatives in forests, bushes, grasslands or roadsides and may consist of seeds, bulbs, cuttings, tubers, corms, whole plants, pollen grains or other in vitro tissues depending on the characteristics of the species, conditions of plant to be collected and its environment, length of trip back after collection, facilities at hand and the intended conservation methods. The aim is to collect maximum genetic diversity and obtain samples that are adequate and general representatives of the genetic resources and making efforts to capture important genes and rare alleles in the population (Witcombe & Erskine, 1984).

During genetic resource acquisition the collecting mission has to obtain appropriate permission from the local indigenous communities and government authorities to gain access to the genetic resources. It is mandatory to observe the relevant local laws and rules laid down by the Convention on Biological Diversity (Biodiversity Convention), Nagoya Protocol on Access and Benefit Sharing, International Treaty on Plant Genetic Resources for Food and Agriculture and Cartagena Protocol as regards acquisition and movement of the genetic resources. The right to benefit sharing by the local communities claiming ownership of the resources has to be upheld. Before taking the collected materials to the ultimate destination the collecting mission needs to obtain phytosanitary certificates declaring the health status of the collected materials and absence of pests, pathogens and noxious weed seeds and other undesirable matter. Before they reach the importing institutions all the introduced germplasm materials are screened at the local quarantine station within the importing country for presence of any diseases and pests.

3.5 Germplasm Characterisation, Documentation, Evaluation, Multiplication and Rejuvenation

The collected plant genetic resources are documented for passport data and all the necessary information for identification and utilisation of the germplasm. Such information includes indigenous knowledge availed by the farmers and the communities who supplied the materials, data about germplasm regarding genus, species, and accession reference number, source of the material, origin, habitat and ecosystem, associated vegetation, soil type, soil moisture conditions, climate, the land terrain and elevation, date of collection and whether it is a wild relative, a weedy relative or a landrace. Other pieces of information for documentation are frequency of distribution in the area, collector's name, height at time of collection, altitude range of the accession, colour, growth habit, important agronomic traits, diseases, pests and pathogens/diseases found on it, any known resistance to pests, weeds and abiotic stress factors in the area of collection, isozyme and DNA analyses and other relevant data. Once the materials are acquired, they are subjected to characterisation by growing them to understand and record highly heritable traits, and take note of both qualitative and quantitative morphological traits. Descriptors given by International Board for Plant Genetic Resources (IBPGR) and International Plant Genetic Resource Institute (IPGRI) or Alliance of Bioversity International-CIAT are used for recording the observations. Plant genetic resources are characterised and evaluated with respect to productivity, quality traits, disease and pest resistance, adaptability, response to abiotic stress factors, agro-morphological trait measurements and genetic assessments entailing field and greenhouse tests to identify accessions with desirable traits. Cytological/karyological, biochemical analyses and genotyping using DNA-sequencing analyses are undertaken for characterisation, taxonomic classification and evaluation. Isozyme and molecular characterisation are more reliable than qualitative and quantitative morphological characters due to lack of environmental influence and that any tissue from a plant in any growth stage can be used to extract DNA for characterisation. Where appropriate facilities are available high-throughput phenotyping and genotyping in conjunction with genomic technologies like genome sequencing and resequencing and DNA molecular marker identification are done to point out molecular markers linked to important traits, distinguish, categorise and cluster the accessions for pre-breeding and plant breeding purposes. DNA markers like random amplified polymorphic DNAs (RAPDs), restriction fragment length polymorphisms (RFLPs), simple sequence repeats (SSR), single nucleotide polymorphisms (SNPs), amplified fragment length polymorphisms (AFLPs), variable number of tandem repeats (VNTR), etc., are employed to examine genetic variation within and among collections and determine inter-relations among the collections (Kwameswara Rao, 2004). Other genomic analyses techniques like genome-wide association studies (GWAS) and diversity array technology sequencing (DArTseq) provide added opportunities for dissecting complex traits to identify genes associated with target traits at high resolution and are efficient in studying, describing diversity and genotyping of large germplasm accession numbers rapidly for conservation and germplasm enhancement, gene and allele discovery, QTL mapping and identification of DNA molecular markers scattered in whole genomes (Halewood et al., 2018; Singh et al., 2020).

Germplasm evaluation which can be done along with characterisation or done for already characterised accessions, is key to germplasm resources utilisation. Evaluation is done along properly defined goals and well-designed procedure and statistically sound experimental designs taking into account numbers and sizes of collections. The activities may require participation of plant scientists of various disciplines, namely, breeders, pathologists, agronomists, weed scientists, entomologists, plant water relations scientists and physiologists, soil scientists and nematologists. Multidisciplinary approach in germplasm evaluation and screening helps in identifying valuable sources of resistance and tolerance to various forms biotic and abiotic stress and combinations of genes for desired morphological/agronomic traits and yield. The germplasm materials are also evaluated for quality traits like oil, fatty acids, proteins, phenols, sugar, amino acids, vitamins, minerals and anti-nutritional factors contents. Evaluation of the germplasm may be done in areas where the stress factors are known to exist like localised disease and pest problems, localised *Striga* and broomrape infestations, localised drought and aridity, salinity, mineral stress and cold condition problems so as to reduce evaluation costs for large numbers of accessions. Evaluation in areas where the production challenges are localised also eliminate spreading crop production problems to where they don't belong, particularly biotic stress factors like diseases, pests and parasitic weeds. Where automated high-throughput phenotyping facilities using sensor technology and automated imaging equipment are available they are employed for precise evaluation, rapid data gathering and processing.

The documentation, characterisation and evaluation data are used to describe the collections, develop documentation systems for conservation, clustering and effective utilisation of the collections, detection of duplications deposited in a gene bank and development of core and mini core collections for managing the germplasm collections. Since characterisation and evaluation of the genetic resources involve several materials and that large amounts of data are gathered over the years, computer technology facilitates the management of the data. Bioinformatics data management for plant genetic resources ensure that the assembled information is well maintained, stored, and allows for addition, removal of some of the data, analysis, presentation, retrieval, sharing, updating of the information and appropriate classification of the data.

Appropriate germplasm rejuvenation is a critical component of germplasm conservation so that the germplasm resource identity and genetic integrity are not lost through successive generations of reproduction, gene flow between reproductively compatible materials, random genetic changes, physical mixing and loss of seed viability in storage. The frequency and methods of rejuvenation are determined by initial sample size, the way the materials are handled, management of the germplasm in storage, storage conditions and length of seed viability. Evaluation plots can be used as the multiplication and rejuvenation plots for some materials and the plots should be adequately isolated depending on reproductive characteristics to avoid genetic contamination through pollen flow from nearby reproductively compatible materials. Though long-distance isolation is not necessary for self-pollinating species, multiplication and rejuvenation of the germplasm for outcrossing species require adequate spatial and temporal isolation. While mechanical mixing of propagules must be avoided adequate population or sample sizes during multiplication and rejuvenation are necessary to prevent any changes that may arise due to drift, particularly in outcrossing species. Some materials may have problems of photoperiodism, vernalization and specific temperature requirements which may lead to germplasm loss unless special conditions are met during evaluation, rejuvenation and multiplication. While multiplying and screening various sorghum germplasm materials obtained through ICRISAT (Ayiecho & Nyabundi, 1994) observed that some exotic sorghum genotypes grew excessively tall and could not flower at Kibwezi in Kenya. Evaluations and rejuvenation exercise take into consideration the suitability of the environment and the site, and where possible such activities should be conducted under conditions similar to the environments where the materials were obtained, especially in regard to soil conditions and photoperiodism or such conditions can be created in the greenhouse.

3.6 Conservation Methods

Conservation of genetic resources takes two complimentary approaches, namely, ex-situ conservation and in-situ conservation, which may involve long term storage, cold storage and even establishment of genetic reserves and on-farm conservation.

3.6.1 In-situ Conservation

In-situ conservation is the maintenance of the germplasm population within its natural habitat in the community to which it is endemic as field gene banks and may be in form of protected forests, nature reserves or national parks for weedy and wild crop relatives and conservation on farmers' fields for landraces in the indigenous agroecosystems (Esquinas-Alcazar, 1993; Quazi et al., 2021). In-situ conservation helps to maintain populations in natural eco-systems, allows the conserved populations to continue undergoing evolution and generating genetic variability naturally and is important for species not easily conserved outside their natural habitats (Salgotra & Chauhan, 2023). Furthermore, in-situ conservation facilities in centres of diversity can be used for agricultural research and allow for introduction of new crops, new varieties, continued exchange of germplasm among farmers and management of crop cultivars by the farmers within their respective agroecosystems with respect to adaptability, resistance and tolerance to various forms of stress. In some regions around the world farmers still retain and manage their traditional landraces and maintain genetic diversity, especially in areas where agriculture is not yet highly commercialised and traditional agriculture still remains the major mode of food production. This form of in-situ germplasm conservation, referred to as on-farm conservation, enhances germplasm diversity by giving opportunities and supporting continued development and maintenance of landrace cultivars and germplasm with broad genetic base and good local adaptation. On-farm conservation is based on the fact that plant genetic resources have been handled by farmers who subjected them to selection through the ages thereby directing the evolution of such resources. On-farm conservation fields complement the institutional gene banks that are supported by governments, Alliance of Bioversity International-CIAT, other CGIAR institutions and other stakeholders concerned with plant genetic resource conservation and utilisation. On-farm conservation gives opportunities to the plant breeders and conservationists to work with traditional farmers and use on-farm conservation fields for community-based participatory breeding.

Though in-situ conservation is an attractive method of conservation it has challenges because it is expensive in terms of space and needs monitoring and special legislation to protect the reserves and national parks, particularly for weedy and wild species. Germplasm conserved in-situ is likely to suffer erosion through diseases, pests, environmental disasters like volcanic eruptions, floods, climate change and associated prolonged droughts, with high temperatures and expansive wildfire incidences and stormy rains leading to destructive flooding and submergence of farms, grasslands and bushes destroying important plant germplasm. They are also vulnerable to political instability, technological and socio-economic changes. Movement of farmers away from their localities due to conflicts will leave the important germplasm to be destroyed and lost. Germplasm resources under in-situ conservation may still suffer destruction due to expanding urbanisation, modern agricultural activities, infrastructure projects like housing, dam and road construction projects and some materials conserved in-situ on roadsides or near residential areas may be lost. Resources in on-farm conservation system may be lost if the farmers change their agricultural and cropping systems or switch to other crop species or more profitable economic activities. However, in-situ conservation helps to address problems associated with ex-situ conservation like recalcitrant seed storage, bulk associated with transportation and storage of vegetative propagules and need for special field conditions where ex-situ gardens are to be established as field gene banks. Creating the right and special environments for some species planted outside their habitats may be difficult, expensive and time consuming. In-situ conservation solves ex-situ conservation problems related to poor field sampling, handling and documentation, possible genetic changes through mutation in storage, gene flow between reproductively compatible but different populations, somaclonal variation, drift, need for rejuvenation and multiplication, seed viability loss, equipment failures and insecurity.

3.6.2 Ex-situ Conservation

Ex situ conservation applies where materials are moved from where they are normally native and taken to other places where they do not belong for conservation in field gene banks as botanical gardens and also subjected to cold storage conservation in international and national gene banks (Esquinas-Alcazar, 1993). Representative samples of the germplasm resources are assembled for maintenance in botanical gardens and greenhouses or as seeds, shoots, tubers, corms, bulbs and other tissues and DNA libraries for storage (Salgotra & Chauhan, 2023). Ex-situ field gene bank conservation like botanic gardens requires creation of suitable ecosystem, land with the right soil, favourable weather and need for water supply if there is no adequate rainfall. Field gene banks are recommended for species producing recalcitrant seeds and vegetatively propagated species. As compared to in-situ conservation, ex-situ methods are cheap and are used to manage the germplasm in small spaces and make the materials more easily accessible for utilisation. The genetic structure of germplasm materials under ex-situ conservation is expected to remain static without any changes because of lack of the natural selection conditions and other forces that are essential for continuous natural adaptation and evolutionary process, hence there are no chances of creating additional variability naturally after being assembled. There are also problems of genetic drift leading to loss of diversity due to small population sizes, sampling errors and undesired selection pressure since multiplication is often done in ecogeographical environments not similar to original natural habitats of the collected germplasm, poor seed germination and loss of seed viability, cost of facilities for storage, possible genetic contamination due outcrossing associated with inadequate isolation during multiplication and regeneration (Esquinas-Alcazar, 1993; Quazi et al., 2021).

Seed Storage

Based on storage characteristics, seeds can be placed in three categories, namely, orthodox seeds, recalcitrant seeds and intermediate seeds. The term orthodox is used for seeds which can tolerate desiccation and freezing temperatures to low levels in storage while recalcitrant seeds do not tolerate desiccation and freezing and are killed if their moisture content is reduced below the critical value of 20–35% (Roberts, 1973). Orthodox seeds can be stored for long periods under dry cool conditions without loss of viability and genetic integrity while recalcitrant seeds like those of sugarcane, cocoa, coconut, citrus, rubber, mango and tea have short life span and cannot withstand dehydration and lose viability if dried (Van der Maesen, 1984). Most agricultural crops including cereals, namely, wheat, maize, barley, oats, sorghum, millets, rice, legumes like field beans, peas, cowpeas, pigeon peas, grams, chickpea, faba bean and vegetables produce orthodox seeds. Some plants such as coffee and palm oil produce seeds that are referred to as intermediate or sub-orthodox seeds with a period of viability not going beyond two years.

Seed storage is the most efficient and convenient ex situ germplasm storage method but its best application is generally for orthodox seeds. Storage of orthodox seeds requires seed maturity with normal size and freedom of damaged seeds, storage pests, diseases and microorganisms. The seeds must not be affected by adverse temperatures and moisture conditions during its development and at maturity. For orthodox seeds, storage for several years in sealed containers can be achieved by reducing seed moisture content to between 3% and 7% and maintaining storage temperatures at −20° to −18°C, depending on the species (Chauhan, 2023; Salgotra et al., 2020). Such conditions prevent loss of genetic diversity, genetic integrity and viability because metabolic processes in the seeds are extremely reduced. However, seeds in storage should be regenerated as often as necessary, especially when viability falls below an acceptable minimum, depending on the species (Van der Maesen, 1984). Low viability leads to loss of genetic integrity arising from high frequency of mutants.

Recalcitrant seeds do not tolerate low temperatures and low water contents and need special conservation methods since they will have reduced viability if there is decreased water contents between 20 and 50% (Priyanka et al., 2021). Some plant species produce recalcitrant seeds that are too large and bulky and therefore difficult and expensive to transport and store, deteriorate quickly and can also be easily infested by pests and diseases. Reduction of moisture content in the recalcitrant seed leads to undesirable chemical processes that kill the seed. While recalcitrant seed longevity varies between a few weeks and a few months with storage under moist conditions, embryos from such seeds can be subjected to low temperature long-term conservation through cryopreservation.

In-Vitro Conservation Methods

Where seed collecting is difficult due to bulk and recalcitrant nature tissues can be collected for in vitro conservation using slow growth and cryopreservation storage methods. In-vitro germplasm storage is most convenient for vegetatively propagated species, species with long juvenile periods and those producing recalcitrant seeds. Vegetatively propagated crops like potatoes, sweet potatoes, sugarcane, pineapple, bananas, sweet potatoes, sisal, yams and other vegetatively reproducing species usually do not produce seeds or if they do the progenies derived are unsuitable for breeding purposes due to unmanageable variation resulting from segregation linked to high heterozygosity levels. Such crops are propagated by corms, stems, stolons, bulbs, bulbils, suckers, tubers, etc., which cannot be stored for long. Germplasm conservation for the vegetatively propagated crops can be done through in situ conservation in the field gene banks or they can be conserved ex-situ in form of botanical gardens and vegetative propagules using special techniques of in-vitro slow-growth and cryopreservation storage. In-vitro biotechnology techniques including embryo culture, cell culture, pollen culture, callus culture, cell suspension culture, etc., are used to solve problems of storage space, transportation, recalcitrant nature of seeds, fast deterioration and challenges associated with field bank conservation. Other advantages of in-vitro conservation are maintenance of genetic integrity, low labour requirements, stored materials are usually free from pathogen and pest attack, low chances of genetic contamination due to unwanted pollination or physical mixing and absence of catastrophic weather conditions. If shoot tips, embryos, cuttings, bulbs, tubers, recalcitrant seeds, meristem cultures, callus cultures, cell cultures, pollen cultures and other tissue cultures are prepared they can be subjected to in-vitro short to medium term or long-term low temperature and humidity-controlled storage conditions that stop growth or reduce growth rate and prevent occurrence of soma-clonal variation employing slow growth, cryopreservation and synthetic seed methods.

Slow Growth

Slow growth germplasm conservation employs physical, chemical and nutrient conditions that reduce mutation rate, cell growth, growth rate of the tissues and where possible stop tissue growth altogether. It eliminates the problem of frequent renewal of culture medium which may lead to errors and losses, thereby raising the degree of safety during conservation, minimising contamination and reduces genetic mutations and somaclonal variation. Slow growth conservation method, also referred to as cold storage, is applicable for storage of explants from tubers, shoots, meristem tips, leaves, seeds (mature and immature embryos) and callus and cell suspension cultures. When a cultured tissue is to be subjected to slow growth conservation by modifying culture medium sugar, mineral and other nutrients concentration, adding osmotically active agents and hormonal inhibitors or other growth retardants to the culture medium and lowering oxygen supply, lighting and storage temperatures, the tissues can be stored for 1 to 15 years (Kwameswara Rao, 2004; Salgotra & Chauhan, 2023). Growth inhibiting chemicals like abscisic acid (ABA), maleic hydrazine and benzyl aminopurine (BAP) are added to the medium and the storage temperatures lowered to between 1 and 5 °C for cold tolerant temperate species and 10–15 °C for tropical and subtropical species which do not tolerate extremely cold conditions (Benelli et al., 2022). Slow growth culture medium is usually changed at intervals of one to two years, depending on the material. Slow growth does not require frequent germination tests and rejuvenation. Slow growth storage has been used for in vitro germplasm conservation for several vegetatively propagated crops like yams, potatoes, strawberry, coffee, bananas, etc.

Cryopreservation

Cryopreservation is a method suitable for long term storage for plant tissues like cells, shoot tips, meristems, ovules, endosperms, somatic and zygotic embryos, callus, pollen, etc. under controlled conditions in liquid or vapour phase nitrogen. It is a cheap, safe, fast and low labour requiring long term conservation method for crops which produce difficult to store recalcitrant seeds or species which do not produce seeds and have to be conserved in form of tissues for indefinite period of time without fear of somaclonal variation, deterioration, any form of genetic changes and genetic erosion in storage (Kaczmarczyk et al., 2008). In cryopreservation cell division and cell metabolism are totally arrested due to application of ultra-low temperatures and cryoprotectant chemicals. Under cryopreservation there is no loss of viability and capability to regenerate. It is suitable for crops like coffee, banana, pineapple, coconut, oil palm, citrus, cassava, potato, sweet potato, strawberry, yam, avocado, mango, palm tree, cocoa.

Successful cryopreservation storage depends on age and type of the tissue, freezing process, storage temperature and concentration of cryoprotectants. Liquid nitrogen at −196 °C or its vapour phase at −160 °C is the cryopreservation storage medium for prolonged periods since it completely arrests cell metabolism and division in materials under storage and there are no fears of genetic instability, somaclonal variation, contamination or human error during subculturing (Salgotra & Chauhan, 2023). In this preservation process the appropriate materials like meristems, shoot apices, root tips, endosperm tissue, leaf tissues, anther, ovules, protoplasts, hypocotyls, callus culture, somatic embryos, zygotic embryos, pollen embryos and cell culture are obtained, washed, sterilised, placed in the culture medium and allowed a brief period of pre-growth to improve the survival rate and eliminate crystallisation in tissues. Growth promoting biochemicals like abscisic acid may also be added and pre-growth is followed by application of cryoprotectants which include glycerol, alcohols, proline, and dimethyl sulphonate to lower the freezing points for the tissues and reduce damage to cells by preventing crystal formation within cells and cooling and subsequently slow cooling is done at a controlled rate to about −40 °C and ultimately the material is frozen by immersing in liquid nitrogen (Quazi et al., 2021). Cryoprotective treatment is more effective for actively growing cells than differentiated tissues because differentiated tissues may suffer damage due to formation of ice crystals which are harmful to cells. This is resolved by vitrification which involves immersion of the cultures in a high concentration of cryoprotectant solution to remove freezable water and prevent crystalised ice formation in the tissues when subjected to cooling (Kwameswara Rao, 2004; Salgotra & Chauhan, 2023). Such treatment is followed by rapid cooling or encapsulation. Through vitrification, solidification of liquid is achieved by elevating viscosity of the liquid during cooling and turning it into a glassy amorphous structure (Fahy et al., 1984). Encapsulation is encasing the explants like shoot tips and somatic embryos in alginate gel to form artificial seed which is dehydrated before cooling for cryopreservation storage. Dehydration is achieved by using compounds like sugars and polyols. Frozen materials for all tissues which have been subjected to pre-growth, cryoprotectant treatment and cooling are stored at temperatures of −70 to −100 °C in liquid nitrogen for a short period in an appropriate container and then the temperature is lowered further to −196 °C for long-term storage in liquid nitrogen or −150°C for long-term storage in vapour form nitrogen (Quazi et al., 2021).

Synthetic seeds and molecular germplasm

Artificial seeds or synthetic seeds is another nonconventional method which has been developed for germplasm conservation, especially for clonally propagated species and those producing recalcitrant seeds or seeds with very poor viability. It is used to conserve forest species, medicinal plants, fruit trees and some vegetable crops. It entails alginate-encapsulation of active shoot tips and somatic embryos in semi-solid material with an artificial seed coat and endosperm to produce disease free propagules referred as beads (Rihan et al., 2017). The beads are packed with the necessary nutrients, hormones and pesticides and are also made desiccation tolerant and are subjected to cold storage and cryopreservation as well. The synthetic seeds can regenerate into plants even after long low temperature storage and are being used for conservation of various plant species like forest species, fruit trees, vegetables and many others.

Another conservation material is molecular germplasm which involves DNA storage. Developments in genomics and genetic engineering has made it possible to conserve genetic resources in form of gene banks having DNA libraries with identified genes and genomic libraries for breeding, genetic and evolutionary studies. These genomic libraries, often referred to as DNA banks or biobanks complement conventional germplasm conservation with the main aim of supporting documentation and DNA sequences for easy access to the conserved materials.

3.7 Types of Germplasm Collections and Gene Banks

Conserved plant germplasm resources are stored under various conditions like working, active, base, core and mini core germplasm collections with durations of short term, medium term and long-term storage. The breeders have working germplasm stocks which are documented, characterised and stored under short term conditions for up to three to five years, and are available for immediate use by the breeders for crop improvement activities (Creech & Reitz, 1971; Esquinas-Alcazar, 1993). Breeders use genes from such materials which include crop wild and weedy relatives, landraces, obsolete cultivars, advanced varieties and any genetically enhanced cultivars to transfer genes into adapted and elite materials for developing improved varieties. Short term storage is at 0–10 °C and relative humidity of 45–50% (Singh et al., 2020). Active collections are under medium term storage conditions of up to 20–25 years and are often used for evaluation, multiplication and distribution. The active germplasm collection orthodox seeds are stored with moisture contents ranging from 7 to 11% or 3 to 8% depending on the species and it is ensured that viability does not go below 65% (Salgotra & Chauhan, 2023). The base germplasm stocks or conserved stocks are under long term storage conditions (Creech and Reitz, 1971), usually duplicated as backup collections in more than one germplasm conservation bank for security reasons. They are stored with low water contents of 3% to 7% under low temperatures of −18 to −20 °C and low humidity for orthodox seeds (Salgotra & Chauhan, 2023; Singh et al., 2020) or stored as tissues under cryopreservation for up to 100 years or more. The backup germplasm collections are supplements to base collections and are insurance against losses due to unexpected catastrophic events. Svalbard Global Seed Vault in Norway is an international backup facility for long term storage for duplicate seeds of genetic resources conserved in various gene banks around the world.

Core collection of a crop species is a condensed germplasm stock having a limited set of accessions adequately representing the genetic diversity of the crop and its wild and weedy relatives with minimum repetitiveness of the genetic diversity and is a subset of accessions, containing most of the genetic diversity (Esquinas-Alcazar, 1993; Egan et al., 2022). A core collection has adequate genetic variation of a crop gene pool such that different accessions in a core collection represent other accessions from particular areas of the world or those with which they share certain characteristics. There should also be genetic and ecological distinctiveness among different accessions in a core collection. The collection has about 70% of the genetic diversity of the base collection and is expected to have 2000–3000 accessions representing about 10% of the size of the base collection (Egan et al., 2022; Jia et al., 2017). Core collections improve the efficiency of germplasm resources management with regard to characterisation, evaluation, maintenance, diversity preservation and make it easy to obtain required information about the diversity in existing collections for access and utilisation of the respective base collections. Construction of a core collection requires teamwork of germplasm curators, plant breeders, molecular scientists, statisticians and other stakeholders and involves clustering the germplasm materials of a given species into groups based on diversity information in regard to origins of accessions, collection sites, protein data, descriptors and biosystematics data, information on observable and non-observable characters, including agronomic data, isozyme data, molecular DNA sequence data and genetic markers and other essential documentation and characterisation information (Egan et al., 2022). Once clustering is achieved a sample of accessions from each cluster is taken for inclusion in the core collection without over-representation of some accessions, with least repetition of the entries and minimum similarities between the entries. Unique accessions in terms of geographic origin, genetic or phenotypic features and rare variants are given special attention for inclusion. Mini core collection, which comprises 1% of the entire base collection is derived from core collection and is constructed to adequately represent the diversity of the base collection (Jia et al., 2017). It is constructed from the core collection following evaluations by selecting 10% of the core collection for more efficient management of the germplasm collection.

Gene banks are facilities for preservation of germplasm resources and may be in form of field gene banks, tissue banks, cryo-banks and seed banks. Genes are conserved in the field gene banks by growing germplasm resources out in the field as botanic gardens for ex-situ conservation or such banks may be protected reserves and parks for in-situ conservation and farmers' fields. In cryo-bank conservation the embryos, cells, synthetic seeds, anther or pollen and other tissues are subjected to cryopreservation storage. Tissue bank is for preservation of seedless plant tissues and others which produce recalcitrant seeds and is used for storage of buds, roots, meristems and other tissues under low temperatures. In seed banks adequately dried seeds having low moisture contents are stored at low temperatures. While orthodox seeds may be stored for long time in seed banks sub-orthodox seeds store in gene banks for short periods. There are also DNA banks or biobanks for DNA storage with single genes and genomic fragments which are helpful for difficult-to-conserve materials or threatened species. In DNA banks the materials can be stored up to two years at −20 °C or at −70 °C using liquid nitrogen for long term storage (Salgotra & Chauhan, 2023). A number of DNA banks have been established in countries like USA, United Kingdom and Australia.

Problems

3.1 Which are the different categories of germplasm resources that are available for breeder use?

3.2 Which activities are responsible for genetic erosion and loss of biodiversity?

3.3 Why are breeders and modern farmers considered as major contributors towards decline in germplasm resources?

3.4 How can genetically modified crops contribute to genetic erosion?

3.5 Why is conservation and maintenance of genetic diversity of crop plants important for food security?

3.6 How does climate change contribute to genetic erosion?

3.7 Design germplasm exploration and collection for crop plants in centres of diversity

3.8 Which features are noted during germplasm characterisation and evaluation?

3.9 Distinguish between core collection mini-core collection.

3.10 How does on-farm conservation help in modern plant breeding?

3.11 Distinguish between slow growth and cryopreservation germplasm conservation.

3.12 What are orthodox and recalcitrant seeds?

3.13 Distinguish between in-situ conservation and ex-situ conservation.

References

Ayiecho, P. O., & Nyabundi, J. O. (1994). Screening of sorghum cultivars for adaptability to dryland conditions of Kenya. In A. Blum, P. O. Ayiecho, J. O. Nyabundi, H. H. Geiger, & B. Haussmann (Eds.), *Sorghum Improvement for Semi-Arid Areas of Kenya*. Third Year Annual Report for Third Year, 1993.

Benelli, C., Tarraf, W., Izgu, T., & De Carlo, A. (2022). In vitro conservation through slow growth storage technique of fruit species: An overview of the last 10 years. *Plants, 11*(23), 3188. https://doi.org/10.3390/plants11233188

Creech, J. L., & Reitz, L. P. (1971). Plant germ plasm now and for tomorrow. *Advances in Agronomy, 23*, 1–49. https://doi.org/10.1016/S0065-2113(08)60149-2

Egan, L. M., Conaty, W. C., & Stiller, W. N. (2022). Core collections: is there any value for cotton breeding? *Frontiers in Plant Science, 13*, 895155. https://doi.org/10.3389/fpls.2022.895155

Esquinas-Alcazar, J. T. (1993). Plant genetic resources. In M. D. Hayward, N. O. Bosemark, & I. Ramagosa (Eds.), *Plant Breeding: Principles and Prospects*. Springer Dordrecht. https://doi.org/10.1007/978-94-011-1524-7_4

Fahy, G. M., MacFarlane, D. R., Angell, C. A., & Meryman, H. T. (1984). Vitrification as an approach to cryopreservation. *Cryobiology, 21*, 407–426. https://doi.org/10.1016/0011-2240(84)90079-8

Halewood, M., Noriega, I. L., Eliis, D., Roa, C., Rouard, M., & Hamilton, R. S. (2018). Using genomic sequence information to increase conservation and sustainable use of crop diversity and benefit sharing. *Biopreservation and Biobanking, 16*, 268–276. https://doi.org/10.1089/bio.2018.0043

Jia J., Li, H., Zhang, X., Li, Z., & Qui, L. (2017). Genomics-based plant germplasm research (GPGR). *The Crop Journal, 5*, 166–174. https://doi.org/10.1016/j.cj.2016.10.006

Kaczmarczyk, A., Shvachko, N., Lupysheva, Y., Hajrezaci, M. R., & Keller, E. R. J. (2008). Influence altering temperature preculture on cryopreservation results for potato shoot tips. *Plant Cell Reports, 27*, 1551–1558. https://doi.org/10.1007/s00299-008-0574-1

Kwameswara Rao, N. (2004). Plant genetic resources: advancing conservation and use of biotechnology. *African Journal of Biotechnology, 3*, 136–145. https://doi.org/10.5897/AJB2004.000-2025

Priyanka, V., Kumar, R., Dhaliwal, I., & Kaushik, P. (2021). Germplasm conservation: instrumental in agricultural biodiversity—a review. *Sustainability, 13*(12), 6743. https://doi.org/10.3390/su13126743

Quazi, S., Golani, T., & Capuzzo, A. M. (2021). Germplasm conservation. In S. Kumar (Ed.), *Endangered plants*. IntechOpen. https://doi.org/10.5772/intechopen.96184

Rihan, H. Z., Kareem, F., El-Mahrouk, M. E., & Fuller, M. P. (2017). Artificial seeds (principles, aspects and applications). *Agronomy, 7*, 71. https://doi.org/10.3390/agronomy7040071

Roberts, R. H. (1973). Predicting the storage life of seeds. *Seed Science and Technology, 1*, 499–514.

Salgotra, R. K., & Chauhan, B. S. (2023). Genetic diversity, conservation and utilisation of plant genetic resources. *Genes, 14*, 174. https://doi.org/10.3390/genes14010174

Singh, K., Gupta, K., Tyagi, V., & Rajkumar, S. (2020). Plant genetic resources in India: Management and utilisation. *Vavilov Journal of Genetics and Breeding, 24*, 306–314. https://doi.org/10.18699/VJ20.622

Van der Maesen, L. J. G. (1984). Seed storage, viability and rejuvenation. In J. R. Witcombe & W. J. Erskine (Eds.), *Genetic resources and their exploitation—chickpeas, faba beans and lentils. Advances in agricultural biotechnology*, Vol. 6. Springer, Dordrecht. https://doi.org/10.1007/978-94-009-6131-9_2

Williams, K. A. (2023). Planning an exploration for plant genetic resources to be incorporated into the USDA National Plant Germplasm System. In G. M. Volk (Ed.), *Fundamentals of Plant Genebanking*. Colorado State University.

Witcombe, J. R., & Erskine, W. J. (Eds) (1984). *Genetic resources and their exploitation—chickpeas, faba beans and lentils. Advances in agricultural biotechnology*, Vol. 6. Springer. https://doi.org/10.1007/978-94-009-6131-9

Chapter 4
Crop Plant Reproductive Systems: Sex Expression and Pollination Control

Successful crop improvement through breeding depends on adequate knowledge of crop plant reproductive systems. The methods of handling breeding materials take into account whether the crop is sexually or asexually reproducing or both and whether it is cross-pollinating or self-pollinating and the occurrence of incompatibility and male sterility. Sexual reproduction or amphimixis involves fusion of male and female gametes formed through normal meiosis giving a zygote within the flower which develops into botanical seed while in apomixis sexual reproduction is replaced by asexual reproduction in the floral parts and there is no fusion of male and female gametes. Vegetative reproduction takes place by means of vegetative organs and tissues.

4.1 Sexual Reproduction

4.1.1 Sex Expression in Plants

Sexual reproduction is the most common mode of propagation in higher plants and it is through sexual reproduction that genetic recombination is achieved and the genes from the female and male parents are brought together in a progeny. Sex expression is an established feature in higher plants and its classification in plants can be based on the individual flowers, individual plants and plant populations.

There may be a situation of dicliny or unisexuality where a plant is either male (androecious) by having only staminate flowers or female (gynoecious) by having only pistillate flowers. The flowers can also be diclinous since a flower may be staminate or androecious by having only stamens or pistillate (gynoecious or carpellate) if it has carpels only. The carpel comprises of the stigma, style and ovules. The unisexual condition is also referred to as dioecy. That is, dioecy is a situation in plant species where a plant is either androecious or gynoecious as in papaya, black pepper, hemp, yam, spinach, hop, Asparagus. There are also cases of andromonoecious plants

P. O. Ayiecho and J. O. Nyabundi, *Conventional and Contemporary Practices of Plant Breeding*, https://doi.org/10.1007/978-3-031-74998-8_4

which are bisexual and each plant carries both hermaphrodite and staminate flowers and gynomonoecious plants which are also bisexual by having both hermaphrodite and pistillate flowers on it. A plant or a flower may be a hermaphrodite by having both stamens and carpel(s) on it. A hermaphrodite plant has hermaphrodite flowers only while a monoecious plant is hermaphrodite and bears on it both staminate and pistilate flowers but in separate positions as in maize. In monoecism all plants of the monoecious plant group are of the same genetic constitution as relates to sex expression. Plant populations can also be classified according to sex expression of the plants in the populations. A hermaphrodite population has only hermaphrodite plants in it while monoecious plant population comprises of monoecious plants only as a maize population. A dioecious plant population has only dioecious plants while androdioecious plant population has both hermaphrodite and androecious plants. A gynodioecious plant population is comprised of both hermaphrodite and gynoecious plants.

Plant sexuality is controlled by both the genotype and environmental conditions like temperatures, day length, light intensity, humidity, nitrogen and water stress (Lai et al., 2018) through balance of growth regulators like auxins and ethylene which can modify sex expression as known in the normally monoecious Cucurbitaceae species (Kocyan et al., 2007). Plants of this family occasionally have hermaphrodite flowers depending on the auxin-gibberellin balance. In cucumber temperatures and photoperiodism impact on sex expression in terms of female/male ratio (Lai et al., 2018) and in squash high temperatures and long days lead to male flower formation while cold conditions and short days promote female flower formation. Ethylene also promotes femaleness at the expense of maleness in plants. In certain plant species there are genes which influence sex expression as for papaya and asparagus in which sex expression is controlled by single gene systems. In some plant species stamen development is suppressed by nuclear genes or cytoplasmic genes leading to gynodioecy where plant populations are comprised of hermaphrodite and gynoecious plants. There are also genes known to suppress the development of pistil or pistillate flowers.

4.1.2 Pollination System and Pollination Control

In sexually reproducing plants pollination and fertilisation within the flower causes union of male and female gametes to form the zygotes which develop into plants following successful seed development and germination. Crop plants are classified as self-pollinating (autogamous) or cross-pollinating (allogamous), based on the amount of pollen flow among plants (Tables 4.1 and 4.2). There are some basically self-pollinating plant species in which cross-pollination takes place reaching about 30%, sometimes getting to 50%, which are referred to as often-cross-pollinated or partially outcrossing. Some often-cross-pollinated crops include cotton, sorghum and grain amaranths (Tables 4.1). The rate of outcrossing in such species is affected by factors like genotype, temperature, humidity, wind, location and abundance of pollination insects or wind (Agong & Ayiecho, 1991). Outcrossing rate is also dependent on flowering time of the populations, genetic compatibility between the populations, pollen production, floral morphology and time of planting (Shivrain et al., 2009). Cross-pollination can take place between compatible self-pollinating populations if the times of their flowering overlap and they are within distances which pollen can reach. Degree of outcrossing determines how easy it is to make crosses between different plants, which is an essential activity in crop plant improvement. All the plants that reproduce by self-pollination have hermaphroditic flowers. Among bisexual plants there may be homogamous or dichogamous plants, a classification based on temporal separation of processes within the sexual organs. In homogamous plants pollen and stigma mature at the same time and pollen is shed when the stigma is ready to receive it, usually leading to autogamy as in wheat, rice, potato and tomato. In dichogamy the stamens and pistils of hermaphrodite flowers mature at different times and therefore, the pollen is shed by the anther at different time when the stigma is mature enough to receive the pollen, necessitating allogamy. There are two forms dichogamy, namely, protogyny and protandry. In protogynous plants the stigma matures and becomes receptive before the anthers are mature enough to shed pollen in the same flower, e.g., pearl millet, while in protandrous plants the anthers mature and shed pollen before the stigma matures in the same flower as in sunflower, rose, carrots, raspberry, red clover, maize and sugar beet. There may be some amount of self-pollination if there is overlap in maturity of anthers and stigma within the flower.

Some plant species are hercogamous by having various morphological and physiological floral features controlled by both the genotype and the environment which either promote autogamy or prevent autogamy in potentially self-fertilising species, suggesting that homogamous plants do not necessarily undergo self-fertilisation. Features which enforce or promote self-fertilisation include cleistogamy (flowers do not open at all) and homogamy while dioecism, heteromorphism, physiological hercogamy, dicliny or unisexuality as in cucumber, water melon, pumpkin, squash and asparagus, enforce allogamy. In cleistogamous plants pollen is shed when the flower is closed and the stigma is also receptive, necessitating autogamy as in wheat, barley, field bean, oats and oat grass, triticale, Rhodes grass, cowpea pea, groundnut, lettuce, rice. In many legume species flowers open but the androecium and gynoecium remain enclosed within the petals. In chasmogamy the pollen is shed as the flower opens when the stigma is also receptive leading to autogamy and chances of allogamy as known in cotton, sorghum, rose and eggplant and some cases of wheat, barley and rice. In certain cases, the mature receptive stigma elongates between the mature stamens enforcing self-pollination. In tomato and brinjal close proximity of the anther and the stigma also promote self-pollination. There is geitonogamy feature in which the pollen from a flower of a particular plant falls on the stigma of other flowers of the same plant as in maize, effecting self-pollination. There are diclinous or unisexual flower conditions which promote cross-pollination but do not preclude self-pollination like situations where the staminate and pistilate flowers are borne separately on the same inflorescence as in banana, mango, cassava, coconut and strawberry or separately on the same plant (monoecism) as for maize, gourds and cucumber. Dioecism enforces complete and obligatory allogamy since male and female flowers are borne on different plants making cross-pollination indispensable, e.g., papaya, spinach, asparagus, yam, date palm, hops and hemp. Heteromorphism, a situation in which flowers of a given plant species are of two or three morphological types, is usually associated with self-incompatibility in some species where it leads to cross-pollination. Flowers may be dimorphic or trimorphic due to heterostyly and heteroanthy which refer to different style lengths and different anther positions respectively. In certain species the stigma projects above the stamens promoting cross-pollination. Physiological hercogamy in plants takes the form of self-incompatibility and male sterility which enforce obligatory cross-pollination.

Autogamous Crops

In autogamous crops there are features which ensure at least 95% self-pollination and less than 5% outcrossing rate. In some autogamous crops like beans, wheat, cowpea, etc., natural outcrossing rate ranges between 0 and 10%. A plant population qualifies as self-pollinating (Table 4.1) if it has outcrossing rate not exceeding 10%. Self-pollination leads to and sustains high level of homozygousity in autogamous crops, restricts genetic recombination and generation of variation. Occasional cross-pollination or occasional mutations occur in such species resulting heterozygousity which disappears rapidly due to continuous natural self-pollination, leading to segregation and exposure of deleterious recessive genes for elimination from the populations by selection. The consequence of continued self-pollination and elimination of any deleterious alleles is homozygous pure-lines with few or no deleterious genes in autogamous species populations, and therefore, self-pollination or any form of inbreeding does not lead to inbreeding depression. In the autogamous crops the amount of natural outcrossing is so low that no long-distance isolation or other form of protection is necessary against cross-pollination in breeding programmes. Breeding methods in autogamous crops are designed for developing of superior pure line varieties.

Partially Cross-Pollinated Crops

In certain crop species cross-pollination often exceeds 10% and may get to 30% and occasionally reaching 50% in some of them. Such species are generally referred to as often cross-pollinated or partially cross-pollinated and includes crops like brinjal, okra, chilli, grain amaranths, cotton, sorghum, sweet pepper. Such crops are enlisted in Table 4.1 along with normally self-pollinated crops.

Table 4.1 Some self-pollinated crops

Common name	Family	Genus	Species	Chromosome number		Propagation means sd/vg
				X	2n	
Wheat	Gramineae/Poaceae	*Triticum*	*aestivum (vulgare)*	7	42	sd
Rice	Gramineae/Poaceae	*Oryza*	*sativa*	12	24	sd
Africa rice	Gramineae/Poaceae	*Oryza*	*glaberrima*	12	24	sd
Barley	Gramineae/Poaceae	*Hordeum*	*vulgare*	7	14	sd
Oats	Gramineae/Poaceae	*Avena*	*sativa*	7	42	sd
Foxtail millet	Gramineae/Poaceae	*Setaria*	*italica*	9	18	sd
Cultivated einkorn	Gramineae/Poaceae	*Triticum*	*monococcum*	7	14	sd
Emmer wheat	Gramineae/Poaceae	*Triticum*	*dicoccum*	7	28	sd
Durum wheat	Gramineae/Poaceae	*Triticum*	*durum (turgidum)*	7	28	sd
Finger millet	Gramineae/Poaceae	*Eleusine*	*coracana*	9	36	sd
Sorghum*	Gramineae/Poaceae	*Sorghum*	*bicolor*	5	20	sd
Garden Pea	Leguminosae/Fabaceae	*Pisum*	*sativum*	7	14	sd
Groundnut	Leguminosae/Fabaceae	*Arachis*	*hypogea*	10	40	sd
Chick pea	Leguminosae/Fabaceae	*Cicer*	*arietinum*	8	16	sd
Green gram/mung bean	Leguminosae/ Fabaceae	*Vigna*	*radiata*	11	22	sd
Cowpea	Leguminosae/Fabaceae	*Vigna*	*unguiculata*	11	22	sd
Soybean	Leguminosae/Fabaceae	*Glycine*	*max*	10	40	sd
Black gram	Leguminosae/Fabaceae	*Phaseolus*	*mungo*	12	24	sd
Lima bean	Leguminosae/Fabaceae	*Phaseolus*	*lunatus*	11	22	sd
Dolichos bean	Leguminosae/Fabaceae	*Dolichos*	*lablab*	11, 12	22, 24	sd
Lentil	Leguminosae/Fabaceae	*Lens*	*esculenta*	7	14	sd
Field bean/French bean/dry bean	Leguminosae/Fabaceae	*Phaseolus*	*vulgaris*	11	22	sd
Broad bean*	Leguminosae/Fabaceae	*Vicia*	*faba*	6	12	sd
Pigeon pea*	Leguminosae/Fabaceae	*Cajanus*	*cajan*	11	22	sd
Coffee*	Rubiaceae	*Coffea*	*arabica*	11	44	sd, also vg
Tomato	Solanaceae	*Lycopersicon*	*esculentum*	12	24	sd
Okra*	Solanaceae	*Abelmoschus (Hibiscus)*	*esculentus*	36, 65	72,144, 130	sd
Lettuce	Compositae	*Lactuca*	*sativa*	9	18	sd
Brinjal/egg plant*	Solanaceae	*Solanum*	*melongena*	12	24	sd
Sweet/chilli pepper*	Solanaceae	*Capsicum*	*annuum*	12	24	sd
Bird chilli/hot pepper*	Solanaceae	*Capsicum*	*frutescens*	12	24	sd

(continued)

Table 4.1 (continued)

Common name	Family	Genus	Species	Chromosome number		Propagation means sd/vg
				X	2n	
Potato	Solanaceae	*Solanum*	*tuberosum*	12	48	vg, also sd
Sesame (simsim)	Pedaliaceae	*Sesamum*	*indicum*	13	26	sd
Linseed	Linaceae	*Linum*	*usitatissimum*	15	30	sd
Apricot	Rosaceae	*Prunus*	*armeniaca*	8	16	sd
Cotton*	Malvaceae	*Gossypium*	*herbaceum*	13	26	sd
Cotton*	Malvaceae	*Gossypium*	*arboreum*	13	26	sd
Cotton*	Malvaceae	*Gossypium*	*barbadense*	13	52	sd
Upland cotton*	Malvaceae	*Gossypium*	*hirsutum*	13	52	sd
Clove	Myrtaceae	*Syzygium*	*aromaticum*	11	22	sd
Grain amaranth*	Amaranthaceae	*Amaranthus.*		21, 23	42, 46	sd
Tobacco*	Solanaceae	*Nicotiana*	*tabacum*	12	48	sd
Safflower*	Compositae	*Carthamus*	*tinctorius*	12	24	sd
Triticale (*X Triticosecale*)*	Gramineae/Poaceae	*Triticale*	*hexaploid*	7	42	sd
Peach	Rosaceae	*Prunus*	*persica*	8	16	sd, also vg

* These crops have some amount of natural outcrossing, up to about 30%, hence partially outcrossing and referred to as often cross-pollinated; sd—seed propagated; vg—vegetatively propagated

Allogamous Crops

In the plant kingdom cross-pollination by which pollen is transferred from one plant flower to the stigma of another plant, either unisexually or bisexually, is more prevalent than self-pollination system. Cross-pollination takes place in both unisexual and bisexual flowers with the former being obligate cross pollinators. The characters enforcing cross-pollination which are both morphological and physiological are genetically and environmentally controlled. Cross-pollination needs agents to transfer pollen from the anther to the stigma and such agents may include wind (anemophily), e.g., rye, spinach, maize, sugar beet; insects (entomophily) e.g., cucurbits, brassicas, mango, apple, citrus, sunflower, cotton, cucumber, peach, grape, onion and carrot and water movement (hydrophily). In anemophilous species the pollen is produced in large amounts and light enough to be carried easily by wind while the stigmas have large surface areas to trap the pollen blown by wind. Since hydrophily is connected with water flow it is rarely found among crop plants.

The insect pollinated species are coadapted with relevant insect pollinators such that they are regularly visited by the insects for nectar and pollen and in the process the sticky and heavy pollen adheres to the insect body, to be deposited on stigmas of other flowers during visits to such flowers. The cross-pollinated crop species (Table 4.2) are

Table 4.2 Some cross-pollinated crops

Common name	Family	Genus	Species	Chromosome number		Propagation means sd/vg
				X	2n	
Maize[b]	Gramineae/Poaceae	*Zea*	*mays*	10	20	sd
Rye[c]	Gramineae/Poaceae	*Secale*	*cereale*	7	14	sd
Bulrush (pearl) millet	Gramineae/ Poaceae	*Pennisetum*	*americanum/ typhoides*	7	14	sd
Field mustard or Turnip rape	Cruciferae	*Brassica*	*campestris*	10	20	sd
Sunflower	Compositae	*Helianthus*	*annuus*	17	34	sd
Castor	Euphorbiaceae	*Ricinus*	*communis*	10	20	sd
Niger	Compositae	*Guizotia*	*abyssinica*	15	30	sd
Alfalfa[d]	Leguminosae/Fabaceae	*Medicago*	*sativa*	8	32	sd, also vg
Red clover[c]	Leguminosae/Fabaceae	*Trifolium*	*pratensis*	7	14	sd
White clover[c]	Leguminosae/Fabaceae	*Trifolium*	*repens*	8	32	sd
Perennial ryegrass[d]	Gramineae/Poaceae	*Lolium*	*perenne*	7	14	sd
Johnson grass	Gramineae/Poaceae	*Sorghum*	*halepense*	5	40	sd
Timothy grass[d]	Gramineae/Poaceae	*Phleum*	*pratense*	7	42	sd
Sugarcane	Gramineae/Poaceae	*Saccharum*	*officinarum*	8	80	vg, also sd
Hemp[e]	Cannabaceae	*Cannabis*	*sativa*	10	20	sd
Pyrethrum	Compositae	*Chrysanthemum*	*cinerariifolium*	9	18	vg, also sd
Cabbage[c]	Cruciferae	*Brassica*	*oleracea* var. *capitata*	9	18	sd
Rape	Cruciferae	*Brassica*	*napus*	19	38	sd
Black mustard	Cruciferae	*Brassica*	*nigra*	8	16	sd
Kale	Cruciferae	*Brassica*	*oleracea* var. *acephala*	9	18	sd
Radish	Cruciferae	*Raphanus*	*sativus*	9	18	sd
Carrot	Umbelliferae	*Daucus*	*carota*	9	18	sd
Broccoli/ Cauliflower[c]	Cruciferae	*Brassica*	*oleracea* var. *botrytis*	9	18	sd
Cucumber[b]	Cucurbitaceae	*Cucumis*	*sativus*	7	14	sd
Onion	Alliaceae	*Allium*	*cepa*	8	16	sd
Pumpkin[b]	Cucurbitaceae	*Cucurbita*	*moschata*	10	40	sd
Watermelon[b]	Cucurbitaceae	*Citrullus*	*lanatus*	11	22	sd
Squash[b]	Cucurbitaceae	*Cucurbita*		20	40	sd
Sweet potato[d]	Convolvulaceae	*Ipomea*	*batatas*	15	90	vg, also sd
Sugar beet[d]	Chenopodiaceae	*Beta*	*vulgaris*	9	18	sd
Cashew nut	Anacardiaceae	*Anacardium*	*occidentale*	21	42	sd

(continued)

Table 4.2 (continued)

Common name	Family	Genus	Species	Chromosome number		Propagation means sd/vg
				X	2n	
Brussels sprouts[c]	Cruciferae	*Brassica*	*oleracea* var. *gemmifera*	9	18	sd
Spinach[e]	Amaranthaceae	*Spinacia*	*oleracea*	6	12	sd
Asparagus[e]	Liliaceae	*Asparagus*	*officinalis*	10	20	vg, also sd
Garlic	Alliaceae	*Allium*	*sativum*	8	16	vg
Apple[d]	Rosaceae	*Malus*	*domestica*	17	34	vg
Avocado[d]	Lauraceae	*Persea*	*americana*	12	24	sd
Mango[b, d]	Anacardiaceae	*Mangifera*	*indica*	10	20	sd
Papaya[e]	Caricaceae	*Carica*	*papaya*	9	18	sd
Coconut[b]	Palmeae/Arecaceae	*Cocos*	*nucifera*	16	32	sd
Macadamia	Proteaceae	*Macadamia*	*integrifolia*	14	28	sd
Macadamia	Proteaceae	*Macadamia*	*ternifolia*	14	28	sd
Macadamia	Proteaceae	*Macadamia*	*tetraphylla*	14	28	sd
Banana[b,g]	Musaceae	*Musa*	*sapientum, acuminata, balbisiana*	11	22, 33, 44	vg
Date palm[e]	Palmeae/Arecaceae	*Phoenix*	*dactylifera*	18	36	sd
Fig[f,g]	Moraceae	*Ficus*	*carica*	13	26	sd
Guava	Myrtaceae	*Psidium*	*guajava*	11	22	sd
Lime	Rutaceae	*Citrus*	*aurantifolia*	9	18	sd
Seville orange	Rutaceae	*Citrus*	*aurantium*	9	18	sd
Pumelo	Rutaceae	*Citrus*	*grandis*	9	18	sd
Grape fruit	Rutaceae	*Citrus*	*paradisi*	9	18	sd
Mandarin (Tangerine)	Rutaceae	*Citrus*	*reticulata*	9	18	sd
Sweet orange	Rutaceae	*Citrus*	*sinensis*	9	18	sd
Lemon	Rutaceae	*Citrus*	*limon*	9	18	sd
Tea	Theaceae	*Camellia*	*sinensis*	15	30	sd
Cocoa	Malvaceae	*Theobroma*	*cacao*	10	20	sd
Grapes[b]	Vitaceae	*Vitis*	*vinifera*	19	38	vg
Plum[d]	Rosaceae	*Prunus*	*divaricate/cerasifera*	8	16	vg, also sd
Almond[c]	Rosaceae	*Prunus*	*dulcis*	8	16	vg, also sd
Pear	Rosaceae	*Pyrus*	*communis*	17	34	vg
Strawberries[b]	Rosaceae	*Fragaria*	*ananassa*	28	56	vg, also sd
Coriander	Apiaceae/ Umbelliferaeae	*Coriandrum*	*sativum*	11	22	sd

[b] monoecious or monoecious cultivars exist
[c] self-incompatibility common
[d] some occurrence of self-incompatibility
[e] dioecious
[f] mainly dioecious
[g] parthenocarpic
sd—seed propagated
vg—vegetatively propagated

highly heterozygous and susceptible to inbreeding depression and loss of vigour if subjected to self-pollination and other forms of inbreeding. Cross-pollination restores and maintains vigour and heterosis in such species. The breeding methods in cross pollinating crop species are designed to develop varieties which exploit heterosis and maintained through development of hybrid and synthetic varieties and other forms of open pollinated varieties.

4.1.3 Self-incompatibility

Self-incompatibility is a physiological mechanism that enforces obligatory cross-pollination in hermaphroditic and some monoecious plants by preventing self-fertilisation or cross-fertilisation of plants having similar incompatibility genes with consequent high levels of genetic heterozygousity. Self-incompatibility is characterised by failure of a hermaphroditic plant having functional gametes to set seeds when self-pollinated or crossed with other plants having similar incompatibility alleles as known in a number of crops, e.g., cabbage, cauliflower, brussels sprouts, almond, white clover, red clover, mustard and some root crops. It is manifested by the failure of viable pollen grains to germinate or the pollen tube to grow and penetrate the style, and therefore self-fertilisation is prevented by the interaction between pollen and the stigma or pollen tube and stylar tissue. Self-incompatibility is a widespread phenomenon in higher plants and is known to occur in more than 100 families of the angiosperm species (Munoz-Sanz et al., 2020). Incompatibility response in some plants like *Brassica* species, especially gametophytic self-incompatibility, is based on recognition of compounds on the pollen grain or pollen tube surface by the stigmatic surface or stylar tissue.

Genetics of Incompatibility

There are two types of incompatibility systems controlled by major genes, homomorphic system and heteromorphic system. In the homomorphic system there is no morphological variation between the incompatible and compatible genotypes while in the heteromorphic incompatibility system there are distinct differences in flower morphology between compatible and incompatible genotypes, particularly the style, stamens and pollen grains. There are two types of homomorphic incompatibility distinguished by genotypes that determine the pollen reaction or incompatibility phenotype, namely, gametophytic self-incompatibility and sporophytic self-incompatibility. In both gametophytic and sporophytic self-incompatibility types the incompatibility phenotypes of the diploid stigma or the diploid style are genotypically determined by stigma/style (sporophyte) genotype, one or more multi-allelic loci are involved and the incompatibility alleles interact in various ways to give complex patterns of incompatibility reactions. The expressivity of the incompatibility genotypes varies with the environment and there may be variable dominance relationships among the alleles with some alleles known to be more dominant than others, especially in the sporophytic case.

Gametophytic Incompatibility

Gametophytic self-incompatibility system is the highly prevalent and is known in several genera like *Lycopersicon, Nicotiana, Solanum, Petunia, Trifolium, Prunus* and *Brassica.* In this system the reaction of the haploid pollen is determined by the multiallelic genes with no recessive-dominance allele relationship and all plants derived from successful fertilisation are heterozygous at the incompatibility locus. Pollen types with distinct incompatibility reactions are produced by the same plant that is heterozygous at the incompatibility locus/loci. Pollen phenotype is determined by the pollen (gamete) incompatibility genotype while that of the stigma/style is sporophytically determined, hence it is haploid tissue-diploid tissue interaction case. In a one-locus self-incompatibility case at least three alleles are necessary for sustained fertility in a given plant population. Possible reactions for the case involving one self-incompatibility locus, S, (S-locus) with three alleles is illustrated in Table 4.3.

The probability of compatibility in a population is increased if there are more than three alleles at the incompatibility locus and a high number of alleles is important where crop yield is a function of fecundity. Homozygousity at the incompatibility locus is not possible under natural conditions. Two-loci gametophytic incompatibility is known to occur in some plant families and is more unstable than the one locus case since there is a lower probability of incompatibility.

In gametophytic incompatibility there is interaction of the haploid pollen and the diploid style/stigma and only when the S-locus allele of the pollen is not similar to any S-locus allele in the in the stigma or style will fertilisation succeed, otherwise the antagonistic reaction between the like alleles of the incompatible gene in the pollen and pistil prohibits fertilisation. The genes responsible for self-incompatibility manifestation encode the phytochemicals which act as male and female specificity determinants and interact to cause self- or non-self-discrimination when pollination occurs leading to fertilisation or no fertilisation (Claessen et al., 2019; Munoz-Sanz et

Table 4.3 Possible interactions between the stigma/style and the pollen grains for a one locus-three allele case*

Stigma/style genotype	Pollen genotype		
	S_a	S_b	S_c
$S_a S_b$	I	I	C
$S_a S_c$	I	C	I
$S_b S_c$	C	I	I

*S_a, S_b and S_c are the alleles at the incompatibility S-locus; C = compatible, I = incompatible

al., 2020). The proteins synthesised in the stigmatic/style tissues act as female determinants while the proteins produced in the anther tissues and deposited on pollen coat acts as male determinants. That is, the S-locus has haplotype system in which there are genes controlling the stigma/style function while other genes control the pollen function. The interaction of the two determinants from the same haplotypes causes rejection of self-pollen and inhibition of pollen germination, hence, fertilisation fails when the haploid pollen S-locus haplotype is similar to either of the haplotypes in the diploid stigma/style with reaction being based on allele specific recognition proteins. Molecular elucidation of gametophytic self-incompatibility expression done in Brassicas, some members of Solanaceae, Rosaceae and Papaveraceae confirmed that the S-locus has haplotype system in which one gene controls the stigma function while one or more genes control the pollen function. For example, in Brassicas serine/threonine kinase synthesised in stigmatic tissues acts as female determinant and a cysteine rich protein produced in the anther tissues and deposited on pollen coat acts as male determinant (Iwano & Takayama, 2012). In Solanaceae the ribonuclease (RNase) synthesised in the style tissues acts to prevent the growth of the pollen tube down the style (Tovar-Mendez & McClure, 2016). Self-pollination triggers the reactions in the incompatible pollen tube so that incompatibility is manifested by degradation of cellular RNA of the germinating pollen tubes in the style by RNase which acts as the female determinant thereby preventing pollen tube growth (Claessen et al., 2019). The S-locus gene expressed in the pistil encodes RNase while the pollen determinant genes which comprise multiple F-box genes at the S-locus, called S-locus F-box brothers, are responsible for encoding recognition proteins (Claessen et al., 2019; Sassa et al., 2007). In addition to the action of RNase on the pollen tube preventing fertilisation, there may also be death of the pollen tube cells (programmed cell death) inhibiting pollen tube growth as has been shown in *Pyrus* by (Liu et al., 2007). The S-locus genes are inherited as a unit without segregation to maintain the functions of the incompatibility mechanism intact (Roalson & McClubbin, 2003).

Sporophytic Incompatibility

There are two types of sporophytic self-incompatibility based on flower morphology differences, homomorphic self-incompatibility and heteromorphic self-incompatibility.

Homomorphic Sporophytic Self-incompatibility

Homomorphic sporophytic self-incompatibility is more common than gametophytic self-incompatibility. It is usually governed by single multiple allelic gene in which the alleles may be independent with no dominant-recessive allele relationships or may

Table 4.4 Possible interactions between the stigma/style and the pollen for a one S-locus case with no dominance-recessive relationship

Stigma/style genotype	Anther tissue (pollen) genotype			
	S_aS_b, (S_a, S_b)	S_aS_c, (S_a, S_c)	S_bS_c, (S_b, S_c)	S_cS_d, (S_c, S_d)
S_aS_b	I	I	I	C
S_aS_c	I	I	I	I
S_bS_d	I	C	I	I
S_cS_d	C	I	I	I

*S_a, S_b, S_c and S_d are the alleles at the incompatibility S-locus; C = compatible, I = incompatible; pollen genotypes are given in parentheses

have dominant-recessive allele relationships in which the behaviour of the gametophyte is controlled by dominant allele of the genotype for heterozygous sporophyte bearing the gamete. The self-incompatibility phenotype of the haploid pollen is determined by the genotype of anther wall tissue, hence determined by the genotype of the sporophyte bearing the anther. The behaviour of the stigma is also determined by the genotype of the sporophyte bearing the stigma. That is, the interaction between the pollen grain and the stigma is a diploid tissue-diploid tissue interaction case because it is based on the genotypes of the sporophytes bearing the stigma and the anther and all the pollen grains from a plant heterozygous at the incompatibility locus react similarly irrespective of the pollen genotype (Table 4.4).

In situations where there are dominant-recessive allele relationships self-incompatibility reactions may be complicated by different allelic pair relations in the pollen and stigma, the presence of several alleles with variable dominance where some alleles may be dominant over others while others may be independent of some other alleles.

Heteromorphic Incompatibility—Heterostyly/Heteroanthy

Heteromorphic self-incompatibility system is well established in *Primula* and *Lythrum.* It is characterised by differences in the morphology of flowers from different plants with dissimilar genotypes for self-incompatibility genes, a condition known as heterostyly because of different style lengths associated with it or heteroanthy where stamens or anthers differ morphologically. Heteromorphic species are mainly dimorphic with a few being trimorphic, where individual plants within a species differ in floral morphological and physiological features. Fertilisation occurs between flowers having different morphologies with different style lengths and different stamen lengths only. There are two known types of heterostyly, distyly and tristyly, which are controlled by major genes.

Table 4.5 Interactions between stigma/style and pollen in *Primula* heteromorphic incompatibility system

Stigma/style genotype	Pollen source: thrum (Ss) pollen, i.e. S and s	Pollen source: pin (ss) pollen, i.e., s
Ss (Thrum)	Incompatible	Compatible
ss (Pin)	Compatible	Incompatible

S and s are the alleles at the incompatibility locus

Distyly

In *Primula* heteromorphic incompatibility is associated with distyly leading to two types of plants, thrum and pin type plants. Thrum type plants have short styles and long stamens and large pollen grains while pin type plants have long styles and short stamens and small pollen grains. These morphological character combinations preclude the chances of self-pollination of pin x pin and thrum x thrum. The morphological and biochemical traits related to incompatibly reactions are associated and combine to prevent self-fertilisation or any pin x pin and thrum x thrum fertilisation while allowing only thrum x pin crosses. The classical belief has been that system is controlled by a single Mendelian gene, with alleles S and s, at the incompatibility locus such that plants with genotype Ss are thrum while ss genotype is for pin type plants with allele S being dominant. Pollen incompatibility phenotype is determined by the sporophyte genotype such that pollen from thrum type plant, Ss, react similarly, i.e., S and s pollen grains from Ss genotype plants behave similarly during fertilisation while all the s pollen from the pin type plants, ss, also behave similarly (Table 4.5).

Recent studies give a more complicated picture for genetic control of distyly and incompatibility in *Primula*. At the incompatibility S-locus in *Primula forbesii* there is a hemizygous supergene having at least five tightly linked genes, one of which controls style elongation and female compatibility reaction and another one is responsible for anther height and pollen morphology (Huu et al., 2016; Jia et al., 2023).

Tristyly

In *Lythrum junceum* where the different incompatibility genotypes have different style lengths three types of flowers exist as determined by stylar length (Dulberger, 1970) which is governed by two epistatic major genes, M and S. Dominant S is epistatic to the M locus alleles and M allele is also a dominant allele. All plants that are homozygous recessive at the two loci have long styles while those homozygous recessive at the S locus but heterozygous or homozygous dominant at M locus have medium style lengths. All other genotypic combinations give short styles. The only compatible combinations are long style × medium style, long style × short style and medium style × short style as determined by morphological and biochemical features.

Overcoming of Self-incompatibility

Where the phenomenon of self-incompatibility exists, it poses a number of challenges in plant breeding and crop production. Self-incompatibility makes inbreeding impossible and consequently, production and maintenance of homozygous lines in hybrid breeding programmes becomes difficult. The presence of self-incompatibility genes in crop populations may lower level of production where crop yield is a function of effective pollination and botanical seed or fruit production. It may also affect crop quality, especially if incompatibility genes are linked to some undesirable genes. In fruit tree production genotypes with different incompatibility alleles have to be inter-planted for effective pollination and productivity.

A number of methods can be employed to create permanent and heritable changes or temporary non-heritable changes to eliminate self-incompatibility. Many self-incompatibility cases are major gene controlled and can be reversed through induced point mutations. Incorporation of alleles and induction of mutations that restore compatibility and induction of polyploidy help to break self-incompatibility by creating permanent heritable changes responsible for self-compatibility phenotype. Induction of polyploidy to overcome self-incompatibility is based on the notion that the tetraploid relatives or artificially induced tetraploids of diploid species having gametophytic mono-factorial self-incompatibility usually display self-compatibility phenotype due to loss of incompatible reaction (Hojsgaard & Horandl, 2019; Nettancourt, 1993). Careful and thorough selection in heterogeneous population can also help identify rare compatibility alleles. The processes which effect temporary non-heritable solutions to self-incompatibility include irradiation of pollen grains prior to pollination, bud pollination in early bud stage, delayed pollination in cases where incompatibility reaction weakens with age, physical or chemical pre-treatment of the pistil like heat treatment or application of chemicals like hexane, surgical removal of stigma where the incompatibility reaction is restricted to the stigmatic surface and mixing of inviable compatible and viable incompatible pollen for use in pollination. If killed compatible pollen grains are used as mentor pollen in a mixture with viable incompatible pollen the dead compatible pollen induces incompatible pollen tube growth to effect fertilisation when such a mixture is applied to the stigma (Knox et al., 1972; Nettancourt, 1993). Bud pollination is applicable where chemicals responsible for self-incompatibility manifestation become active in the mature flower but not in bud stage. Treatment of the pistil with high temperatures, radiations, or appropriate chemicals before pollination overcomes incompatibility by interfering with physiological processes like protein synthesis linked to incompatibility reaction.

4.1.4 Male Sterility

Male sterility is the condition in which either no pollen is formed or the formed pollen is inviable, the plant may lack the stamens or anthers and some plants may have functional pollen grains which are not released because the anthers do not release them. Therefore, in a monoecious or in a hermaphrodite plant male sterility is attributable to irregularities in the androecium which may take the form of pollen sterility due to lack of viable pollen grains, structural or staminal male sterility where the androecium is sexually reversed, malformed, non-functional or absent and situations where the plant is male sterile yet it has functional pollen grains which remain enclosed in indehiscent anthers. In situations where the anthers have functional pollen but their walls do not dehisce to release the pollen grains the plant is effectively male sterile.

There are situations in which there is aberrant androecium development and the plant ends up having no stamen or no anther or it may have only rudimentary stamens and anthers. Faulty stamen differentiation and development in sorghum, rice, pea and cucumber is known to be influenced by single Mendelian gene in each case with the sterility alleles being recessive. There are cases of sex reversion where the anthers develop into stigma and style with ovules that can develop into seeds on fertilisation. Plasmon-genome interactions are known to cause irregularities in floral development resulting in sex reversion, particularly in cases of interspecific crosses. Breakdown in microsporogenesis is characterised by faulty timing of the various stages and abnormal or malfunctional tapetum are other mechanisms responsible for male sterility, particularly pollen sterility. Processes that affect any stage of normal cell division will interfere with microsporogenesis leading to male sterility. Asynapsis or desynapsis of the homologous chromosomes during meiosis I due to sticky chromosomes prevents terminalisation of the chiasmata, interferes with microsporogenesis and prevents formation of cell metaphase plate and the spindle fibre. There may also be spindle divergence in meiosis I or meiosis II and cytokinesis failure.

Genetics of Male Sterility

Male sterility can be classified as nuclear (genetic or genic) male sterility, cytoplasmic male sterility and cytoplasmic-genic male sterility. Male sterility can also be chemically induced by application of gametocides also referred to as chemical hybridising agents (CHA). In some cases, certain environmental factors like temperature, photoperiodism and humidity also have influences in the extent of male sterility expression.

Genic Male Sterility (GMS)

Genic male sterility (GMS) is strictly of nuclear genetic nature and may be due to single locus or multi-loci major genes with recessive alleles being responsible for sterility in most cases. It arises from spontaneously mutations or induced using mutagens like radiations and chemical mutagens and can also be obtained through hybridisation. It is a common natural feature among many crop plants like maize, sorghum, soybean, rice, wheat, beans, cotton, tomato, Lima beans, sesame, barley and several other plants. Dominant male sterility alleles are also known in a number of crops. Genic male sterility may not be under the influence of environmental conditions or may environment-sensitive by virtue of being responsive to changes in the environmental conditions.

Genetic male sterility is usually of sporophytic nature where all the gametes from plants heterozygous for fertility and sterility genes are fertile or sterile depending on which allele is dominant. In major gene sporophytic male sterility, fertility/sterility genotypes of the progenies depend on the genotype of the parents. For example, if an allele denoted as Rf is for male fertility while the allele rf is for sterility with Rf being the dominant allele, male fertility genotypes and phenotypes of the progenies from various crosses will be as follows.

Cross $♀rfrfx♂RfRf$ gives $Rfrf$ F_1 progenies which are all male fertile and F_2 progenies arising from Rfrf x Rfrf mating have a 1 sterile : 3 fertile segregation ratio, i.e., $1rfrfand2Rfrf + 1RfRf$.

There are cases of gametophytic genic male sterility where the pollen phenotype is determined by the pollen genotype making plants heterozygous for sterility genes have 50% male fertility and 50% male sterility. Genic male sterility is expressed through precocious or delayed programmed cell death in the tapetum, asynapsis, desynapsis or failure of homologous chromosomes to separate in anaphase I resulting in failure in microsporogenesis (Farinati et al., 2023). Control of genic male sterility genes involves transcription factors which bind directly to promoters of target genes for expression of the sterility genes (Abbas et al., 2021) and in some cases there is increased methylation on certain gene promoters leading to reduced transcription and precocious programmed cell death in developing anthers (De Guzman & Oard, 2019). There may also be coordinated regulation where different transcription factors interact with each other and synergistically regulate the expression of target sterility genes or some transcription factors may indirectly regulate other transcription factors which in turn activate the expression of downstream sterility genes (Abbas et al., 2021). The sterility genes may also interfere with the plant male reproductive tissues through disruption of lipid and sugar metabolism during anther and pollen development leading to failure of proper anther cuticle and pollen formation. Genetic male sterility manifestation may also be in other forms like absence of stamens and anthers or rudimentary forms of these structures and sex reversion.

Cytoplasmic Male Sterility (CMS)

In cytoplasmic male sterility there are cytoplasmic genes associated with the mitochondria which are responsible for sterility and crosses between a male sterile plant and a male fertile give male sterile progeny only. These mitochondrial genes are not paired as alleles for nuclear genes and are inherited through the female parent only. If S is the cytoplasmic factor responsible for fertility while s is the corresponding cytoplasmic factor responsible for sterility, there are only two cytoplasmic genotypes with the corresponding phenotypic expression of fertile cytoplasm and sterile cytoplasm respectively. The cross between male sterile and male fertile parents, i.e., ♀ s x ♂ S, gives progenies which are all sterile because all of them have s in the cytoplasm which they have inherited from the male-sterile female parent. The sterile s line can be maintained by having its fertile version to which it is crossed. This kind of sterility is commercially exploitable in vegetable crops where there is no need for seed yield in the commercial hybrid as in onions, cauliflower, cabbage or ornamental crops. Cytoplasmic male sterility is well established in several crops like maize, sorghum, pearl millet, sunflower, rice, wheat, soybean, onion, tomato, sugar beet, pepper and others. Some known cytoplasm associated with male sterility include Texas cytoplasm (T-cytoplasm), C-cytoplasm and S-cytoplasm in maize, ogu-cytoplasm in rapeseed, PET1-cytoplasm in sunflower, S-cytoplasm in onion and *timophevi*-cytoplasm (T-cytoplasm) in wheat (Zurek, 2023). Others are D1-ctoplasm and BT-cytoplasm in rice, milo-cytoplasm, VZM-cytoplasm and A3cytoplsm in sorghum, A2 and A3 cytoplasm in pigeon pea. Cytoplasmic male sterility may be autoplasmic or allospamic where autoplasmic male sterility is used for sterility arising due to mutations in cytoplasmic factors within a species cytoplasm while allospamic cytoplasmic male sterility refers to sterility arising from interspecific hybridisation and caused by unfavourable interaction between the nuclear genome of one species and cytoplasm of the other species interfering with the development of anthers. Cytoplasmic male sterility in products of protoplast fusion also fall in allospamic category. In some plants cytoplasmic male sterility may be associated with certain types of cytoplasm having undesirable factors. In maize some cytoplasmic male sterility genes are associated with T-cytoplasm which also has susceptibility factors to southern corn leaf blight (*Cochliobolus heterostrophus/Bipolaris maydis*).

Cytoplasmic male sterility is usually associated with the synthesis of undesired toxic proteins as a result of the expression of aberrant genes in the mitochondria. The expression of the aberrant mitochondrial genes responsible for cytoplasmic sterility is linked to unusual open reading frames (orf) which are chimeric in structure and are responsible for production of the proteins that disrupt pollen development and production (Kotchoni et al., 2010; Zurek, 2023). Cytoplasmic male sterility orf genes are known in many crop species including maize, rapeseed, sunflower, rice, wheat, cotton, sorghum, soybean, barley, tomato, onion, sugar beet and many other crops. In plant genotypes having this aberrant reproductive condition the mitochondrial genes responsible for male sterility encode the toxic proteins which adversely impact on the plant's male reproductive tissues, particularly the tapetum, by causing mitochondrial malfunction and death of cells resulting in the abortion of the developing gametes (Farinati et al., 2023; Zurek, 2023). The accumulation of the toxic proteins and reactive oxygen species (ROS) disrupts mitochondrial membrane functions and mitochondrial operations, suppresses adenine triphosphate (ATP) synthesis resulting in ATP deficiency and leads to abnormal cell death in the anther tissues (Melonek et al., 2021). The pollen mother cell development may be disrupted during meiosis or after meiosis resulting lack of pollen or inviable pollen production. The tapetum is disrupted, undergoes precocious programmed cell death and autolyses with consequent failure to nourish the developing pollen (Farinati et al., 2023; Liu et al., 2023). Maize cultivars with C and T sterility cytoplasm experience the collapse of tapetum cells preventing pollen release.

Cytoplasmic-Genic Sterility (CGMS)

Cytoplasmic-genic male sterility (CGMS) is a consequence of combined action of male sterility genes located in the mitochondria within the cytoplasm and nuclear genes making a plant have male sterility or male fertility phenotype. In the CGMS situation the nuclear gene, Rf, often referred to as the restorer gene, suppresses or downgrades the cytoplasmic male sterility genes to restore viable pollen production and release despite the aberrant mitochondrial male sterility genome. Usually, the restorer allele, Rf, is dominant to the non-restorer recessive allelic form, rf, and fertility restoration is in most cases monogenic, though there are cases of oligogenic or multigenic control. There is also involvement of quantitative trait loci in partial fertility restoration known in some crops like maize.

In monogenic inheritance, progenies of male sterile plants can be male fertile or sterile depending on the fertility genotypes of plants used as pollinators. For illustration S is taken as the cytoplasmic factor responsible for fertility while s is the corresponding cytoplasmic factor responsible for sterility and the nuclear genome alleles being Rf and rf. The sterility/fertility genotypes and phenotypes of the progenies obtained by crossing male sterile parents with various fertility genotypes are as follows.

(i) The cross ♀ s.rfrf x ♂ s.RfRf gives the F1 that are s.Rfrf and are all male fertile. The F2 progenies are in a 3-male fertile: 1 male sterile ratio which comprise of 1 s.RfRf and 2 s. Rfrf which are all fertile and 1 s.rfrf which is sterile.
(ii) The cross ♀ s.rfrf x ♂ s.Rfrf gives progenies that have a 1 male fertile: 1 male sterile ratio. These are 1 s.Rfrf and 1 s.rfrf respectively.
(iii) The cross ♀ s.rfrf x ♂ S.rfrf gives s.rfrf progenies which are all male sterile.
(iv) The cross ♀ s.rfrf x ♂ S.RfRf gives s.Rfrf progenies which are all male fertile

The restorer genes, Rf, encode proteins which restore male fertility by preventing the accumulation of products of CMS open reading frames (orf). The Rf encoded proteins are RNA-binding pentatricopeptide proteins (PPR) which are moved to the mitochondria and protoplasts after their synthesis (Melonek et al., 2021). The RNA-binding pentatricopeptide proteins modify mitochondrial CMS gene expression through post-transcription processing which may include editing, splicing and nuclease cleavage (Zurek, 2023). They restore male fertility by capturing and binding to the transcripts of mitochondrial orf genes and inducing the cleavage of such transcripts (Zurek, 2023). That is, Rf encoded pentatricopeptide proteins overcome mitochondrial associated male sterility and allows normal metabolism in the male reproductive tissues and microsporogenesis by binding and modifying RNA transcribed by orf gene, resulting in reduction or non-production of the deleterious CMS associated proteins. The Rf genes are known in both cross-pollinating and self-pollinating crops, e.g., wheat, tomato, cotton, maize, tomato, rapeseed, sunflower, sorghum, soybean, barley, onion, sugar beet, rice and other crops. In some maize genotypes having T-cytoplasm there are at least two Rf genes, e.g., Rf1 and Rf2 (Kotchoni et al., 2010). The two genes work in conjunction to restore fertility. The protein coded by RF2 is an aldehyde dehydrogenase (ALDH) which helps to compensate for the defects caused by CMS protein URF13, allowing for anther development while RF1 protein causes the reduction of CMS toxic protein T-URF13 by altering the T-URF13 transcript (Kotchoni et al., 2010). In wheat cultivars having CMS T-cytoplasm the restorer genes, Rf1 and Rf3, restore male fertility by encoding proteins which bind to the transcripts of orf genes, cleaving the RNAs to prevent expression of male sterility and allowing the development of anther and pollen formation (Melonek et al., 2021).

Environment-Sensitive Genic Male Sterility

Genetic male sterility may or may not be influenced by environmental conditions, namely, photoperiod, temperature and humidity. Genetic male sterility dependent on the environment is usually referred to as environment-sensitive genic male sterility (EGMS) and may be in form of photoperiod-sensitive genic male sterility (PGMS), thermo-sensitive or temperature-sensitive genic male sterility (TGMS), photoperiod thermos-sensitive genic male sterility (PTGMS) and humidity-sensitive genic male sterility (HGMS) (Abbas et al., 2021; Liu et al., 2023). PGMS, TGMS and HGMS are responsive to changes in photoperiod day-length, temperature and humidity respectively so that plant male sterility status is converted to male fertility status and vice versa, depending on the changes in these environmental variables. PTGMS sex expression through male sterility-male fertility conversion is simultaneously influenced by both day-length and temperature conditions. Conversions from male sterile status to male fertile status and vice versa arise because of the influence of these factors on male gametogenesis resulting in promotion or disruption of microspore development (Abbas et al., 2021). Environment-sensitive genic male sterility may be of monogenic or oligogenic control or of multigenic and more complex nature with multiple biochemical pathways influencing pollen development process. Photoperiod-sensitive genic male sterile genotypes have no pollen under long day conditions (day-length of at least 14 h) but are male fertile if grown under short day conditions (less than 10 h day-length) (De Guzman & Oard, 2019; Liu et al., 2023). In TGMS cultivars there is male fertility at low temperatures (less than 24 °C) while the same cultivars are devoid of pollen at high temperatures (28 °C and above) irrespective of day-length (De Guzman & Oard, 2019; Liu et al., 2023). In case of PTGMS condition the plants show male fertility under long days (at least 14 h of daylight) and low temperatures (below 24 °C) or short-day conditions of less than 10 h of light and high temperature (at least 28 °C) conditions but the same plants are devoid of pollen under long day photoperiod conditions and high temperatures (De Guzman & Oard, 2019; He et al., 1999). The actual temperatures (low or high) and day-lengths (long day or short day) at which conversion to male fertility or male sterility happens depends of the plant species, the genes and the alleles involved.

Male sterility and fertility expression arising from environmental condition changes have independent control mechanisms attributed to differentially expressed genes (DEGs) and differentially expressed proteins (DEPs) related to environment sensitivity as regards sex expression, particularly DEGs and DEPs for long-day sensitivity response and DEGs and DEPs for short-day sensitivity response (Liu et al., 2023). It is also established that low temperatures, low light intensity and short-day length reduce the rate of anther and pollen development while favouring viable pollen production in photoperiod- and temperature-sensitive male sterile plants, converting such plants to male fertile status, suggesting that specific genes respond to these environmental factors as regards transcription and translation (Liu et al., 2023). Receptor-like kinases (RLK) and leucine-rich repeat (LRR)-RLK (i.e., LRR-RLK) are involved in tapetum initiation, development and cell death and pollen development in response to changes in photoperiod day-length, temperature and humidity (Abbas et al., 2021; Liu et al., 2023). Transcription factors like plant homeodomain (PHD) and MYB are also important in the regulation of genes that control anther development in response to influential environmental factors. That is, some male sterility related genes encode transcription factors e.g., MYB (myeloblastosis)-like DNA binding domain which has influence on transcription associated with plant response to changes in day-length and temperature (Abbas et al., 2021; De Guzman & Oard, 2019). The levels of certain mRNAs also change under varied photoperiod and temperature conditions leading to variation in male sterility/fertility condition.

Since EGMS plants respond to changes in environmental conditions, control of day-length, temperature and humidity can be employed to manipulate plant male fertility-sterility conversion with the aim of conducting self-pollination and cross-pollination for breeding exercise and hybrid seed production (De Guzman & Oard, 2019). Such conditions can be changed in the greenhouse or growth chamber for the desired purpose. The crop can also be planted in the field at the right time/season and the right location so as to take advantage of appropriate conditions for expression of male sterility or male fertility, whichever is desired.

Use of Male Sterility in Plant Breeding

Effective commercial exploitation of heterosis requires economically feasible production of the F_1 hybrid seed in sufficient quantities for distribution to the farmers and the hybrid variety cultivation must be economically viable to the farmers who buy the hybrid seed to plant. The value and the profit from the hybrid crop to the farmer must be much higher than that of the conventional crop. Production and distribution of the hybrid seed by the seed company must also be economically viable, self-sustaining and sufficiently profitable. In case of maize the breeder can easily emasculate the plant by removing the tassel without damaging the female flowers. Other than maize, commercial hybrid varieties are successfully produced in sorghum, sugar beet, sunflower, rice, tomato, carrot, onion, Brassicas and several ornamental crops and other vegetable and field crops. For most crops hand emasculation is difficult, tedious, labour intensive, expensive and not economically viable for hybrid seed production and use of male sterility is the only viable alternative. Cytoplasmic-genetic male sterility is the sterility system most commercially used for hybrid seed production in a number of crops like maize, sorghum, pearl millet, sunflower, rice, pepper, cotton, sugar beet, various *Brassica* species and other crops where hybrid production is commercialised because it solves the problems of hand emasculation, unreliable use of hot water, vacuum suction of pollen, use of gametocides and helps to save on costs. Male sterility is applied to improve both self-pollinated and cross-pollinated crops through recurrent selection to enhance genetic recombination and increase the frequencies of favourable genes and gene combinations. Dominant male sterility genes are usually not used for commercial hybrid seed production. They are mainly used to facilitate crossing in recurrent selection programs, especially in self-pollinated crops. In case of recessive male sterility, if sterile male plants are fertilised by homozygous male fertile plants the crop in farmers' field will be fertile and may give a good yield. However, in case of dominant male sterility the hybrid obtained between a male fertile plant and a male sterile plant will be male sterile with no pollen production and no crop yield unless the harvestable produce is vegetative.

Cytoplasmic male sterility alone is economically useful in hybrid seed production for vegetable crops, forage and ornamental crops in which productivity is not a function of botanical seed production or fruits and that the commercial product is purely based on vegetative parts. Otherwise, cytoplasmic male sterility is most useful in seed production if it is combined with nuclear male sterility system having dominant restorer (Rf) genes. The female lines having a combination of sterile cytoplasm and homozygousity for the recessive sterility gene (s,rfrf) are often referred to as A lines in hybrid seed production programmes. A version of such a line with fertile cytoplasm and homozygousity for the recessive nuclear sterility alleles, S.rfrf, is called maintainer or B line since male sterile A line produces a uniform male sterile offspring when pollinated by plants of B line. The maintainer is used for development and maintenance of the female A parent. A genotype with dominant nuclear fertility Rf allele suppresses the expression of cytoplasmic male sterility trait and when used as a pollinator on a plant with cytoplasmic male and genic sterility it restores pollen fertility for the derived hybrid. A genotype homozygous for dominant nuclear fertility, S.RfRf or s.RfRf is a restorer or C line. In a breeding programme and commercial seed production where cytoplasm-nuclear sterility system is employed there is the cytoplasm male sterile A (s,rfrf) line, the maintainer or the B (S,rfrf) line and restorer or C (S.RfRf or s.RfRf) line. Since the restorer genotype is homozygous dominant for the fertility nuclear allele it is used as the male parent for final hybridisation in seed production to restore fertility of the commercial hybrid because a cross of the type ♀ s.rfrf x ♂ S.RfRf or ♀ s.rfrf x ♂ s.RfRf gives progenies that are all male fertile and can therefore be used as the hybrid for commercial production by the farmers.

Male sterility genes and the respective fertility restorer genes can be sourced from cultivars in production, obsolete cultivars, landraces and wild and weedy crop relatives. Male sterility can also be obtained through induced mutagenesis, genetic modification and genome editing. While genetic transformation can be used to transfer difficult to transfer male sterility genes genome editing can be employed to modify the functions of the genes connected with regulation of male sterility and fertility for better exploitation of such genes in plant breeding and hybrid seed production.

Problems Associated with Male Sterility

Use of male sterility in breeding programmes has the handicaps of modifier genes affecting the expression of male sterility and in some cases a small amount of cytoplasm may be carried by pollen grain, resulting in partial loss or breakdown of sterility or fertility. Multigenic control of male sterility and involvement of quantitative genes in some cases makes male sterility genes challenging to transfer from the various sources. Sometimes male sterility genes may not be available in the primary gene pool of the crop in question making it necessary to source the genes form wild and weedy crop relatives enforcing interspecific hybridisations which are difficult to make and if they are successful there are problems of undesired alien genes tightly linked to the sterility genes. Since male sterility genes are expressed at flowering, they also difficult to select for and require tightly linked flanking molecular markers. Male sterility may also be associated with some undesirable characters like disease susceptibility as the case of T-cytoplasm which is linked to susceptibility to southern corn leaf blight, *Bipolaris maydis* (*Cochliobolus heterostrophus* in teleomorph state) in maize. Environmental conditions also affect the expressivity and penetrance of expression of male sterility such that some genetically male sterile plants will produce viable pollen under certain conditions. Environment-sensitive genic male sterility may also not have 100% conversion. In self-pollinating crops like wheat, barley, sorghum, rice, beans, cowpea, etc., there may be insufficient pollination mechanisms because of the structure and inadequate opening of the flower leading to incomplete pollination of male sterile plants. Use of male sterility in breeding programmes necessitates the development and maintenance of the male sterile lines, adding to costs of breeding. Despite these challenges the advantages of male sterility phenomenon in crop plants overrides its disadvantages in that it saves on time and cost of breeding and development of hybrid varieties and production of hybrid variety seed. There is no need for gametocides and human labour to conduct emasculation to make crosses.

4.2 Asexual Reproduction

Asexual reproduction is a regular mode of propagation in crops which are poor in seed production or produce no seeds at all and have long reproductive cycles. Plants may reproduce asexually in several ways without involvement of meiosis and fertilisation and no union of gametes to form zygotes. In asexual reproduction vegetative tissues and organs of the plant are used as propagules in place of botanical seed or the reproductive organs may give new plants without fertilisation. Tissues and organs used as propagules may be in form of modified underground and sub-aerial roots and shoots, stem cuttings, bulbils and suckers. Modified underground and sub-aerial stems take the form of suckers (bananas, sisal, roses), bulbs (onion, garlic), rhizomes (ginger, hops, turmeric, banana), tubers (potato), corms (Colocasia, yam), stolons (strawberry) for use in propagation. The modified stems serve as storage organs and have buds which develop into shoots and produce plants under appropriate conditions. Stem cutting propagation is commonly used in sugarcane, cassava and some forage grasses where nodal portions of stems are planted. Some plants like sugar beets, alfalfa and sweet potatoes have succulent stems from which cuttings are obtained for planting. Mango, tea and coffee are also propagated using cuttings from tender portions of shoots. Tuberous roots are used for sweet potato and dahlia propagation. Bulbils which are modified inflorescences that develop into plants without fertilisation and seed formation are used for propagation in sisal, onions, garlic and lilies. Leaf bulbils also form in *Bryophyllum* where they can be used as propagules as well.

Significance of vegetative reproduction is that it helps in perpetuation of superior cultivars and vegetative production of large numbers of propagules for desired clones. Vegetatively propagated species (Table 4.6) are generally polyploids and have high levels of heterozygousity which is a serious challenge in sexual reproduction because of segregation leading to a high number of variants that are difficult to manage. A major advantage of vegetative reproduction is that a desirable genotype can be used as a commercial variety for long whether it is homozygous or heterozygous and newly identified superior genotypes can simply be multiplied and released as new varieties. Vegetative reproduction helps in propagation of self-incompatible varieties as in avocado, fixation of heterozygotes and hybrid vigour in crops like roses, coffee and other tree crops and production of pathogen free clones. Apart from above forms of vegetative propagation layering and grafting can also be used for vegetative propagation.

Table 4.6 Some vegetatively propagated crops

Common name	Family	Genus	Species	Chromosome number		Propagation means **vg/sd**[i]
				X	2n	
Potato	Solanaceae	*Solanum*	*tuberosum*	12	48	vg, also sd
Sweet potato	Convolvulaceae	*Ipomea*	*batatas*	15	90	vg, also sd
Cassava	Euphorbiaceae	*Manihot*	*esculenta*	9	36	vg, also sd
Jerusalem artichoke	Compositae	*Helianthus*	*tuberosus*	17	34	vg
Yam	Dioscoreaceae	*Dioscorea*	*alata*	10	40, 60, 80	vg
Taro (cocoyam)	Araceae	*Colocasia*	*esculenta*	14	28, 42	vg
Banana	Musaceae	*Musa*	*sapientum, acuminata, balbisiana*	11	22, 33, 44	vg
Pineapple	Bromeliaceae	*Ananas*	*cosmos*	25	50	vg
Arrow root	Marantaceae	*Maranta*	*arundinacea*	6	48	vg
Sisal	Agavaceae	*Agave*	*sisalana*	30	150	vg
Pyrethrum	Compositae	*Chrysanthemum*	*cinerariifolium*	9	18	vg, also sd
Strawberries	Rosaceae	*Fragaria*	*x ananassa*	28	56	vg, also sd
Sugarcane	Gramineae/Poaceae	*Saccharum*	*officinarum*	8	80	vg, also sd
Alfalfa	Leguminosae/Fabaceae	*Medicago*	*sativa*	8	32	vg, also sd
Asparagus	Liliaceae	*Asparagus*	*officinalis*	10	20	vg, also sd
Apple	Rosaceae	*Malus*	*domestica*	17	34	vg, also sd
Grapes	Vitaceae	*Vitis*	*vinifera*	19	38	vg
Plum	Rosaceae	*Prunus*	*divaricata*	8	16	vg, also sd
Almond	Rosaceae	*Prunus*	*dulcis*	8	16	vg, also sd
Ginger	Zingiberaceae	*Zingiber*	*officinale*	11	22	vg
Coffee	Rubiaceae	*Coffea*	*arabica*	11	44	vg, also sd
Coffee	Rubiaceae	*Coffea*	*robusta*	11	22	vg, also sd

vg—vegetatively propagated;
sd—seed propagated

4.2.1 *Apomixis*

Apomixis is reproduction in which the plant sexual organs or related structures are involved in seed production without the union of gametes and seedlings thus produced are known as apomictics. In most cases of apomixis the embryo is formed without reduction of the chromosome number during meiosis. This reproduction phenomenon, which is the replacement of sexual reproduction by asexual reproduction in the floral parts, is common in plants and is known in at least 300 plant species belonging to about 30–35 families (Fiaz et al., 2021; Marshall & Brown, 1981). It occurs in both diploids and polyploids and is known in *Paspalum, Panicum, Eragrostis*, *Citrus*, *Rubus*, sugar beet, apple, wheat, sorghum, *Pennisetum glaucum*, Kentucky blue grass (*Poa pratensis*), buffel grass (*Cenchrus ciliaris*), maize and maize wild relatives (Bashaw, 1980; Fiaz et al., 2021). Apomixis may be obligate if the plants produce apomictic embryos only or facultative where the plants produce both apomictic embryos and embryos arising from regular union of haploid gametes (amphimixis). Obligatory apomixis is known in sisal while facultative apomixis is found in sugarcane, citrus, mango, Kentucky blue grass and guinea grass. A plant of facultative apomixis nature usually has a mixture of embryos comprising of 2n apomictic embryos, 2n embryos produced through meiosis and union of haploid gametes and haploid (n) embryos produced parthenogenically from haploid egg cells.

Mechanisms of Apomixis

Apomixis takes several forms namely, agamospermy, parthenogenesis, apogamy, apospory, recurrent apomixis, non-recurrent apomixis, adventitious embryony, vegetative apomixis, diplospory and pseudogamy. Other forms are apomeiosis, generative apospory, somatic apospory, polyembryony, androgenesis, semigamy, androgamy and parthenocarpy. Among these, four types, namely, apospory, diplospory, adventitious embryony and parthenogenesis are mechanisms of apomixis which lead to asexual seed development and are grouped as agamospermy. Adventitious embryony applies to situations where the embryo develops directly from the nucellus somatic cells.

Apospory is used to define the situation when the embryo develops from an unreduced egg in the embryo sac while diplospory refers to the development of the embryo through mitosis of the megaspore mother cell. This is a common apomixis mechanism in higher plants and is responsible for the occurrence of multiple embryo sacs in some species like sorghum, buffelgrass, Kentucky bluegrass, dallis grass, apple, *Crepis*, citrus and mango. Apospory may also occur along with apogamy.

Agamospermy refers to apomictic reproduction in which embryos and seeds are formed by asexual means e.g., diplospory, apospory and adventitious embryony. It is a type of apogamy where plants reproduce by seed without normal sexual cycle since meiosis or union of gametes or both do not take place and the embryo develops from the unreduced female gametophyte or from the diploid sporophytic cells of the embryo sac like nucellus or integument cells. Agamospermy also results from haploid parthenogenesis by the development of a haploid embryo from the egg cell (gynogenesis) or microspore (androgenesis).

In parthenogenesis embryos arise from gametic cells, usually the reduced female egg cell in the embryo sac without fertilisation. The egg cells from which such embryos develop are products of regular meiotic process and are haploid. Parthenogenesis, therefore, gives rise to haploid embryos and plants developing from such embryos are haploid. If haploid embryos are diploidised the resulting diploid plants are 100% homozygous. There are rare cases of failure of chromosome number reduction in some species arising from abnormalities during meiosis leading to diploid egg cells that give embryos through parthenogenesis.Therefore, there are two types of parthenogenesis, haploid parthenogenesis and diploid parthenogenesis. In haploid parthenogenesis, also referred to as non-recurrent apomixis, the embryo develops from the reduced egg cell or any other haploid cell in the embryo sac without fertilisation. This type of apomixis, also referred to as generative or haploid apospory is known to occur in early purple orchid. Rare incidences of haploid parthenogenesis are also known in maize and some *Solanum* and *Nicotiana* species.

Diploid parthenogenesis which is also referred to as somatic apospory or diploid apospory involves the development of the embryo from unreduced egg cell as meiosis fails to take place in megaspore mother cell. The embryo forms directly from unreduced egg cell or an unreduced diploid somatic cell, without formation of gametes and fertilisation. Nucellus or integument develops into a diploid embryo sac along the normal embryo sacs which may also function in embryo formation. This is a form of recurrent apomixis.Though parthenogenesis is believed to be rare and spontaneous it is genetically controlled as known in maize and cotton (Bashaw, 1980). Incidences of diploid parthenogenesis are known in *Allium and Crepis* species.

In diplospory the megaspore mother cell develops into an ovule without undergoing meiosis thereby producing a diploid embryo sac through mitosis. In the initial stages the megaspore mother cell may undergo an irregular meiosis I in which asynapsis takes place and its restitution nucleus ends with 2n chromosomes. This may be followed by mitosis leading to the unreduced egg from which the embryo develops. That is, the egg cell develops parthenogenetically into an embryo, or another cell of the embryo-sac divides and develops into an embryo. Examples of plants in which diplospory is known are *Paspalum. Eragrostis, Rubus, Allium tuberosum* (Hojsgaard & Horandl, 2019).

In apogamy, also called apogamety, the embryo develops from any one of the synergid or antipodal cells other than the egg cell within the embryo sac without fertilisation. The derived embryo is haploid if it develops from any haploid cell in the haploid embryo sac formed through generative apospory as in certain *Lilium* species. This is non-recurrent type of apomixis. If the embryo develops from a cell in a diploid embryo sac developed by somatic apospory it will be diploid. This is diploid apogamy which is a recurrent apomixis type. Synergids and antipodals may also form supernumerary embryos. Apogamy is known in some *Allium* and *Antennaria* species.

Adventitious embryony is another form of apospory where an embryo develops apomictically from a diploid nucellus or integuments along the normally sexually produced embryo or the latter may be eliminated. It also referred to as nucellar budding or nucellar embryony. There may be one or two apomictic embryos, making the mature seed contain one or two or more embryos, a feature known as polyembryony. Only one among such embryos is the true sexual embryo while the others are nucellar embryos. This is common in citrus and mango.

In some plants there is vivipary, also known as vegetative apomixis, in which vegetative bulbils form as accessory organs in the inflorescence axils or have replaced the flowers as found in *Agave*, spp. *Allium spp.* and some grass species. In pseudogamy a diploid embryo is formed following pollination without fertilisation. In this type of apomixis the development of pollen tubes is necessary for apomictic seed production so that the pollen stimulates embryo and endosperm development without fertilisation (Hojsgaard & Horandl, 2019). Pseudogamy is known to occur in mango, citrus, raspberry, apple, and strawberry. Sometimes haploid embryos arise through androgenesis when the embryo develops from the sperm cell which is already within the embryo sac following pollination. This kind of non-recurrent apomixis is called androgamy. Another form of apomixis is semigamy which is a type of parthenogenesis where fertilisation takes place but the male gamete nucleus and the egg nucleus do not unite for normal zygote formation. Instead, the two gametic cells form a heterogeneous haploid embryo with chimeral sectors in which one sector is of male gamete origin while the other sector is of female gamete origin. This rare form of apomixis has been reported in cotton. In certain cases, pollination without fertilisation induces the development of the ovary into a fruit without seeds, a process called parthenocarpy. In this situation the ovary develops and becomes enlarged as in pineapples and citrus. Such stimulation can be caused by incompatible pollen.

Genetic Control of Apomixis

Genetic control of apomixis reproduction in higher plants is well established in many plant species and the presence of genes responsible for its manifestation has been detected through segregation in crosses between amphimictic and apomictic plants of the same species. In a number of plants, it has been established that apomixis is simply inherited with single nuclear gene control and usually influenced by recessive Mendelian genes as in wheat, rice, maize, sorghum, potato, sugar beet (Savidan, 1999). Multigenic control for apomixis feature is also known in some plant species. (Powers, 1945) suggested that at least three loci control apomixis reproduction in rubber plant (*Parthenium argenteum*) where the alleles responsible for apomixis expression were found to be recessive. (Taliaferro & Bashaw, 1966) observed that obligate apomixis in *Cenchrus ciliaris* was under the control of two epistatic loci. Apart from being controlled by simple qualitative genes apomixis is believed to be under the influence of transposable elements and epigenetic processes in some cases (Fiaz et al., 2021) and environmental factors like temperature, day length and light intensity influence the degree of apomixis expression as well (Hoojsgaard & Horandl, 2019). Apomictic plants are often highly heterozygous and apomictically derived offspring remain genetically uniform and identical to the mother plants since there is no segregation, especially for obligate apomictics (Bashaw, 1980). In facultative apomictics there is some genetic variation among the progenies since those arising from sexual reproduction have some segregation and may differ from their apomictic sibs and parents.

Genes for apomixis phenomenon are known to exist in some wild relatives of crop plants like *Elymus rectisetus*, a relative of wheat, *Tripsacum dactyloides*, a relative of maize and *Penissetum* species related to pearl millet (Hanna & Bashaw, 1987). Apomixis genes are also known in some species within *Panicum, Eragrostis, Cenchrus* genera and wild relatives of sugar beet and mango. Apomixis genes can be transferred from such species to their crop relatives through interspecific hybridisation as has been demonstrated for maize (Fiaz et al., 2021). Apart from interspecific hybridisation apomictic feature can be induced in crop plants through induced mutations and biotechnology methods. Genome editing to substitute mitosis for meiosis (mitosis instead of meiosis or MiMe) is a technology for possible generation of apomictics (Fiaz et al., 2021; Mieulet et al., 2012) in crop plants.

Significance of Apomixis in Plant Breeding and Crop Production

Recurrent apomixis mechanisms like diploid forms of apospory, parthenogenesis, apogamy, adventive embryony and vegetative propagation are useful in plant breeding programmes because they give viable diploid embryos which can be used for propagation and derivation of clones. Where possible hybrid vigour can be fixed, maintained and exploited in F_1 obtained from facultative apomictics which are then propagated indefinitely for agronomic production. In obligate apomictics the progenies of a particular genotype breed true, are genetically similar to all other progenies derived apomictically from the same plant and are exact replicas of the parental plant making apomixis an excellent method of cloning and a tool in plant breeding. Therefore, within a clone there is genetic purity and no genetic variation while clones derived from different cultivars are genetically different. Despite heterozygousity, the genotypes and the vigour of the superior clones will remain stable, uniform and can be sustained by apomictic seed production for prolonged periods. Since apomixis helps to conserve the genetic constitution and maintains maternal effects it can be employed for rapid and continued multiplication of genetically uniform genotypes without segregation and without loss of hybrid vigour and maternal effects in clones obtained by hybridising superior clonal parents. In facultative apomictics the breeder has the opportunity to select superior clones with good combining ability and cross them for favourable gene combinations. It is also possible to develop superior hybrids in a short duration since there is no need to produce inbreed lines as selected good parental clones are hybridised and the arising best hybrids multiplied for commercial production. Selections from apomictically obtained progenies, especially in obligate apomictics, can be also multiplied and used as superior varieties with high levels of genetic uniformity and stability. Occasional sexual reproduction in obligate apomictics and regular sexual reproduction in facultative apomictics create variability in the apomictic crop species for exploitation by identifying good genotypes for use as superior clones or as parents in hybrids for commercial production. Induced mutagenesis, tissue culture and genetic engineering methods are applicable for creating genetic variability in plants of apomictic nature. New morphological, physiological and biochemical variants obtained through induced mutations, somaclonal variation and genetic engineering give opportunities for isolating superior novel variants with favourable traits.

Seed production of superior genotypes is cheap because no emasculation, male sterility and isolation are needed and that a particular planting is used to produce propagules year after year, especially for obligate apomictics, thereby saving the farmers the expenses of buying the expensive hybrid seed every season for planting. Apomixis and parthenocarpy are applied in horticulture for production of seedless fruits which are preferred by consumers and are more marketable because they have increased portions of edible parts. Apomixis and parthenocarpy are applied to propagate and maintain superior genotypes for prolonged periods. The genotype of the variety can be kept intact as long as desired while ensuring that there is no degeneration through disease infection, mutation and physical mixing of planting materials. Haploid apomictics are useful for genetic analyses and research purposes and are used to produce 100% homozygous plants through diploidisation.

Apomixis, particularly obligate apomixis, has been viewed previously as not being adequately flexible with regard to crop evolution by providing very limited opportunities for recombination to create genetic variation required for evolution. However, apomictic plants also have avenues of natural generation of genetic variability through irregular sexual reproduction, referred to as leaky apomixis and facultative reproduction giving room for recombination and mutations (Fiaz et al., 2021). Apomixis is widespread in the plant kingdom and is known in at least 35 families and over 300 species (Fiaz et al., 2021; Mieulet et al., 2012) most of which have undergone extensive and successful evolution. Furthermore, apomixis is a common method of reproduction in many crop species some of which are of high commercial value or important staple crops in various parts of the world or important forage crops.

Problems

4.1 Which morphological features enforce the following?

(a) Autogamy
(b) Allogamy

4.2 Which determinants are responsible for self-incompatibility phenotype and how do they operate?

4.3 How are self-incompatibility challenges resolved in plant breeding programmes for purpose of conducting inbreeding?

4.4 Distinguish between non-recurrent apomixis and recurrent apomixis.

4.5 What morphological features are associated with self-incompatibility in *Primula* plants?

4.6 A plant breeder attempted to make crosses between black mustard plants having sporophytic self-incompatibility system. The crosses were made using plants having incompatibility alleles as follows:

(a) ♀S_1S_1 x♂ S_1S_2
(b) ♀S_2S_3 x♂ S_1S_4

Give the genotypes of the progenies obtained from the crosses

4.7 Indicate any three problems associated with male sterility in crop plants.

References

Abbas, A., Yu, P., Yang, Z., Chen, D., Cheng, S., & Cao, L. (2021). Exploiting genic male sterility in rice: From molecular dissection to breeding applications. *Frontiers in Plant Science, 12*, 629314. https://doi.org/10.3389/fpls.2021.629314

Agong, S. G. & Ayiecho, P. O. (1991). The rate of outcrossing in grain amaranths. *Plant Breeding, 107*, 156–160. https://doi.org/10.1111/j.1439-0523.1991.tb00544.x

Bashaw, E. C. (1980). Apomixis and its applications in crop improvement. In W. R. Fehr & H. Hadley (Eds.), *Hybridisation in crop plants* (pp. 45–63). American Society of Agronomy and Crop Science Society of America. https://doi.org/10.2135/1980.hybridizationofcrops.c3

Claessen, H., Keulemans, W., de Poel, B. V., & de Storme, N. (2019). Finding a compatible partner: Self-incompatibility in pear (*Pyrus communis*); molecular control, genetic determination and impact on fertilisation and fruit set. *Frontiers in Plant Science, 10*, Article 407. https://doi.org/10.3389/fpls.2019.00407

De Guzman, C., & Oard, J. (2019). Genetics and breeding system for cytoplasmic and genetic male sterility in rice. In Y. Jia (Ed.), *Protecting rice grain in the post-genomic era*. IntechOpen. https://doi.org/10.5772/intechopen.85191

Dulberger, R. (1970). Tristyly in lythrum. *New Phytologist, 69*, 751–759. https://doi.org/10.1111/j.1469-8137.1970.t602460.x

Farinati, S., Draga, S., Betto, A., Palumbo, F., Vanozzi, A., Licchin, M., & Barcaccia, G. (2023). Current insights and advances into plant male sterility: New precision breeding technology based on genome editing applications. *Frontiers in Plant Science, 14*, 122386. https://doi.org/10.3389/fpls.2023.1223861

Fiaz, S., Wang, X., Younas, A., Alharthi, B., Riaz, A., & Ali, H. (2021). Apomixis strategies to induce apomixis to preserve hybrid Vigour for multiple generations. *GM Crops and Food, 12*, 57–70. https://doi.org/10.1080/21645698.2020.1808423

Hanna, W. W., & Bashaw, E. C. (1987). Apomixis: Its identification and use in plant breeding. *Crop Science, 27*, 1138–1139. https://doi.org/10.2135/cropsci1987.0011183X002700060010x

He, Y. Q., Yang, C. G., Zhang, Z. G., & Zhang, Q. (1999). Genetic bases of instability of male sterility and fertility reversibility in photoperiod-sensitive genic male-sterile rice. *Theoretical and Applied Genetics, 99*, 683–693. https://doi.org/10.1007/s001220051285

Hojsgaard, D., & Horandl, E. (2019). The rise of apomixis in natural plant populations. *Frontiers in Plant Science, 10*, Article 358. https://doi.org/10.3389/FPLS.2019.00358

Huu, C. N., Kappel, C., Keller, B., Sicard, A. et al. (2016). Presence versus absence of CYP734A50 underlies the style-length dimorphism in primroses. *eLife, 5*, e17956. https://doi.org/10.7554/eLife17956

Iwano, M., & Takayama, S. (2012). Self/non-self-discrimination in angiosperm self-incompatibility. *Current Opinion in Plant Biology, 15*, 78–83. https://doi.org/10.1016/j.pbi.2011.09.003

Jia, Y., Liu, C., Li, Y., Xiang, Y., Pan, Y., Liu, Q., Gao, S., Yin, X., & Wang, Z. (2023). Inheritance of Distyly and homostyly in self-incompatible *Primula forbesii*. *Heredity, 130*, 259–268. https://doi.org/10.1038/s41437-023-00598-6

Knox, R. B., Willing, R., & Ashford, E. A. (1972). Role of pollen-wall proteins as recognition substances in interspecific incompatibility in poplars. *Nature, 237*, 381–383. https://doi.org/10.1007/BF02981926

Kocyan, A., Zhang, L. B., Schaefer, H., & Renner, S. S. (2007). A multilocus chloroplast phylogeny for Cucurbitaceae and its implications for character evolution and classification. *Molecular Phylogenetics and Evolution, 44*, 553–577. https://doi.org/10.1016/j.ympev.2006.12.022

Kotchoni, S. O., Jimenez-Lopez, J. C., Gachomo, E. W., & Suefferheld, M. J. A. (2010). A new and unified nomenclature for male sterility restorer (Rf) proteins in higher plants. *PLoS ONE, 5*(12), e15906. https://doi.org/10.1371/journal.pone.0015906

Lai, Y. S., Zhang, W., Zhang, X., Wang, H. Dou, X., Li, S., Wu, Y., Song, J., Ji, G., & Li, X. (2018). Temperature and photoperiod changes affect cucumber sex expression by different epigenetic regulations. *BMC Plant Biology, 18*, 268. https://doi.org/10.1186/s12870-018-1490-3

Liu, D., Shi, J., Liang, W., & Zhang. (2023). Molecular mechanisms underlying plant environment-sensitive genic male sterility and fertility restoration. *Seed Biology, 2*, 13. https://doi.org/10.48130/SeedBio-2023-0013

Liu. Z., Xu, G., & Zhang, S. (2007). *Pyrus pyrifolia* stylar S-RNase induces alterations in the actin cytoskeleton in self-pollen and tubes in vitro. *Protoplasma, 232*, 61–67. https://doi.org/10.1007/s00709-0269-4

Marshall, D. R., & Brown, A. D. H. (1981). The evolution of apomixis. *Heredity, 47*, 1–15. https://doi.org/10.1038/hdy.198154

Melonek, J., Duarte, J., Martin, J., Beuf, L., Murigneux, A., et al. (2021). The genetic basis of cytoplasmic male sterility and fertility restoration in wheat. *Nature Communications, 12*, 1036. https://doi.org/10.1038/s41467-021-21225-0

Mieulet, D., Jolivet, S., Rivard, M., Cromer, L., Vernet, A., Mayonove, P., Pereira, L., Droc, G., Courtois, B., & Guiderdoni, E. (2016). Turning rice meiosis into mitosis. *Cell Research, 26*, 1242–1254. https://doi.org/10.1038/cr.2016.117

Munoz-Sanz, J. V., Zuriaga, E., Cruz-Garcia, F., McClure, B., & Romero, C. (2020). Self-(in)compatibility systems: target traits for crop production, plant breeding and biotechnology. *Plant Science, 11*, Article 195. https://doi.org/10.3389/fpls.2020.00196

de Nettancourt, D. (1993). Self- and cross-incompatibility systems. In M. D. Hayward, N. O. Bosemark, I. Ramagosa (Eds.), *Plant breeding: Principles and prospects*. Springer Dordrecht. https://doi.org/10.1007/978-94-011-1524-7_14

Powers, L. (1945). Fertilization without reduction in guayule (*Parthenium argenteum* Gray) and a hypothesis as to the evolution of apomixis and polyploidy. *Genetics, 30*, 323–346. https://doi.org/10.1093/genetics/30.4.323

Roalson, E. H., & McClubbin, A. G. (2003). S-RNase and sexual incompatibility: Structure, functions and evolutionary perspectives. *Molecular Phylogenetics and Evolution, 29*, 490–506. https://doi.org/10.1016/S1055-7903(03)00195-7

Sassa, H., Kakui, H., Miyamoto, M., Suzuki, Y., Hanada, T., Ushijima, K. et al. (2007). S-locus F-box brothers: Multiple and pollen specific F-box genes with S-haplotype-specific polymorphisms in apple and Japanese pear. *Genetics, 175*, 1869–1881. https://doi.org/10.1534/genetics.106.068858

Savidan, Y. (1999). Apomixis. In J. Janick (Ed.), *Plant breeding reviews*. https://doi.org/10.1002/9780470650.158.ch2

Shivrain, V. K., Burgos, N. R., Sales, M. A., Mauromoustakos, A., Gealy, D. R., Smith, K. L., Black, H. L., & Jia, M. (2009). Factors affecting the outcrossing rate between Clearfield^{TM}rice and red rice (*Oryza sativa*). *Weed Science, 57*, 394–403. https://doi.org/10.1614/WS-09-003.1

Taliaferro, C. M. and Bashaw, E. C. 1966. Inheritance and control of obligate apomixis in breeding buffelgrass, *Pennisetum ciliare*. *Crop Science, 6*, 473–476. https://doi.org/10.2135/cropsci1966.0011183X000600050025x

Tovar-Mendez, A., & McClure, B. (2016). Plant reproduction: Self-incompatibility to go. *Current Biology, 26*, R115–R117. https://doi.org/10.1016/j.cub.2015.12.001

Zurek, M. 2023. Genetic basis of the phenomenon of male sterility and fertility restoration in maize (*Zea mays* L.)—a review. *Agronomy Science, 78*(3), 89–99. https://doi.org/10.24326/as.2023.5120

Chapter 5
Population Genetic Variation, Genetic Structure of Crop Plant Populations, Inbreeding Depression and Heterosis

5.1 Population Genetic Variation

Population genetics relate to the behaviour of genes in populations and genetic structure of populations in terms of frequencies of alleles (genes), genotypes and phenotypes and the forces that affect them. Changes in the gene frequencies affect the genotypic frequencies which in turn influence phenotypic frequencies. One of the key aspects that affects and helps in understanding population genetic structure is the population breeding system. A population may be said to be random mating or non-random mating. A random mating population is one in which the probability of mating between individuals is independent of their genotypes, phenotypes and relationships among them and that each individual has equal chances of mating with others in the population. In such a population mating between genotypes is determined by their proportions in the population. If the population is non-random there may be assortative mating and the pairing of individuals to mate is affected by genotype and phenotype such that partners are chosen because they are phenotypically similar or because they have certain characters in common. Non-random mating also involves mating between related individuals and may take the form of self-fertilisation, mating of brothers and sisters (sib-mating), parents and their offspring and mating between cousins. Self-fertilisation is the most extreme form of non-random mating and is possible and takes place only in plants. In animals and dioecious plants self-fertilisation is not possible. Most theories of population genetics have been developed with assumptions of large population size, random mating populations, absence of linkage, diploid segregation and absence of selection.

P. O. Ayiecho and J. O. Nyabundi, *Conventional and Contemporary Practices of Plant Breeding*, https://doi.org/10.1007/978-3-031-74998-8_5

5.1.1 Hardy-Weinberg Principle

In 1908 G. H. Hardy and W. Weinberg developed the Hardy-Weinberg law which is used to define and understand population genetic structure in terms of allelic and genotypic frequencies. The law states that for a diallelic single autosomal locus the genotypic frequencies are given by a binomial function of allelic frequencies after one generation of random mating, i.e.,

$$(p+q)^2 = p^2,\ 2pq,\ q^2 \text{ for } A_1A_1,\ A_1A_2,\ A_2A_2,$$

where frequency of A_1, $f(A_1) = p$ and $f(A_2) = q$; and for a multi-allelic single autosomal locus the genotypic frequencies, are also given by a multinomial function of allelic frequencies after one generation of random mating, i.e.,

$$(p+q+r+s+\ldots)^2 = p^2, 2pq, q^2, 2pr, 2qr, r^2, 2ps, 2qs, 2rs, s^2, \ldots.$$

For $A_1A_1,\ A_1A_2,\ A_2A_2,\ A_1A_3,\ A_2A_3,\ A_3A_3,\ A_1A_4,\ A_2A_4,\ A_3A_4,\ A_4A_4\ldots.$ where $f(A_1) = p,\ f(A_2) = q,\ f(A_3) = r,\ f(A_4) = s.$

The genotypic frequencies of the progenies depend on the gene frequencies of the parental population and not the genotypic frequencies.

As an example, there is a population of 1000 individuals among whom 500 are A_1A_1, 250 are A_1A_2 and 250 are A_2A_2 and that the heterozygotes A_1A_2 are distinguishable from the two distinct homozygotes.

Number of A_1 allele from $A_1A_1 = 500 \times 2 = 1000$ and number of A_1 allele in $A_1A_2 = 250$. Total A_1 in the population = 1250.

Number of A_2 allele from $A_2A_2 = 250 \times 2 = 500$ and number of A_2 allele in $A_1A_2 = 250$. Total A_2 in the population =750. Total number of A_1 and A_2 alleles in the population $= 1250 + 750 = 2000$.

Frequency (proportion) of $A_1 = f(A_1) = \frac{1250}{2000} = 0.625 = p$.

Frequency of $A_2 = f(A_2) = 1 - 0.625 = 0.375 = q$.

This can be generalised by taking the observed numbers of the three genotypes as A, B and C for A_1A_1, A_1A_2 and A_2A_2 respectively with the total for the population being N,

$f(A_1) = p = \frac{(2A+B)}{2N}$ and $f(A_2) = q = 1.00 - p$.

These gene frequencies can also be given in terms of genotypic frequencies as follows.

$$p = P + \frac{1}{2}H = 1.00 - q \text{ and } q = Q + \frac{1}{2}H = 1.00 - p.$$

The values P, H, and Q are the observed genotypic frequencies in the population.

In this example $P = f(A_1A_1) = \frac{500}{1000} = 0.50$, $H = f(A_1A_2) = \frac{250}{1000} = 0.25$, and $Q = f(A_2A_2) = \frac{250}{1000} = 0.25$.

H is the heterozygosity (proportion of heterozygotes) in the population.

In the population the frequency of the gametes with A_1 allele is p and the gametes with A_2 allele is q. If the population undergoes random mating the two gametes unite at random to give the three genotypes whose frequencies can be given by the binomial function $(p+q)^2$ leading to p^2, $2pq$ and q^2 for A_1A_1, A_1A_2 and A_2A_2 with the actual values being $(0.625)^2 = 0.390625$, $2(0.625)(0.375) = 0.46875$ and $(0.375)^2 = 0.140625$ respectively. The three genotypic frequencies must sum up to 1.00, that is, $p^2 + 2pq + q^2 = 1.00$.

Also, $p^2 = 1.00 - 2pq - q^2$, $q^2 = 1.00 - 2pq - p^2$ and $2pq = 1.00 - p^2 - q^2$.

If the population is large and is undergoing random mating, diploid segregation with no selection, no mutation, no migration the frequencies of genotypes are given by $(p+q)^2 = p^2, 2pq$ and q^2 for a two-allele case and $(p+q+r+\ldots)^2 = p^2$, $2pq, q^2, 2pr, 2qr, r^2, \ldots$ for multi-allelic case after one generation according to Hardy-Weinberg principle. This is expected to remain the same from generation to generation so long as there is sustenance of random mating in a large population with no selection, no migration and no mutation. If these conditions hold the population attains an equilibrium situation, usually referred to as Hardy-Weinberg equilibrium, for genotypic frequencies after one generation of such mating. Open pollinated plant populations of cross-fertilising crops like synthetic and composite maize and sunflower cultivars are expected to attain such equilibrium because natural cross-pollination is equivalent to random mating. A population in which no conditions violate Hardy-Weinberg law, and is also in Hardy-Weinberg equilibrium, is often referred to as a Hardy-Weinberg population.

If the Hardy-Weinberg conditions hold in a population, gene frequencies can be estimated from the genotypic frequencies observed in the population since P and Q will be assumed to have no deviations from p^2 and q^2 respectively, hence, $p = \sqrt{P}$ and $q = \sqrt{Q}$. A set of data from a population can be subjected to statistical analysis to verify whether the data are from a population obeying Hardy-Weinberg conditions. This is done by conducting a χ^2 test as follows;

$\chi^2 = \sum \frac{(O-E)^2}{E}$ with $n-k-1$ degrees of freedom, where n is the number of categories or classes (genotypes), k is the number of parameters estimated, O and E are respective observed and expected numbers for a given genotype in the population.

The expected numbers are based on the estimated gene frequencies. A significant calculated χ^2 suggests that the population does not obey Hardy-Weinberg law and that at least one of the conditions mentioned, namely, random mating, large population size, absence of selection, absence migration and absence mutation does not hold.

5.1.2 Factors that Violate Hardy-Weinberg Principle

The factors which will a make population not behave as a Hardy-Weinberg population are mutation, migration, selection, non-random mating and drift.

Mutation

Mutation is the primary source of variation in populations of all organisms since changes in alleles due to mutation alter the numbers and proportions of different alleles. The process of mutation may involve a single DNA base, several DNA bases, a segment of a chromosome, a whole chromosome or sets of chromosomes. Mutation may be natural (spontaneous) or artificially induced using mutagens. Compared to other forces like selection, migration, mating system and population size (drift) that affect population genetic structure making populations deviate from Hardy-Weinberg conditions, spontaneous mutation is the least significant and has the smallest effect on gene frequencies per generation because it occurs at very low rates in the order of 1×10^{-8} to 1×10^{-5}, that is 1 in 100000000 to 1 in 100000 gametes carry a newly mutated allele at any particular locus in the population per generation. It takes several generations for its effect to be noticed in populations. Spontaneous forward and backward (reverse) mutation may involve alleles A_1 and A_2 such that A_1 changes to A_2 at rate u while A_2 changes to A_1 at rate v, i.e., in the population the fraction of A_1 that mutates to A_2 is u and the fraction of A_2 that mutates back to A_1 is v per generation. If the frequencies of alleles A_1 and A_2 are p and q respectively the proportion of A_1 that changes to A_2 per generation is pu and the proportion of A_2 that changes to A_1 per generation is qv in the population.

If the reverse mutation rate is 0 the allelic frequencies, p_1 and q_1, after one generation of mutation are given as

$$f(A_1)_1 = p_1 = p - pu \text{ and } f(A_2)_1 = q_1 = 1.00 - p_1 = q + pu.$$

However, if each of forward and reverse mutation rates is greater than 0 the allelic frequencies after one generation of mutation are

$$p_1 = p - pu + qv \text{ and } q_1 = 1.00 - p_1 = q + pu - qv.$$

Change in gene frequency due to one generation of mutation is given by

$$\Delta p = p_1 - p = p - pu + qv - p = qv - pu \text{ and}$$
$$\Delta q = q_1 - q = q + pu - qv - q = pu - qv.$$

The change in gene frequency is the same in both directions except for the sign. The change in gene frequencies will change the genotypic frequencies to p_1^2, $2p_1q_1$ and q_1^2 for A_1A_1, A_1A_2 and A_2A_2 respectively. As both forward and backward mutations take place a point is reached when the mutation forces balance and $\Delta q = 0$, when Δq due to $u = \Delta q$ due to v. The point at which forward and backward mutations balance is an equilibrium for the mutation forces.

Migration

Migration of genes, also referred to as geneflow, through mating alters gene frequencies of the recipient population. Cross-pollination causes geneflow where pollen flow from one plant population to another one leads to changes in gene frequencies in the population receiving the foreign pollen. Migration or geneflow nullifies lack of genetic connectedness or genetic distinctness between populations. The degree of genetic connectedness between populations depends on the amount of effective migration or geneflow that takes place between populations or subpopulations. Effective migration refers to involvement of the migrants in successful mating with the members of host population leading to production of progenies leading to changes in gene frequency in the recipient population. Migration is also ineffective if the two populations are not genetically distinct, that is, the two populations have to differ in gene frequencies for changes to arise in gene and genotypic frequencies in recipient population as a consequent of geneflow. Ineffective migration is where there are seasonal migration of animals or birds that go to a place but get back to their original habitat without mating with the native hosts. They may migrate for food supplies but do not get involved in gene flow.

If a donor population has $f(A_1) = p_m$, $f(A_2) = q_m$ and the recipient population has $f(A_1) = p$, $f(A_2) = q$ and the migration rate from the donor population to recipient population is m where m is proportion of migrants in the recipient population while $1.00 - m$ makes up the proportion of original residents of the recipient population, the new gene frequencies in the recipient population including the migrants are

$$q_1 = mq_m + q(1.00 - m) = m(q_m - q) + q \text{ and } p_1 = 1.00 - q_1.$$

The migrants contribute mq while the natives contribute $q(1.00 - m)$ towards the overall gene frequency in the new population comprising of the migrants and the natives. Gene frequency change due to migration in one generation is the difference between the new gene frequency, q_1, and the original recipient population gene frequency, q, and is obtained as follows.

$$\Delta q = q_1 - q = m(q_m - q) + q - q = m(q_m - q) = mq_m - mq = -m(q - q_m)$$

If $q < q_m$ allelic frequency change will increase since Δq will be positive. On the other hand, if $q > q_m$ allelic frequency decreases since Δq will be negative. New genotypic frequencies among the zygotes arising after one generation of random mating in the mixed population are p_1^2, $2p_1q_1$ and q_1^2 for A_1A_1, A_1A_2 and A_2A_2 respectively.

Selection

Naturally, populations vary in different characteristics some of which are important for the organism survival and fecundity. Genotypic composition of a population and the environment will determine the capacity of the population to survive, increase in numbers and therefore the ability of individuals within the population to leave progenies behind and pass their genes to the progenies. The relative survival ability of different genotypes is determined by natural selection forces that affect the population by acting directly on phenotypes which are manifestations of genetic effects. Selection on organism populations is due to the fact that individuals in the population differ in viability, fertility and survival rates and in plant populations individual genotypes may differ in seed production, seedling emergence and survival to maturity, flower production, fertility, pollination and effective fertilisation, predation (being eaten up by grazers or browsers or seed consumers), diseases, pests and other forms of biotic and abiotic stresses. These fitness components may affect the plants at various stages of the life cycle as determined by extent of genetic variation for them and reflected in the phenotypes.

The primary effect of selection at the various stages of organism life cycle is change in gene frequencies which impacts on genotypic frequencies and ultimately phenotypic frequencies. Selection effect on gene frequencies depends on the magnitude of selection quantified as selection coefficient for different genotypes or fitness of the various genotypes in the population. Selection coefficient measures strength of selection in terms of relative reproductive and survival disadvantages of different genotypes by taking into account the proportion of individuals eliminated by selection before mating or proportion that fails to get involved in effective mating because of reduced fitness. Selection includes directional selection, stabilising selection and disruptive selection. If s is considered as the selection coefficient, that is proportion of the population eliminated by natural selection, the three genotypes A_1A_1, A_1A_2 and A_2A_2 have selection coefficients of s_1, s_2 and s_3 respectively with each value ranging between 0.00 and 1.00 as dictated by the pattern of selection.

In directional selection against a recessive allele s_1 and s_2 are both 0 and s_3 has value greater than 0. However, if each of s_2 and s_3 has a value greater than 0 and s_1 has a value of 0 there is directional selection against a dominant allele. Directional selection in either pattern eliminates one allele by favouring the genotype or genotypes having a particular allele. The favoured allele will end up being fixed in the population with frequency being 1.00 when the deleterious allele is totally eliminated and the population will have only one genotype, the favoured homozygote. In case of selection against a recessive allele gene frequencies and various population parameters change as follows. Weighted contribution of each genotype is the product between genotypic frequency before selection and relative fitness.

The population mean fitness after selection which is the sum of the weighted contributions of the three genotypes = $p^2 + 2pq + q^2(1 - s_3) = p^2 + 2pq + q^2 - q^2s_3 = 1 - q^2s_3$.

Genotype	A_1A_1	A_1A_2	A_2A_2
Frequency before selection	p^2	$2pq$	q^2
s value	0	0	s_3
Relative fitness	1	1	$1 - s_3$
Weighted contribution after selection	p^2	$2pq$	$q^2(1 - s_3)$

Normalised frequencies after selection are $\frac{p^2}{(1-q^2s_3)}$ for A_1A_1, $\frac{2pq}{(1-q^2s_3)}$, for A_1A_2 and $\frac{q^2(1-s_3)}{(1-q^2s_3)}$ for A_2A_2.

Gene frequency after one generation of selection is given by

$$\begin{aligned} q_1 &= \frac{q^2(1-s_3)}{(1-q^2s_3)} + \frac{1}{2}\left[\frac{2pq}{(1-q^2s_3)}\right] \\ &= \frac{[q^2(1-s_3)+pq]}{(1-q^2s_3)} \\ &= \frac{q(1-qs_3)}{(1-q^2s_3)} \text{ and} \\ p_1 &= 1.00 - q_1 \\ &= 1.00 - \left[\frac{(q-s_3q^2)}{(1-s_3q^2)}\right]. \end{aligned}$$

The frequency of the deleterious allele, A_2, is decreased by selection while the favourable allele, A_1, has its frequency increased as selection progresses.

In other directional selection patterns, the selection coefficients take the form $s_2 \le s_3$, ranging from partial to complete dominance. In case of selection against complete dominant allele, $s_3 > 0$, $s_2 > 0$, $s_1 = 0$ and $s_3 = s_2$, with A_2 being detrimental and dominant while the favoured allele A_1 is recessive. The changes in population genetic structure parameters after one generation of selection are as follows.

Genotype	A_1A_1	A_1A_2	A_2A_2
Frequency before selection	p^2	$2pq$	q^2
s value	0	s_3	s_3
Relative fitness	1	$1 - s_3$	$1 - s_3$
Weighted contribution	p^2	$2pq(1 - s_3)$	$q^2(1 - s_3)$

As in the above case weighted contribution of each genotype is the product between genotypic frequency before selection and relative fitness while population mean fitness after selection is the sum of the weighted contributions of the three genotypes.

$$\begin{aligned} \text{The mean fitness of the population after selection} &= p^2 + 2pq(1-s_3) + q^2(1-s_3) \\ &= 1 - (2pq + q^2)s_3 \\ &= 1 - (1 - p^2)s_3 \end{aligned}$$

Normalised frequencies after selection are $\frac{p^2}{[1-(1-p^2)s_3]}$ for A_1A_1, $\frac{2pq(1-s_3)}{[1-(1-p^2)s_3]}$ for A_1A_2 and $\frac{q^2(1-s_3)}{[1-(1-p^2)s_3]}$ for A_2A_2.

After one generation of selection the allelic frequencies are

$$q_1 = \frac{\{\frac{1}{2}[2pq(1-s_3)] + q^2(1-s_3)\}}{[1-(1-p^2)s_3]}$$
$$= \frac{\{pq(1-s_3) + q^2(1-s_3)\}}{[1-(1-p^2)s_3]},$$
$$p_1 = 1.00 - q_1 = 1.00 - \frac{\{pq(1-s_3) + q^2(1-s_3)\}}{[1-(1-p^2)s_3]}.$$

In this selection regime elimination of unfavoured dominant allele from the population is very fast, leading to $p = 1.00$ and $q = 0.00$ in a few generations. If the dominant allele is lethal q goes to 0.00 in one generation of selection.

Stabilising (balancing) selection favours the heterozygote, a condition of heterozygote advantage similar to heterotic situations. This leads to maintenance of both alleles in a state of stable equilibrium in which all the homozygous and heterozygous genotypes exist in the population in equilibrium frequencies. In stabilising selection $s_1 > s_2 < s_3$, s_1 and s_3 may be unequal or have equal magnitudes. The selection coefficients can be taken as $s_1 > 0$, $s_2 = 0$, $s_3 > 0$, in which case, s_1 and s_3 are positive and the heterozygote has the highest fitness since it has the lowest selection coefficient. If $s_2 = 0$, fitness value for A_1A_2 is 1.00 and the gene frequencies and other population parameters are obtained as follows after one generation of such selection.

	A_1A_1	A_1A_2	A_2A_2
Freq before selection	p^2	$2pq$	q^2
Relative fitness	$1-s_1$	1	$1-s_3$
Weighted contribution	$p^2(1-s_1)$	$2pq$	$q^2(1-s_3)$

$$\text{The mean fitness of the population} = p^2(1-s_1) + 2pq(1) + q^2(1-s_3)$$
$$= p^2 - p^2s_1 + 2pq + q^2 - q^2s_3$$
$$= 1 - p^2s_1 - q^2s_3$$

Normalised frequencies after selection are $\frac{p^2(1-s_1)}{(1-p^2s_1-q^2s_3)}$ for A_1A_1,

$$\frac{2pq}{(1-p^2s_1-q^2s_3)} \text{ for } A_1A_2 \text{ and } \frac{q^2(1-s_3)}{(1-p^2s_1-q^2s_3)} \text{ for } A_2A_2.$$

After one generation of selection the allelic frequencies will be

$$\begin{aligned} q_1 &= \frac{[\frac{1}{2}(2pq) + q^2(1 - s_{3)}]}{(1 - p^2 s_1 - q^2 s_3)} \\ &= \frac{[pq + q^2(1 - s_{3)}]}{(1 - p^2 s_1 - q^2 s_3)} \\ &= \frac{[(1 - q)q + q^2 - q^2 s_3]}{(1 - p^2 s_1 - q^2 s_3)} \\ &= \frac{(q - q^2 + q^2 - q^2 s_3)}{(1 - p^2 s_1 - q^2 s_3)} \\ &= \frac{(q - q^2 s_3)}{(1 - p^2 s_1 - q^2 s_3)}, \\ p_1 &= 1.00 - q_1 = 1.00 - \frac{(q - q^2 s_3)}{(1 - p^2 s_1 - q^2 s_3)}. \end{aligned}$$

This selection pattern does not lead to fixation of any allele but leads to an equilibrium situation where both alleles exist in the population. At equilibrium gene frequency is given by $q_e = \frac{s_1}{(s_1+s_3)}$. This a stable equilibrium and the population cannot be fixed for any one allele or any one homozygote. It is the point of maximum fitness for the population under stabilising selection.

Disruptive selection is a case of heterozygote disadvantage favouring the homozygous phenotypes at the expense of the heterozygotes. The heterozygotes are the ones eliminated through selection by virtue of having the highest selection coefficient i.e., $s_1 < s_2 > s_3$ and $s_1 = s_3$ or $s_1 \neq s_3$. If the fitness of the heterozygote is set equal to 1.00 and s_1 and s_3 are given negative values, the fitness of each homozygote will be greater than 1.00 since $1 - (-s_1)$ and $1 - (-s_3)$ for A_1A_1 and A_2A_2 respectively is each greater than 1.00. Population genetic structure parameters after one generation of selection are obtained as follows.

Genotype	A_1A_1	A_1A_2	A_2A_2
Freq before selection	p^2	$2pq$	q^2
Relative fitness	$1 - (-s_1)$	1	$1 - (-s_3)$
Weighted contribution	$p^2(1 + s_1)$	$2pq$	$q^2(1 + s_3)$

The mean fitness of the population after selection $= p^2(1 + s_1) + 2pq(1) + q^2(1 + s_3)$

$$\begin{aligned} &= p^2 + p^2 s_1 + 2pq + q^2 + q^2 s_3 \\ &= 1 + p^2 s_1 + q^2 s_3. \end{aligned}$$

Normalised frequencies after selection are $\frac{p^2(1+s_1)}{(1+p^2s_1+q^2s_3)}$ for A_1A_1, $\frac{2pq}{(1+p^2s_1+q^2s_3)}$ for A_1A_2 and $\frac{q^2(1+s_3)}{(1+p^2s_1+q^2s_3)}$ for A_2A_2.

After one generation of selection

$$\begin{aligned} q_1 &= \frac{1}{2}\frac{(2pq)}{(1+p^2s_1+q^2s_3)} + q^2\frac{(1+s_3)}{(1+p^2s_1+q^2s_3)} \\ &= \frac{[pq+q^2(1+s_3)]}{(1+p^2s_1+q^2s_3)} \\ &= \frac{(q+q^2s_3)}{(1+p^2s_1+q^2s_3)}; \\ p_1 &= 1.00 - q_1 \end{aligned}$$

In this selection pattern there is an equilibrium at which the two alleles are both in the population. This equilibrium point is the point of least fitness for the population making the equilibrium unstable for allelic frequencies since any allelic frequency lower or higher than the equilibrium gene frequency forces the gene frequency to drift to the nearest fixation point where the population is fixed for one of the alleles. This type of selection will ultimately lead to the elimination of one allele or maintain both alleles depending on the original gene frequencies.

Non-random mating

Non-random mating usually refers to assortative mating or consanguineous mating (mating of relatives who have one or more ancestors in common). This leads to inbreeding and the level of inbreeding is influenced by amount of ancestry shared by the common parents. The primary effect of inbreeding is change in genotypic frequencies while the primary effects of mutation, migration and selection are changes in gene frequencies which lead to genotypic frequency changes. The change in genotypic frequencies due to non-random mating is reflected in increased homozygote frequencies and reduced frequencies of heterozygotes. Effects of inbreeding affects all loci in the same genome. The magnitude of the effect of inbreeding depends on the rate of inbreeding, which is the extent of mating of related individuals in the population. Self-fertilisation is the most extreme form of inbreeding leading to the fastest rate of inbreeding. Other forms of inbreeding like parent-progeny mating, full sib mating and half sib mating are less severe and have lower inbreeding rates than self-fertilisation. In dioecious plant species and in situations of where there is self-incompatibility and male sterility in plants self-fertilisation is impossible and the only inbreeding options are sib-mating or backcrossing

Inbreeding coefficient (f) defines the proportion by which homozygotes are increased due to inbreeding and is the probability that two similar alleles at the same locus in an individual are identical by descent by virtue of having come from a particular allele as a result of common ancestry (Wright, 1931). Since genes identical by descent are responsible for inbreeding the coefficient of inbreeding (f) is the probability that a homozygous individual received two identical alleles which originated from the same ancestral gene. High levels of inbreeding as reflected by high f values lead to high reduction of heterozygotes frequencies and increased homozygotes frequencies. Based on (Wright, 1931) derivations, mating of various types of relatives has different levels of inbreeding coefficient after one generation of inbreeding as follows. Genotypic frequencies are defined in terms of inbreeding

Mating type	Inbreeding coefficient, f
Self-fertilisation	$\frac{1}{2} = 0.5$
Parent-offspring mating	$\frac{1}{4} = 0.25$
Full sib mating	$\frac{1}{4} = 0.25$
Half sib mating	$\frac{1}{8} = 0.125$

coefficient (Wright, 1931) such that after one generation of non-random mating the genotypic frequencies are as follows.

A_1A_1	A_1A_2	A_2A_2
$p^2 + pqf$	$2pq - 2pqf$	$q^2 + pqf$

Self-fertilisation reduces heterozygosity, H, by $\frac{1}{2}$ while increasing frequency of each homozygote by $\frac{1}{4}$ of the heterozygote frequency as follows for a population in which the original genotypic frequencies were P, H and Q for A_1A_1, A_1A_2 and A_2A_2 respectively.

Generation of self-fertilisation	A_1A_1	A_1A_2	A_2A_2
0	P	H	Q
1	$P + \frac{1}{4}H$	$\frac{1}{2}H$	$Q + \frac{1}{4}H$
2	$P + \frac{3}{8}H$	$\frac{1}{4}H$	$Q + \frac{3}{8}H$
At limit	$P + \frac{1}{2}H$	0	$Q + \frac{1}{2}H$

Since heterozygousity is reduced by $\frac{1}{2}$ each generation the amount of heterozygosity left after t generations of self-fertilisation at a single locus is derived by the equation

$$H_t = \left(\frac{1}{2}\right)^t.$$

For the case of two alleles, A_1 and A_2, having frequencies p and q the homozygotes at fixation will have frequencies p for A_1A_1 and q for A_2A_2, reflecting the frequencies of pure lines or inbred lines with respect to the locus under consideration. A segregating population of a self-pollinating crop like wheat subjected to bulking advancement without selection every generation ends up with this genetic structure at each locus after several generations of self-pollination and bulking advancement.

Genetic drift

Small populations experience unpredictable random changes in gene frequencies arising from dispersive processes due to sampling of gametes from generation to generation leading to genetic drift. For example, if a number of small populations with same population size and similar allelic frequencies are allowed to go through a number of generations and sizes kept the same across generations as well, they will undergo unpredictable random changes independently and will diverge and differ with respect to gene and genotype frequencies after some generations. Due to such genetic differentiation some populations may even be fixed for certain alleles and become genetically uniform. Random changes in gene frequencies which can be in either direction, are also responsible for reduction in heterozygosity and increased homozygosity leading to inbreeding in small populations. Inbreeding in a small population is due to the fact that the limited number of individuals leads to consanguineous mating in the population. Even though there may be random mating in the population, inbreeding and fixation of alleles occur due to small population size and not necessarily due to mode of breeding. Inbreeding coefficient, f, arising from drift effects for one generation in a population of size N, is the probability that any two gametes uniting to give a zygote have alleles at a particular locus that are autozygous by virtue of having similar ancestral descent. This probability is given by $f = \frac{1}{2}N$ for one generation and the change in the population genetic structure in terms of heterozygosity, H, attributable to small population size after t generations of drift is $H_t = (1 - \frac{1}{2}N)^t H_0$ if heterozygousity started at a level of H_0.

Effective population size

Normally, the number of breeding individuals in a population does not remain constant generation after generation or season after season. Occasionally, small population sizes arise due to natural bottlenecks in form periodical occurrence of disease and pest epidemics, seasonal floods or droughts which reduce numbers of breeding individuals in population of many species and occasional increase in populations of predators and grazers/browsers. The unfavourable situations reduce population sizes such that only a few individuals mate to produce the next generation of the population. Organisms in which male individuals are different from female individuals the number of the two sexes are rarely equal and constant. To deal with such deviations from an ideal population with a constant size from generation to generation and in which the number of male and female individuals are the same the population size is expressed in terms of number of breeding individuals, or effective population size, N_e. The effective population size, N_e, is substituted for the actual number N for estimating f and gene frequency variance. Variation in numbers of sexes is common in animals and is also found in dioecious plant species where male plants are different from female plants like pawpaw, spinach, cannabis, light carnation, yam, asparagus and date palm. The effective population size where there are differences in numbers of the two sexes is estimated according to expression given by (Falconer & Mackay, 1996) as follows.

$\frac{1}{N_e} = \frac{1}{4}N_m + \frac{1}{4}N_f$ and

$N_e = \frac{4N_m N_f}{(N_m + N_f)}$, where N_m = number of males, N_f = number of females.

In situations where generations have unequal numbers the generation with the smallest number has the greatest impact on the rate of inbreeding, gene frequency variance and genotypic frequencies. For unequal population sizes over t generations effective population size, N_e, is obtained using expression given by (Falconer & Mackay, 1996),

$$\frac{1}{N_e} = \frac{1}{t}\left(\frac{1}{N_1} + \frac{1}{N_2} + \frac{1}{N_3} + \cdots \frac{1}{N_t}\right)$$

$$N_e = \frac{t}{\Sigma\left(\frac{1}{N_i}\right)}$$

where $N_1, N_2, N_3, \ldots. N_t$ refer to population sizes in generations $1, 2, 3, \ldots\ldots t$.

5.2 Genetic Structure of Crop Plant Populations

Success in crop improvement is conditioned on adequate understanding of genetic structure of the plant populations being handled. This is necessary because understanding of selection response patterns and breeding strategies are closely linked to the population genetic structure and dynamics and desired type of variety. Genetic structure of plant populations is affected by plant reproductive system, i.e., whether the crop being handled is self-pollinating, cross-pollinating or often cross-pollinated and presence of pollination control mechanisms.

Table 5.1 Changes in genotypic frequencies in % due to progressive self-pollination without selection

Selfing of generation	Genotypes and their % proportions			% Heterozygousity
	A_1A_1	A_1A_2	A_2A_2	
S_0	0	100%	0	100
S_1	25	50	25	50
S_2	37.50	25	37.50	25
S_3	43.75	12.50	43.75	12.50
S_4	46.875	6.25	46.875	6.25
S_5	48.4375	3.125	48.4375	3.125
S_6	49.21875	1.5625	49.21875	1.5625
S_7	49.609375	0.78125	49.609375	0.78125

5.2.1 Autogamous Situations

Crops that are normally self-pollinated differ in population genetic structure from the normally cross-pollinated crops because they are comprised of pure lines or homozygous plants while cross pollinating crops have highly heterozygous populations. In cultivars of self-pollinating crops like beans, wheat, barley, soybean, field peas, cowpeas, rice and finger millet there are mixtures of pure lines, unless there had been previous selection for one pure line genotype. The landrace populations of such crops are comprised of pure lines due to natural continuous self-pollination making any heterozygotes present segregate into homozygotes that breed true and produce respective homozygous progenies generation after generation. For example, if a cultivar is initiated using a group of heterozygous plants or a plant heterozygous at a two-allele locus, A_1A_2, first generation of natural self-fertilisation (S_1) produces 25%A_1A_1, 50%A_1A_2, and 25%A_2A_2 among the derived progenies. If each of these progenies is undergo natural self-fertilisation A_1A_1 and A_2A_2 will breed true to produce only homozygous progenies, A_1A_1 and A_2A_2 respectively. On the other hand, A_1A_2 will produce segregants among their progenies in the proportions of 25%A_1A_1, 50%A_1A_2 and 25%A_2A_2 of the 50% heterozygousity obtained in S_1 generation, leading to a total of 37.50%, 25%, and 37.50% for A_1A_1, A_1A_2 and A_2A_2 respectively in S_2 generation (Table 5.1). In each generation the cultivar will increase in homozygousity by 50% of the previous heterozygote proportions and have the heterozygote component reduced by 50% of the previous heterozygote proportions.

By the fifth generation, each of A_1A_1 and A_2A_2 genotypes comprises 48.4375% of the cultivar while the heterozygous type, A_1A_2, comprises only 3.125% of the cultivar. In the seventh generation homozygousity in the population is over 99%. In a plant population with initial heterozygousity level of H_0 continued self-fertilisation reduces heterozygosity by half each generation to a level of $(\frac{1}{2})^t H_0$ at a diallelic locus in generation t.

If self-pollination continues without selection for indefinite number of generations the population will finally be comprised of two pure line genotypes only, A_1A_1 and A_2A_2, which will appear in equal proportions of 50% each on condition that the original population was started with all plants being heterozygous at the diallelic locus. Alternatively, the homozygotes will be in proportions equal to the respective gene frequencies, i.e., the frequency of A_1A_1 pure line will equal p where p is the frequency of allele A_1 and the frequency for A_2A_2 pure line will be equal to q where q is the frequency of allele A_2. Where there are more than one locus segregating for two alleles at each locus the number of individuals that are completely homozygous at all the loci will be determined by the number of segregating loci and the number of generations of self-fertilisation. Self-pollination in a population, therefore, results in genetic variability because different alleles are concentrated in different lines and consequently reflected in wide phenotypic variability. However, if selection is done for a particular homozygous genotype the population will be comprised of a single line with no genetic variation. Natural selection also favours lines with desirable genes or non-deleterious genes while the lines with deleterious or lethal genes are eliminated with the consequence that under natural conditions the surviving lines may not harbour harmful alleles.

Heterozygotes often arise in self-pollinating crops due to occasional cross-pollination between different genotypes and rare spontaneous mutations. When heterozygotes arise due to occasional outcrossing and mutations, the heterozygotes segregate fast into true breeding homozygous lines. Accidental or intentional mixing of cultivars, change in gene frequencies due to natural selection among the genotypes therein and genetic drift also determine the composition of self-pollinating crop cultivars.

A self-pollinating cultivar has no genetic variability if it has only one pure-line like commercial pure line varieties or it may have substantial variability if it is comprised of a mixture of pure-lines. Landrace cultivars of self-pollinating crops often comprise of mixtures of pure-lines. If individual pure lines in such a mixture are artificially isolated and multiplied each line will produce progenies that are similar to the parental line and among themselves, hence, true breeding. This is the basis for pure-line theory of breeding developed by Johannsen at the beginning of the 20th century. Pedigree breeding and most breeding strategies for self-pollinating crops are designed based on this theory since they lead to development of pure-line varieties.

5.2.2 *Allogamous Situations*

In naturally cross-pollinated crop plants like maize, sunflower, rye, castor, hemp, cabbage, carrot and rapeseed the populations are comprised of highly heterozygous plants since in such crops self-fertilisation does not normally take place unless enforced as in case of production of inbred lines.

In populations of cross-pollinated crops natural cross-pollination, also referred to as open pollination, is equated to random mating and the population genetic structure is explainable using Hardy-Weinberg genetic equilibrium conditions. Genetic equilibrium in open pollinated plant populations is achieved in one generation of random mating on condition that there is no selection, no mutation, no pollen migration and no drift and the equilibrium will be maintained from generation to generation. Such a population with alleles A_1 and A_2 will be comprised of A_1A_1, A_1A_2 and A_2A_2 genotypes in frequencies p^2, $2pq$ and q^2 respectively. Unlike self-pollinating crop populations, the open pollinated crop populations usually have deleterious or lethal recessive alleles in relatively high frequencies because such alleles remain hidden in heterozygous genotypes. Enforcing of inbreeding through self-fertilisation or other form of inbreeding leads to segregation which makes plants homozygous for undesired alleles show up with lowered vigour, reduced plant size, fertility, yield and lowered fitness in populations and some plants may die altogether. Such reduction in vigour and low general performance affects both characters of qualitative and quantitative nature. Self-fertilisation is difficult or impossible to achieve in certain crops in which self-incompatibility genes are present like Brassicas and in crops where the male and female flowers are borne on different plants like pawpaw.

In crops like sorghum, *Amaranthus*, pigeon peas and cotton there are varying amounts of mixed cross-pollination and self-pollination. In populations of such species having mixed mating system the proportions of heterozygotes and homozygotes depend on the extent of mixed mating.The extent of mixed mating is highly influenced by the environment, presence of pollination agents and genotypic variation.

Outcrossing gives chances for recombination, helps to create new gene combinations, new genotypes and therefore creates genetic variation and enhances selection process. Inbreeding does not allow for recombination, maintains genetic linkages intact and restricts creation of variation. In crop improvement sustained response to selection requires the breakage of linkages and recombination for creation of variability. This is the rationale in intermating among selected genotypes followed by selection in improvement of both self-pollinated and cross-pollinated crops.

5.2.3 Vegetative Propagation Situation

A number of crops are propagated vegetatively such as bananas, sisal, potatoes, coffee, tea, pineapples, pyrethrum, sweet potatoes, sugarcane and cassava. In vegetatively propagated crops lack of union of male and female gametes keeps genotypes the same and unchanged from one generation to another unless there are mutational events. This is advantageous in plant breeding because once a desirable genotype has been developed it can be propagated indefinitely and its genetic purity will be easy to maintain unless there is mechanical mixing of propagules or some rare mutations arise.

Vegetative reproduction restricts the formation of new genotypes, especially in crops where zygote formation is impossible because of obligate vegetative reproduction. Furthermore, vegetatively propagated crops have the problem of restricted or narrow genetic base for cultivars under cultivation and are likely to be vulnerable in case of disease and pest or other form new stress outbreaks. However, such problems can easily be solved by induced mutagenesis and methods of modern crop biotechnology and genetic modification followed by clonal selection. In some plant species there is mixed reproduction of both asexual reproduction by apomixis and sexual reproduction leading to zygotes arising from apomixis and sexual reproduction. For a plant population having such mixed reproduction a proportion of the progenies will arise through apomixis and another proportion through sexual reproduction.

5.3 Inbreeding Depression and Heterosis

5.3.1 Inbreeding Depression

In naturally cross-pollinated crops like maize, self-fertilisation does not normally occur unless enforced by controlled mating in breeding programmes. Inbreeding may take the form of self-fertilisation, mating of relatives like half sibs and full sibs and parent-offspring mating with self-fertilisation being the most extreme form of inbreeding. Inbreeding leads to increased homozygosity, a situation that is not part of the normal genetic structure of allogamous crops. As inbreeding takes place deleterious and lethal recessive alleles usually hidden in heterozygous conditions segregate and become homozygous leading to dramatic reduction of vigour, size, and fertility resulting in lowered productiveness, decline in resistance to biotic and abiotic stress factors, general weakness and some homozygous plants may simply die.

Inbreeding leads to appearance of many undesirable and harmful characters controlled by deleterious recessive alleles, e.g., chlorophyll deficiency, plants with diminutive leaves, defective roots, defective floral parts, reduced height, weakness, reduced adaptability, etc., which are not expressed in heterozygous conditions because the responsible recessive alleles are masked by the respective dominant alleles. This phenomenon of reduced performance and loss of vigour and fertility due to inbreeding is inbreeding depression. It is accompanied by phenotypic uniformity within lines as homozygosity progresses due to continued inbreeding. The progress of inbreeding depression through self-pollination is very rapid and after about five generations of inbreeding the inbred progenies will have significant reduction in yield, general performance and survival. Reproductive ability in some lines will be extremely low and only a small number of them will be fertile enough for use in plant breeding. Through inbreeding the population differentiates into genetically and phenotypically distinct lines which are homozygous and fixed for various alleles resulting in increased variability among the lines and low variability within lines. After about seven generations of self-pollination most lines are almost 100% homozygous and further inbreeding and maintenance of the inbred lines can only be done by sib-mating. The inbred lines will have highly reduced yields as compared to parental open pollinated populations from which they are derived as commonly observed in maize, sunflower, alfalfa, carrots, onions, cucurbits and other cross-pollinated crops

The magnitude of inbreeding depression is a function of inbreeding coefficient, gene frequencies and gene effects and its quantification in terms of these parameters for a diallelic locus is defined as

$$M_f = a(p - q) + 2pqd(1 - f) = M_0 - 2pqdf;$$

where p and q are mean allelic frequencies in a population with $+a$, $-a$ and d being genotypic values for A_1A_1, A_1A_2 and A_2A_2 respectively; M_f is mean of the inbred population with inbreeding coefficient f (Falconer and Mackay, 1996). M_0 is mean of the outbred population before inbreeding. The inbreeding depression, I_d, is the difference between the inbred population and the outbred population means.

$$I_d = M_f - M_0 = M_0 - 2pqdf - M_0 = -2pqdf.$$

A locus will contribute to inbreeding depression for a trait only if d is not zero, and the value of the heterozygote differs from average value of the homozygotes, i.e., there is presence of dominance. This quantification of inbreeding depression can be extended for a locus with multiple alleles. If the trait is influenced by more than one locus the combined effects of all the loci towards inbreeding depression is the summation over the loci involved for the effects of inbreeding, i.e.,

$$M_f = \sum a(p - q) + 2\left(\sum pqd\right)(1 - f) = M_0 - 2f\sum pqd \text{ and}$$
$$I_d = M_f - M_0 = M_0 - 2f\sum pqd - M_0 = -2f\sum pqd \text{ (Falconer \& Mackay, 1996).}$$

These expressions indicate that a character will show inbreeding depression if it has a net directional dominance. Interpretation of inbreeding depression for quantitative characters is explainable in terms of dominance and over-dominance which are reduced as alternative alleles segregate into homozygous state and deleterious alleles are fixed in homozygous lines.

5.3.2 Heterosis

When inbred lines are crossed vigour and fertility are restored in the hybrid progeny to a level comparable to the performance of the original open pollinated population from which the lines are derived. The phenomenon by which vigour, fertility and general good performance of the hybrid are restored when inbred lines are crossed in the allogamous species, whether they are from the same parental open pollinated population or from diverse open pollinated populations, is referred to as heterosis.

Heterosis is the superiority of the F_1 hybrid or any other form of hybrid as compared to the mid-parent or the better parent, particularly with respect to yield and other traits related to vigour, adaptability and survival. In F_2 there is a sharp decline in heterosis, especially for hybrids between two inbred lines. Heterosis is reduced by half and maintained at that level in subsequent generations of random mating within the derived population of the cross for outcrossing species.

Though it was originally defined for inbred lines for allogamous crops, heterosis is not confined to crosses between inbred lines of cross-pollinated species only. It is a phenomenon observed in crosses between pure lines of autogamous species and open pollinated populations of diverse origin. Heterosis is more dramatic in allogamous species than is autogamous species because in the latter, through evolution, self-pollination and homozygous nature helped eliminate most of deleterious and lethal alleles which have remained hidden in heterozygotes in allogamous species. In allogamous species inbred lines have different genes in recessive homozygous or dominant homozygous states in various combinations. When such lines are crossed, they will complement each other in hybrids because favourable alleles in different loci are brought together in the F_1 hybrid individuals. For example, if there are five loci with desirable dominant alleles M, N, R, S and T and the respective undesirable recessive alleles m, n, r, s and t which are distributed in two poor performing inbred homozygous genotypes, namely, mmNNRRsstt and MMnnrrSSTT, a hybrid between the two lines gives a progeny heterozygous at all the loci, i.e., MmNnRrSsTt. This hybrid performs better than either of the parents, hence heterotic. Heterosis is manifested in form of increase in yield, fertility, plant size, tiller number, node number and node length, number and size of leaves, faster growth rate, earlier flowering and maturity, larger seeds in cereals, larger fruits in crops like tomatoes and fruit crops, larger maize cob size, larger heads in crops like sunflower and sorghum and larger boll size in cotton. Hybrids show heterosis for resistance and tolerance to biotic and abiotic stress factors and survive better under weed infestation. Hybrids also have better quality in terms of shelf life, higher levels of desired protein, sugars and vitamins than their inbred parents. Hybrid vigour depends on genetic divergence of the parents, and less diverse parents give less heterotic hybrids while greater heterosis is obtained when more diverse parents are crossed (East, 1936; Singh & Sharma, 1989).

Physiological and molecular basis of heterosis

Heterosis is associated with better physiological processes like more efficient soil water and nutrient uptake, improved enzyme activities, photosynthesis, better and improved sink capacity to store metabolites/photosynthates leading to faster and more luxuriant growth, better flowering for good seed set or fruit set as known in most crops like maize, tomato, onion, cucurbits, rice, sorghum, sunflower, Brassicas, fruit crops, and other cross-pollinated crops. Increased number and size of fruiting bodies and grain lead to greater sink for assimilates.

As compared to their parental inbred lines hybrids have more balanced metabolic processes, better mitochondrial function through enhanced and improved activities of enzymes like adenosine triphosphatase (ATPase) activity and more efficient mitochondrial absorption of oxygen. Some enzymes in heterozygotes have unique conformations which may reflect hybrid enzymes and certain biochemicals in hybrids may be of heterobeltiosis content (Yu et al., 2021). Because of larger cells and higher cell number, improved cellular metabolic activities and regulatory mechanisms, hybrids also have faster cell division processes, increased rate of DNA replication, transcription and translation for increased protein synthesis, improved efficiency of enzymatic activities, better regulation of metabolic processes and morphogenesis resulting in faster development and improved partitioning of metabolites for improved vigour, growth and yield performance. Assembling of different favourable alleles at various loci in the hybrid leads to genetic interactions and altered regulation of genes that control biological functions affecting traits associated with hybrid vigour making the hybrid performance surpass those of the parents. Epigenetic changes related to DNA methylation, small RNAs and histone modification also influence and stabilise heterosis due to their effects on gene expression patterns (Shiraki et al., 2023; Yu et al., 2021).

Genetic basis of heterosis

There are two schools of thought regarding explanation of genetic basis of heterosis. One school supports the views of scientists like Davenport and Bruce that dominance plus other genetical complications are responsible for heterosis while the other school supports the proposal of (Shull, 1908) which considers over-dominance, hence heterozygousity per se along with allelic interaction and physiological stimulation, to be responsible for heterosis (Shiraki et al., 2023; Yu et al., 2021). According to dominance hypothesis heterosis is expressed so long as there are gene frequency differences between the parents and there is presence of dominance. Heterosis is also believed to be due to assemblage of dominant favourable alleles and is a product of complementation action of such alleles from both parents at different loci. Dominance hypothesis suggests that at each locus there is a favourable dominant allele and an unfavourable recessive allele influencing the character. Due to heterozygous condition in the hybrids, the desirable characters are expressed because of the dominance allele relationships resulting in heterosis. There is the masking of deleterious effects of recessive alleles by their dominant counterparts. Therefore, this hypothesis works on assumptions that the dominant alleles are beneficial and recessive alleles are deleterious and undesirable, with the loci showing additive effects, no epistasis and that there is free recombination between the genes. An example of the hypothetical case of five genes M, N, R, S, and T given above serves as an illustration here.

For dominance hypothesis, it should be possible by selection and inbreeding, to get true breeding homozygous lines with all the desirable dominant alleles together that are comparable to their vigorous F_1 hybrids in terms of performance, resulting in corresponding decrease in heterosis. However, this is not possible because of linkages involving the polygenic loci, some of which being so closely linked that they are not easy to recombine. Furthermore, decline in heterosis does not necessarily accompany improvement of inbred lines. To the contrary, heterosis is usually improved along with enhanced inbred parents' performances. Another expectation for dominance-recessive phenotypes is that there should be asymmetrical 3:1 segregation in F_2 with skewness towards the dominant phenotype for quantitative characters. Instead, there is symmetrical distribution which is explained in terms of the number of polygenes and magnitudes of their effects, random distribution of the polygenes on chromosomes and tight linkages in blocks for the polygenes within the genomes.

Overdominance hypothesis arises from heterozygosity at individual loci because of strong intra-locus interactions of the alleles. Heterozygotes are superior to their inbred parents because of overdominance arising at various loci, presence of different divergent alleles at various loci leading to high the heterosis attributable to intra-loci interactions. (East, 1936) proposed the idea of increasing divergence among multiple alleles at a locus to explain heterosis. In situations of multiple allelism where alleles at a locus, A_1, A_2, A_3, A_4, etc., have divergence in order of $A_1 > A_2 > A_3 > A_4$ so that A_1 and A_4 are the most divergent the heterozygote A_1A_4 is more heterotic than any other heterozygous genotype involving these alleles. That is, in cases of multiple alleles there is higher overdominance interaction between more divergent alleles than less divergent alleles. Despite overdominance hypothesis explanation for heterosis, there are cases where hybrid vigour is also attributable to epistasis involving several loci and there are cases where homozygotes surpass the heterozygotes, making it difficult to apply this hypothesis alone to explain heterosis.

Reconciliation for the two hypotheses arises from the fact that there is general acceptance that heterosis is due to dominance, overdominance and epistatic gene effects. That is, heterosis is attributable to complimentary interactions of additive, dominant or recessive genes and epistasis and that heterosis is a product of heterozygosity at the various loci controlling a particular trait. Involvement of epistasis implies that genes at different loci interact making the combined effects of the genes in the hybrids higher than expected based on parental performance. Heterosis is also attributable to additive effects between active genes. In a homozygote it is hypothesised that only one allele is active while in a heterozygote both alleles are considered to be active (Yu et al., 2021).

It is an established fact that higher heterosis is obtained by crossing genetically diverse parents, be they inbred lines or populations and there is increase in heterosis as genetic diversity between the parents increases. However, the effect of genetic divergence on heterosis has limits and heterosis increase occurs only within a restricted range of divergence since extremely divergent parents give hybrids with limited heterosis. This is because extremely divergent parents do not nick well due to lack of harmonious allele functions and inefficient physiological functions in the hybrids. Remote hybridisation involving extremely distant parents like different species and genera fails to exhibit any heterosis, especially in regard to reproductivity, survival and fitness related traits. Such hybrids may be sterile because of interspecific barriers between the species.

Quantification of heterosis

Heterosis is influenced by gene frequencies, dominance effects and the difference between the gene frequencies of the two parental populations P_1 and P_2. This relates to all loci affecting a particular trait. If the parental population means are M_{P1} for P_1 and M_{P2} for P_2, the mid-parent mean, mean of the two parents, is

$$M_{\dot{P}} = \frac{1}{2}(M_{P1} + M_{P2}).$$

When the two populations are crossed to produce F_1 by taking individuals at random from each population to be used as parents and the F_1 genotypic value (mean) is taken as M_{F1}, heterosis expressed as the difference between F_1 mean and mid-parent value and in terms of d and y is

$H_{F1} = M_{F1} - M_{\dot{P}} = dy^2$, where M_{F1} is mean performance of F_1 hybrid.

Heterosis depends on the presence of dominance, d, at each locus and extent of diversity between the parents quantified in terms of difference of gene frequencies between the parents, y. Heterosis is maximum when one parent is homozygous for one allele and other parent is homozygous for another allele and the absolute value of allele frequency difference between the two parents, y, is highest.

Heterosis is also expressed as deviation from the mean of the superior parent, often referred to as heterobeltiosis, H_{F1b}, and given as

$$H_{F_{1b}} = M_{F_1} - \dot{P}_b \text{ or as percentage, } \%H_{F_{1b}} = \left[\frac{(M_{F_1} - \dot{P}_b)}{\dot{P}_b}\right] \times 100,$$

where $\dot{P}_b$ is mean of better parent.

Heterosis as percentage of the mid-parent value is given as

$$\%H_{F1} = \left[\frac{(M_{F1} - M_{\dot{P}})}{M_{\dot{P}}}\right] \times 100,$$

For economic reasons the magnitude of heterosis can also be given in comparison to the best local commercial variety and pre-existing hybrids.

Problems

5.1 An important medicinal plant in a riverbed is suffering from a high mutation rate of 0.01 due to a nearby industrial construction. The frequency of petiole pigmentation allele, c, affected by mutation is q = 0.2. Only allele c mutates to allele C. What is the allelic and genotypic frequencies after one generation of mutation followed by random mating?

5.2 Consider a plant population in which it is known that some individuals are self-pollinated and that inbreeding is the only departure from Hardy-Weinberg conditions. At a particular locus in the plant population the genotypic frequencies are as follows. Calculate the inbreeding coefficient of this population.

AA	Aa	aa
0.42	0.36	0.22

5.3 On a papaya farm 50 male and 500 female plants are used to produce seeds for distribution. On another farm only one male papaya plant is used to provide pollen for 400 female papaya plants.

(i) Calculate the effective population size, N_e, in each case.

(ii) Give your comments on the differences of the two effective population size estimates.

5.4 A population of a cross-pollinated plant growing in a grassland is subjected to inflow of genes from a neighbouring donor population at a migration rate of $m = 0.4$. The recipient population has gene frequencies of $P = 0.6$ and $p = 0.4$ affecting pod colour. The frequencies of the same alleles in the donor population are $P = 0.8$ and $p = 0.2$.

(i) What are the genotypic frequencies of the recipient population before migration?

(ii) What are the gene frequencies of the recipient population after one generation of migration?

(iii) Give the frequencies for the genotypes PP, Pp and pp after one generation of migration followed by random mating in the recipient population.

5.5 An insect population native to a riverbed habitat is known to undergo half-sib matting. A scientist created a heterogeneous population of this insect through controlled hybridisation giving the following genotypes for antenna colour.

Genotype	BB	Bb	bb
Number	700	500	600

(i) Give the gene frequencies in this population.

(ii) The scientist allowed the insect population to undergo its regular half-sib mating. Give the frequencies of the three genotypes among the progenies obtained after one generation of half-sib mating.

5.6 Write a brief comment on differences between genetic structure of self-pollinated cowpea landrace and genetic structure of maize double-cross hybrid cultivar

References

Agong, S. G. & Ayiecho, P. O. (1991). The rate of outcrossing in grain amaranths. *Plant Breeding, 107*, 156–160. https://doi.org/10.1111/j.1439-0523.1991.tb00544.x

East, E. M. (1936). Heterosis. *Genetics, 21*, 375–397. https://doi.org/10.1093/genetics/21.4.375

Falconer, D. S. & Mackay, T. F. C. (1996). *Introduction to quantitative genetics* 4th *edn*. Oxford University Press

Shiraki, S., Fujiwara, K., Kamiya, Y., Akter, M. A., Dennis, E. S., Fujimoto, R. & Mehraj, H. (2023). Studies on molecular basis of heterosis in Arabidopsis thaliana and vegetable crops. *Horticulturae, 9*, 366. https://doi.org/10.3390/horticulturae9030366

Shull, G. H. (1908). The composition of a field of maize. *The American Breeders Association Report, 4*, 296. https://doi.org/10.1093/jhered/os-4.1.296

Singh, S. P. & Sharma, J. R. (1989). Genetic improvement of pyrethrum. *Theoretical Application Genetics, 78*, 841–846. https://doi.org/10.1007/BF00266668

Wright, S. (1931). Evolution in Mendelian populations. *Genetics, 16*, 97–159.

Yu, D., Gu, X., & Zhang, S. (2021). Molecular basis of heterosis and related breeding strategies reveal its importance in vegetable breeding. *Horticulture Research, 8*, 120. https://doi.org/10.1038/s41438-021-00552-9

Chapter 6
Quantitative Genetic Variation, Theory of Selection and Marker Assisted Selection

6.1 Quantitative Genetic Variation

6.1.1 Gene Effects, Genotypic and Phenotypic Values

Changes in gene and genotypic frequencies lead to genetic variation within populations and among populations. Individuals within a population are different among themselves because of different alleles they carry at different loci and populations also differ among themselves because of variation in alleles and allele frequencies they have at different loci. Some alleles may be specific and characteristic of certain populations. Genetic variation may be easily detected for some characters while it may not be that obvious for other characters.

Differences in some characters may be attributable to genes with small effects on the characters they control, often referred as minor genes, quantitative genes or polygenes. Most characters in plants, particularly those of economic importance in crop plants, are of quantitative nature and are called quantitative characters, metric or polygenic characters. Usually, several polygenes affect a particular trait with each polygene contributing only a small effect towards the expression of the trait it influences which makes it difficult to detect the changes in the effects of individual polygenes. Small effects of the many genes affecting a trait lead to continuous distribution for the trait as the genes are progressively changed, hence, they show continuous variation in segregating populations. Characters controlled by polygenes do not segregate into discontinuous phenotypic classes, but instead have continuous variation in segregating populations, a feature that is not observed for traits under simple inheritance of Mendelian nature. Therefore, quantitative characters are assessed by quantitative measurements like height, weight, colour intensity, seed yield, biomass yield, leaf size, tuber yield, stem diameter, days to flowering, days to maturity, etc. and expressed in terms of means, variances and standard deviations. Though there may be a few cases of cytoplasmic influence, polygenes controlling these traits are located on chromosomes within the nucleus and polygenes for a given

P. O. Ayiecho and J. O. Nyabundi, *Conventional and Contemporary Practices of Plant Breeding*, https://doi.org/10.1007/978-3-031-74998-8_6

character tend to be linked in blocks on chromosomes. This is the basis for quantitative trait loci (QTLs) analysis for quantitative traits which has made it easy to conduct detailed estimate of the number of genes or blocks of genes controlling quantitative traits. A quantitative trait locus is a segment of the chromosome associated with the expression of a quantitative trait since the genes controlling that trait are considered to be located in the segment.

The quantitative traits are also highly influenced by environmental effects which complicate the distribution so that there is a high degree of continuity in the distribution. Simultaneous segregation of the several genes affecting a quantitative trait and superimposition of environmental effects are responsible for such pattern of variation. Qualitative or Mendelian traits controlled by major or oligogenes are not as highly influenced by environmental conditions since the effects of qualitative genes are large and variation and distribution of such traits are easily quantified in terms of ratios, gene, genotypic and phenotypic frequencies. Some qualitative traits include sex expression, colour, shape, quality, disease resistance in many cases, pigmentation, vigour, synchrony of tillering, leaf senescence, dwarfness versus normal height, spinnability of lint cotton, grain colour, texture, etc. They are simply inherited and are under the control of one, two or a few major genes and are highly heritable and therefore amenable to improvement by simple plant breeding methods like selection, crossing and backcrossing and induced mutagenesis. Though metric traits differ from qualitative traits in terms of phenotypic variation, superimposition of environmental effects on genetic effects and quantification of the variation, the polygenes controlling them behave like Mendelian genes with respect to the mode of gene function, as they are also located on nuclear chromosomes, have additive, dominance, pleiotropic and epistatic action, gene frequency variation in populations and undergo segregation for diploid organisms or those polyploids that behave like diploid. At each polygene locus there is 1:2:1 genotypic segregation for the heterozygote but the effects may be unnoticeable phenotypically due to small effects of each allele, the number of other contributing loci and confounding environmental effects.

In practice metric traits variation is given in terms of means, variances, and standard deviations and covariances which can also be used to quantify resemblance of relatives but without distinct classification. Though the metric traits cannot practically be studied in terms of gene, genotypic and phenotypic frequencies because of the polygenic nature and environmental effects the behaviour of genes for quantitative traits can be investigated with the assumptions of additive and dominance gene action.

When a quantitative trait is measured on an individual a phenotypic value is obtained, e.g., a given height measurement in cm or m, or biomass or seed yield in g or kg, etc. Parameters like means, variances, covariances and standard deviations are based on measurements of phenotypic value determined by genes and environmental factors. The environmental component (E) of the phenotype comprises all the non-genetic factors affecting the phenotype. Therefore, the phenotypic value P of an individual is given by

$$P = G + E, \text{ Were } G \text{ is the genotypic value (or genotype).}$$

The genotype imparts some value on an individual while the environmental factors cause positive or negative deviations from the value attributable to the genotype.

If several individuals in a population like a crop cultivar are measured the values are summed up to get the population total and mean and the positive and negative environmental deviations from the mean cancel out so that the environmental deviations sum to 0. Under such a situation the population mean, which is the mean phenotype, can be equated to the genotypic mean of the population.

In addition to the genotypic effects and environmental deviations, different genotypes may respond differently to different environments leading to different performances for different genotypes under different environmental conditions. The variation in performance attributable to differential response of various genotypes exposed to different environments is called genotype-environment interaction (G × E) which is another component of the phenotype. The genotype and environment are interdependent and interact with each other to produce the phenotypic value of an individual for any trait as follows.

$$P = G + E + G \times E.$$

The genotypic value is comprised of two major effects, namely, the additive effects or breeding value due to additive genes and dominance effects attributable to dominance deviation. The additive effects are due to the average effects of genes that individuals possess and pass to their progenies which determine the performance and mean genotypic value of the progenies.

The average effects of genes of an individual influence the mean genotypic value of its progeny which in turn gives the breeding value or additive genetic merit of the parent. Breeding value of an individual mated at random to others in the population is twice the mean deviation of its the progeny from the population mean (Falconer & Mackay, 1996). This forms the basis for progeny testing and estimation of combining ability for individuals and families to be selected for crop population improvement

The overall breeding value of a particular genotype is obtained by the summing up the values across the loci involved in control of the target trait. For a large population in H-W equilibrium the mean breeding value can be equated to the mean genotypic value and the mean phenotypic value. Breeding value can be designated as A and comprises the additive gene component of the genotype. Dominance deviation which is due to dominance interaction between the alleles at a locus can be designated as D. The two components are the main components of genotypic value G of an individual.

$$G = A + D.$$

In addition to additive and dominance components of the genotypic value there is epistatic component, I, attributable to epistasis across the loci. This makes the total genotypic value become

$$G = A + D + I.$$

6.1.2 The Variance Components

Apart from mean values the phenotypic measurements are also used to derive other parameters such as variances, standard deviations and covariances. Variance components are used to measure variation among the individuals within a population or among populations. The total variance in a population is the phenotypic variance (σ_P^2) which is comprised of genetic variance component (σ_G^2), environmental variance component (σ_E^2), and variation due to genotype-environment interaction (σ_{GE}^2):

$$\sigma_P^2 = \sigma_G^2 + \sigma_{GE}^2 + \sigma_E^2$$

Genetic variance component also has two major components, the additive variance component attributable to additive gene effects (σ_A^2) and dominance variance component arising from variation of dominance deviations (σ_D^2). Under normal circumstances for any trait, there is also variance due to epistasis, σ_I^2, which is sometimes ignored and lumped together with σ_D^2 but may be important in hybrid breeding programs, hence the genotypic variance has three components as follows.

$$\sigma^2 G = \sigma_A^2 + \sigma_D^2 + \sigma_I^2.$$

If various components of epistasis are taken into account epistatic variance, according to Kempthorne (1955) is given as

$$\sigma_I^2 = \sigma_{AA}^2 + \sigma_{DD}^2 + \sigma_{AD}^2 + \sigma_{AAA}^2 + \sigma_{DDD}^2 + \sigma_{AAD}^2 + \sigma_{ADD}^2 + \text{ etc.},$$

with the definitions of the components in this partitioning being

$\sigma_{AA}^2, \sigma_{AAA}^2$, etc. = epistatic variances due to interaction of additive effects
$\sigma_{DD}^2, \sigma_{DDD}^2$, etc. = epistatic variances due to dominance deviation interactions
$\sigma_{AD}^2, \sigma_{AAD}^2$, etc. = epistatic variances due to interaction of additive and dominance effects.

The total genetic variance, based on Kempthorne (1955), for a diploid population in H-W equilibrium is

$$\sigma_G^2 = \sigma_A^2 + \sigma_D^2 + \sigma_{AA}^2 + \sigma_{DD}^2 + \sigma_{AD}^2 + \sigma_{AAA}^2 + \sigma_{DDD}^2 + \sigma_{AAD}^2 + \sigma_{ADD}^2 + \text{etc.}$$

Genotypic variance components are determined by gene frequencies and genotypic values, implying that additive and dominance genetic variances and the interaction components in any population are influenced by the gene frequencies in the population.

Environmental variance component includes all the non-genetic variance components attributed to biotic and abiotic factors like climatic conditions, soil moisture and nutrients, pests, diseases, etc. which affect phenotypic expression and error term. The environmental variance also has two components, σ_E^2 and σ_{GE}^2 as given above.

The variance components are useful in estimating heritability and other parameters in breeding programmes. Heritability gives the proportion of phenotypic variance that is attributable to genetic component. Heritability is given as broad sense heritability (degree of genetic determination) or narrow sense heritability (or heritability) as follows.

$$\text{Broad sense heritability} = h_{BS}^2 = \frac{\sigma_G^2}{\sigma_P^2}$$

$$\text{Narrow sense heritability} = \text{heritability} = h_{NS}^2 = h^2 = \frac{\sigma_A^2}{\sigma_P^2}$$

Heritability measurements are important in breeding programs because they are used in predicting selection gain.

6.1.3 Covariance Between Relatives

Relationships and resemblance between relatives are often quantified in terms of covariance and correlation between relatives. Resemblance between relatives is due to the fact that relatives have genes in common which they inherit from their parents leading to genetic similarities between them. Based on the pedigree relationships the covariances between relatives are expressed in terms of components of genetic variance (Kempthorne, 1955) for reference populations. In appropriately designed breeding experiments the covariance information is used for estimation of genetic variance components. The variances, taking epistasis into account, based on family pedigrees and covariance of relatives for random mating populations are in Table 6.1.

Covariance information for different family structures is important in breeding programmes since selection progress depends on the degree of resemblance and

Table 6.1 Family pedigrees and the corresponding covariances and genetic variances (Based on Kempthorne (1955))

Relatives	Covariance	Genetic variance among relatives
Parent-offspring	Parent-offspring regression	$\sigma_G^2 = \frac{1}{2}\sigma_A^2 + \frac{1}{4}\sigma_{AA}^2 + \cdots$.
Full sibs	COV_{FS}	$\sigma_G^2 = \frac{1}{2}\sigma_A^2 + \frac{1}{4}\sigma_D^2 + \frac{1}{4}\sigma_{AA}^2 + \frac{1}{8}\sigma_{AD}^2 + \frac{1}{16}\sigma_{DD}^2 + \cdots$
Half sibs	COV_{HS}	$\sigma_G^2 = \frac{1}{4}\sigma_A^2 + \frac{1}{16}\sigma_{AA}^2 + \cdots$

covariance between relatives used in selection, be they individuals or families. Environmental effects are usually considered uncorrelated between relatives hence the covariance of relatives deals only with genotypic values in plants, especially if appropriate mating designs and experimental designs are applied. Appropriate breeding and experimental designs, like various diallel designs, North Carolina designs, generation mean analysis, etc., are used to generate relatives which are evaluated and variance components estimated which are in turn expressed in terms of covariances of relatives and therefore genetic variance components. The assumptions necessary for interpretations of genetic composition of the covariance of relatives obtained from different mating designs are normal Mendelian diploid inheritance, absence of environmental correlations with relatives, lack of maternal effects and linkage disequilibrium in the population sampled, non-inbred relatives and relatives used are a random sample of some specified population.

6.2 Response to Artificial Selection

The primary aim of the plant breeder is to develop improved crop varieties. This is done through changing the genetic properties of populations by enforcing artificial selection and controlling the reproduction of the selected parents which may involve inbreeding, pairwise crosses, random mating or some other form of structured cross-mating of the individuals. Selection involves using the most superior individuals from the population as parents to give the next generation where superiority may mean highest or lowest value, depending on the trait under selection. Selection is used to discriminate among individuals within a population or select the best population among a set of different related or unrelated populations. Products of selection are used either directly for commercial production by farmers or can be used for further breeding activities to get superior varieties. The presence of genetic variability accompanied by appreciable heritability is mandatory for selection response. Selection primarily changes gene frequencies resulting in changes in genotypic frequencies and consequently changed mean phenotype or performance of the population. Since the individuals used as parents for the next generation will have a restricted number, selection may also lead to inbreeding due to small population size or the population being fixed for the wrong alleles leading to unexpected limited response.

In quantitative traits it is difficult to assess the effects of selection in terms of gene and genotype frequency changes because it is not easy to identify individual polygenes and quantify the changes in terms of their frequencies. The effects of selection on quantitative traits are viewed in terms of changes in the mean phenotypes, variances and standard deviations by comparing populations derived through selection and the respective parental populations.

When artificial selection is conducted on a variable population there is likely to be a change in the population mean performance, the change being referred to as selection response, genetic gain or selection gain, symbolised as R, and given as

$R = \tilde{Y}_s - \tilde{Y}_0$; where $\tilde{Y}_0$ is the mean performance of the original population before selection, $\tilde{Y}_s$ is the mean performance of the population derived after selection.

The difference between the original population mean performance before selection, $\tilde{Y}_0$, and the mean performance of the selected parents to be used to derive the next population, $\tilde{Y}_P$, is selection differential which can be given as

$$s = \tilde{Y}_P - \tilde{Y}_0.$$

Heritability, h^2, of the trait under selection can be used to predict selection response as follows

$$R = sh^2 \text{ (Falconer and Mackay 1996) or } R = s\left(\frac{\sigma_A^2}{\sigma_P^2}\right).$$

Heritability is also considered as the regression of selected individuals' breeding values on their phenotypes. Predicted response to selection, R, is valid for only one generation and for every generation of selection the predicted response must be estimated afresh because of changes in population genetic structure arising from selection. Since heritability depends on gene frequencies of the population it will change as gene frequencies, genotypic frequencies and variance component magnitudes change with selection. Therefore, predicted response must be estimated afresh for each population as each population or generation has its unique genetic properties.

Predicted selection differential can be obtained if the trait subjected to artificial truncation selection has normal distribution and the best individuals are chosen based on their phenotypic performances (Falconer and Mackay 1996). Selection differential, s, depends on selection pressure and the phenotypic standard deviation, σ_P, and can be expressed as a fraction of phenotypic standard deviation, i.e.,

$$i = \frac{s}{\sigma_P} \quad \text{(Falconer \& Mackay, 1996).}$$

This standardised selection differential, i, referred to as intensity of selection, is determined by the percentage of population selected and its values are presented in the tables of Fisher and Yates (1963). Some i values for situations where selection is assumed to be conducted on large populations are given in Table 6.2.

Intensity of selection is also employed to predict selection differential, s, and response to selection, R, if the variance components are known (Falconer & Mackay, 1996) as follows.

$$s = i\sigma_P$$

$$R = ih^2\sigma_P = \frac{i(\sigma_A^2)\sigma_P}{\sigma_P^2} = \frac{i\sigma_A^2}{\sigma_P}$$

Table 6.2 Selection intensity, i, values for corresponding selection pressure % (Based on Fisher and Yates (1963))

% Selected	Selection intensity i	% Selected	Selection intensity i
0.01	3.960	7.0	1.91
0.1	3.367	8.0	1.858
0.5	2.892	9.0	1.804
1.0	2.665	10	1.755
2.0	2.421	15	1.554
3.0	2.268	20	1.400
4.0	2.154	25	1.271
5.0	2.063	30	1.159
6.0	1.985	35	1.058

Realised heritability, h_r^2 is one way of describing the relationship between selection response and selection differential, that is,

$$h_r^2 = \frac{R}{S}.$$

6.3 Marker Assisted Breeding

6.3.1 *Use of Yield Components and Morphological Characters for Indirect Selection*

In conventional plant breeding selection is normally based on phenotypic expression of individual plants or mean phenotypes of families or populations for the target characters. Though plant breeders sometimes recommend the use of component characters for improvement of yield, direct or indirect selection for yield and related metric characters based on phenotypic expression has always had limited progress. In crop plants many characters of economic importance are in one way or another related to each other. Crop yield has also been defined in terms of what are believed to be its component characters for selection purposes. Most crop yield component characters are quantitative and are equally as complex as yield in inheritance with low heritability and therefore, inappropriate and inefficient to use as selector variables to improve one another. However, some of component characters have been proposed as criteria for indirect selection in form of marker assisted selection to improve traits like yield, quality aspects and resistance and tolerance to various stress factors. This has been based on the phenotypic relationships of various characteristics which may be due to physiological and genetic associations.

Indirect selection of desired superior genotypes using markers simplify selection for the target characters making selection cheaper and easier. Employment of indirect selection requires that a selector trait should have simple inheritance with high heritability, easy to evaluate, distinct and genetically associated with the target trait. Many yield components that are of quantitative nature rarely meet such requirements and are difficult to select for due to continuous variation resulting from polygenic control and environmental influence. Some of them are either morphological traits or are associated with morphological characters making them undesirable selection markers. A number of morphological markers like height, fruit and flower colour, fruit and seed shape and size, pubescence, leaf shape, waxiness, leaf colouration and others can be used as markers in selection for a number of traits they may be associated with like disease and pest resistance, tolerance and resistance to abiotic stress factors, male sterility and other traits. However, use of most morphological or physiological markers for indirect selection in breeding is undesirable because they may be influenced by polygenes, dominance, pleiotropy and epistasis and some may have late growth stage expression and even affect the adaptability of the crop.

6.3.2 Phenotyping, Genotyping and Bioinformatics

Phenotyping

Phenotyping, which is the analysis and evaluation of complex and multi-trait growth and development characters (e.g., plant growth, development, morphology, architecture, various functional and physiological aspects), is important for genetic variation analysis in breeding programmes. Phenotypic measurements for both qualitative and quantitative traits in germplasm resources and other breeding materials makes it possible to identify lines and individuals with the desired genotypes for crop improvement. Since multi-trait complex quantitative characters are controlled by minor genes and highly influenced by the environment appropriate methods have to be designed for precise data gathering and predictive accuracy in phenotypic assessments. This calls for the use of modern phenotyping approaches and appropriate facilities. Precise phenotyping requires use of high resolution high-throughput facilities that capture the required information without destruction of the plant. The availability of unmanned automated and computerised equipment with high resolution imagery and sensors for high throughput non-destructive phenotyping allows accurate and repeatable measurements, data gathering several times in a day and throughout the season, generating large amounts of data, evaluation and screening of germplasm with respect to plant architecture traits and early detection of desirable genotypes (Varshney et al., 2018; Razzaq et al., 2021). Qualitative and quantitative traits including agronomic traits like seedling vigour, flowering and maturity time, flower and fruit counts, biomass, grain yield, height, leaf erectness, tillering, branching, canopy structure, etc., and resistance and tolerance to biotic and abiotic stress factor scan be measured with high degree of accuracy and reliability with the help of such facilities.

Phenotyping can be done under speed breeding system where generation time is significantly reduced for faster growth and development, especially for the complex quantitative traits. In some cases, phenotypic assessment under speed breeding protocols correlates well with field phenotypic measurements and the traits are expressed distinctly because speed breeding conditions help to eliminate variations that are common with field-based measurements, especially if high-throughput facilities are employed (Alahmad et al., 2018). High-throughput phenotyping generates large amounts of data for analysis and storage which need to be efficiently handled and managed professionally employing bioinformatics tools.

Omics technologies, namely, genomics, transcriptomics, proteomics, metabolomics, phenomics and epigenomics assays, along with bioinformatics tools improve the breeders' ability to accurately evaluate, identify candidate genes, understand aspects of plant growth and development, predict and select best materials for breeding and fast-track development of appropriate crop varieties with high yield potential, stability, quality and better tolerance/resistance to biotic and abiotic stresses (Janni et al., 2020). The assays assist in identification of stress responsive proteins and other related metabolites and their functions under stress conditions. Omics technologies elucidate the complex molecular mechanisms pertaining to plant growth and development, their interactions with the environment, plant response to stress and expression and regulation of stress response genes under stress conditions (Janni et al., 2020). Phenomics which helps to envisage the plant traits and characterise phenotypes to be used in selection can be integrated with genomic analysis using next-generation sequencing technologies for genotyping to enhance plant breeding.

Genotyping

Genotyping of lines, individuals and samples in populations through mapping is an important activity in genetic analysis for identifying alleles/genes/QTLs and genotypes with the desired characters for use in breeding. Genomics is useful in genetic mapping and gene/DNA sequencing, helps in determining the structure and function of genomes, understanding the functions of genes and their interaction and molecular basis of biological processes. To deal with large populations in a short time in this respect the breeder relies on use of high-throughput next-generation sequencing techniques to reveal genetic variation in the germplasm resources and breeding materials and identification of target genes and QTLs. High-throughput next-generation sequencing technologies, techniques like genome-wide association studies (GWAS) and genotype-by-sequencing (GBS) help in efficient identification of markers like single nucleotide polymorphisms (SNPs) which in turn enhance phenotypic and genotypic diversity investigations, genotyping and screening of germplasm and understanding of the genetics of traits under study (Gomez-Casati et al.,2018). Several genomic markers are available which can be used in DNA mapping and tagging of genes and QTLs that are of importance in breeding and genetic analysis. With the use of GWAS and GBS, SNPs has stood out as one of the most useful markers that have been employed by plant scientists and geneticists for genotyping in a number of crops like maize, wheat and rice (Bakala et al., 2020). Availability of genetic maps and molecular information about various traits has made it possible to associate candidate genes in model plant species with corresponding loci in crop plants for conventional and molecular breeding.

Genomic analyses, DNA sequencing and phenotyping data facilitate identification of molecular markers linked to genes/QTLs controlling target traits for use in regular breeding, molecular marker assisted breeding and genomic selection. Such data are employed in distinguishing different lines, cultivars and their relationships for purpose of designing hybridisation between selected genotypes/lines to maximise heterosis and other selection procedure. Integration of genome sequencing data with other omics technologies analyses and bioinformatics can be employed to fast-track the development of desired crop varieties that are high yielding, climate change resilient, tolerant or resistant to various abiotic and biotic stresses and have the required produce quality and nutraceutical properties for food crops.

Bioinformatics

Bioinformatics is a multidisciplinary scientific field merging biology with mathematics, information technology, statistics and computer technology. It helps in managing biological information for understanding of biological phenomena and includes the collection of biological information, storage, management of the data and the development of software for analysis of large data. It utilises mathematics and computer technology to manage, analyse, store and retrieve biological data. In plant sciences some information covered by bioinformatics databases pertain to phenotyping, genotyping, genome sequences, DNA mapping and sequencing, RNA sequencing, metabolomics, genomics, phenomics, transcriptomics, proteomics, gene expression, protein functions, gene and protein structure analyses, protein-protein interactions, amino acid sequences, aligning different DNA and protein sequences and genetic interaction pathways (Alemu, 2015); (Kushwaha et al., 2017). The data also relate to modelling of transcriptional and translational control, metabolic pathway analyses, evolutionary relationship investigations, identification of target genes of qualitative and quantitative traits, markers for breeding purposes, host-pathogen interactions and other aspects of molecular biology. The technology is also useful in analysing and understanding of phylogenic relationships and various biological functions and networks. In the sequence analyses bioinformatics identifies and distinguishes DNA nucleotide sequences from different organisms and develops the relationships between the organisms. Large amounts of data obtained in these areas of inquiry require application of bioinformatics tools for efficient data management, maintenance, storage, retrieval, analyses, addition, updating, annotation of data and modelling, visualisation and prediction of results (Tiwari et al., 2020). The bioinformatics databases are categorised into primary and secondary databases. Primary (archived) database is information and annotation of DNA and protein sequences, structure and expression profiles while secondary database (derived database) comprises of results of analysis on the primary datasets including information on sequence patterns or motifs, variants and mutation and evolutionary relationships (Alemu, 2015).

Model plants like *Arabidopsis thaliana* are used for genetic studies and applying the information gathered to understand other plant species and food crops. Plant scientists have constructed databases for *Arabidopsis* and various crops like groundnut, coffee, pawpaw, pigeon pea, soybean, cotton, barley, cassava, beans, potato, rice, African rice, wheat, maize and Chinese cabbage, which can be used to for improvement of the crops (Kushwaha et al., 2017). The genomics datasets on genome sequences and the markers that are managed through bioinformatics are highly beneficial to the field of agriculture, especially plant breeding and genetics. The data on proteins, metabolites and transcriptomes make it easy to understand genetic and molecular basis of biological processes and metabolic pathways, thereby facilitating planning and designing efficient and effective programmes for using plant genetic resources in crop improvement programmes.

Bioinformatics plays a major role in germplasm resource characterisation, evaluation and conservation and plant breeding through judicious application of bioinformatics tools to handle datasets obtained for germplasm resources. Availability of the sequence data for the relevant genes and QTLs and their markers and, in many cases genome wide analyses, make bioinformatics play a key role in germplasm conservation and plant breeding. Bioinformatics acts as guide to plant scientists as regards the use of genetic resources and helps identify which crop relatives are close enough for ease of transfer of target genes to desired crop cultivars, e.g., genes or QTLs for high yield, adaptability and tolerance to abiotic stresses, climate change resilience, resistance and tolerance to pests and diseases, and quality attributes like genes for elevating nutraceutical qualities for food crops. Bioinformatics data are utilised to distinguish plants with respect to differences in gene expression or protein profiles under different environmental conditions like drought, heat, mineral stress, disease and insect pest attack and can reveal genes expressed in high levels under such conditions for better plant survival. Such genes are targeted by selection for development of better adapted cultivars or incorporated into cultivars which lack them. Bioinformatics together with next generation sequencing technology and 'omics' (genomics, phenomics, proteomics, transcriptomics and metabolomics) analyses provide opportunities for plant breeding programmes to develop crop plants with desired characters for improved agricultural production and sustenance of food security in the light of climate change challenges.

Through bioinformatics tools and 'omics' technologies plant scientists are able to understand better the genetic architecture of plant pest and pathogen virulence and the molecular basis of interaction with the hosts and to strategize the development of disease and pest resistant cultivars. Genomics and proteomics techniques and bioinformatics have made it possible for the scientists to map the entire or partial genomes and protein functions of pathogens and their plant hosts facilitating molecular understanding of the genetics of pathogenicity and host resistance to diseases (Alemu, 2015).

Though bioinformatics plays key roles in modern crop improvement programmes and genetic analyses through efficient and professional management of the vast phenotyping and genotyping data and other related information for the benefit of breeding programmes, it has some challenges. The technology has the challenge of large amounts of data which must be professionally managed with easy access calling for appropriate computer tools to handle, store the large voluminous data and making them available whenever required.

6.3.3 Genotypic Selection Using Biochemical and Molecular Markers

Use of biochemical and molecular markers resolves the limitations of breeding for quantitative traits in crop plants by integrating them in the breeding programmes as selector traits by virtue of being linked to the genes controlling the expression of particular target traits. Marker assisted selection using isozyme and DNA molecular markers improves response to selection for difficult to select yield and related agronomic traits, selection for tolerance to abiotic and biotic stress conditions and quality aspects. These markers also solve the problem of need for special selection conditions and facilities which are needed for expression and detection of certain characters like resistance to abiotic and biotic stress factors, e.g., drought and mineral stresses, insect pests, diseases and parasitic weeds.

Apart from being abundant in the genome, it is a requirement that an ideal marker be expressed early to allow for selection in early stages of plant life to save time and resources and the genes controlling the marker traits should be codominant so that the homozygous and heterozygous genotypes at the marker locus are easily and distinctly identifiable. The best markers are those which are tightly linked to the genes controlling the target trait to help precisely tag a gene or genes controlling the trait for effective and simultaneous selection of genes controlling the particular trait without unnecessary and undesirable genes being included. An ideal marker should be neutral with respect to expression of agronomic characters, performance and adaptability of the crop and have no pleiotropic or epistatic effects on the target trait or other agronomic traits. These requirements make morphological and phenological traits and resistance to biotic and abiotic stresses undesirable markers for selection. Isozyme and DNA markers are ideal by virtue of meeting these requirements in addition to being polymorphic and having high heritability, making them attractive markers for gene mapping and use in selection.

Markers are most useful when the target characters have low heritability like most metric traits or the trait is controlled by recessive genes. Markers are important where the crop under improvement has long juvenile periods and selection is for characters expressed late in the life cycle like flower characteristics, fruit quality attributes and fruit/seed yield. Markers are also appropriate for early indirect selection in field crops for genes which are expressed late in plant development stage, such as high lysine and tryptophan genes in maize.

Isozyme makers

Isozymes are different forms of an enzyme with different molecular weights and electrophoretic characteristics coded by different alleles of the same gene but having a similar function. The isozymes can be separated on gel electrophoretic column and stained to reflect functions of different alleles of a particular gene at a specific locus on a chromosome. These products of allelic expression are co-dominant so that enzyme extracts from each homozygote produce one band and the heterozygote extract gives two distinct bands where each of the two bands is attributable to the enzyme product of each allele. The isozyme banding patterns are easy to interpret and several different isozyme loci can be studied simultaneously from a sample. They are handy selection markers for target traits like quality attributes, yield and yield related traits, resistance and tolerance to diseases, pests, parasitic weeds and abiotic stress factors if it is ascertained that they are associated with the target trait through genetic linkage.

DNA molecular markers for breeding

Though they have good features as ideal markers, isozymes have limitations in that different isozymes need different electrophoretic conditions and staining requirements, their expression is dependent on developmental stages of the tissues from which they are extracted and they are also not adequately polymorphic and abundant as compared to DNA based molecular markers. DNA molecular markers which are small polymorphic segments of DNA sequences are the best markers for molecular breeding (DNA marker assisted breeding) because they meet most requirements for ideal markers for use in marker assisted selection, marker assisted recurrent selection and marker assisted backcrossing. Furthermore, DNA markers are not tissue specific, not affected by tissue development stage and environment, have high levels of polymorphism, have no adverse effects on the phenotype, can easily be applied as selection markers for both major gene and polygene-controlled characters, are evenly distributed in the genome and have codominant and distinct allelic expression with no pleiotropy (Jiang, 2013). DNA markers are both nuclear genome based and cytoplasmic organelle genome based and can be identified in the genomes of chloroplasts and mitochondria. They make it possible to study more loci and more alleles than isozyme analysis. Use of DNA markers require dense molecular maps with uniformly distributed markers for locating the target genes or quantitative trait loci with which they are closely associated by virtue of close linkage or flanking the target genes/QTLs or being located within the target genes/QTLs. The various genomic regions controlling a particular quantitative trait can be tagged using tightly linked flanking molecular markers, one on either side of the QTL with the consequence of improving the efficiency and precision of selection and improvement of the target metric trait.

Some of the commonly used DNA molecular markers which have been developed by various molecular geneticists are restriction fragment length polymorphisms

(RFLPs), variable number of tandem repeats (VNTR), single nucleotide polymorphisms (SNPs), random amplified polymorphic DNAs (RAPDs), amplified fragment length polymorphisms (AFLPs) and micro-satellite loci or simple sequence repeats (SSRs) also referred to as short tandem repeats (STRs) or sequence-tagged micro-satellites (STMS) (Jiang, 2013; Amiteye, 2021). The RFLP marker system is a Southern blotting-based marker method for genome mapping which is simple, reproducible but requires large amounts of DNA and not easy to use while SNPs, SSRs, AFLP, and RAPDs are polymerase chain reaction (PCR) based marker systems which need only small amounts of DNA for amplification, are easy to use and are highly polymorphic in plant genomes with high marker indices (Amiteye, 2021). The SSRs and SNPs makers can be used for all purposes while AFLPs and RAPDs are useful for diversity studies.

The DNA markers have heritability of 1.00 because the genotype is always fully reflected in the phenotype and are used as indirect selector traits with extremely high efficiency. Once the linkage of a molecular marker to a trait of interest has been ascertained selection can be initiated for the marker since such selection will always bring with it the desired trait. Flanking markers tightly linked to the target gene/QTL or markers located within the target genes/QTLs improve the effectiveness of indirect selection and the probability of getting superior genotypes with the desired traits through marker assisted selection, marker assisted pedigree breeding, marker assisted recurrent selection, marker assisted backcrossing and gene pyramiding. Use of molecular markers like RFLPs, RAPDs, RFLPs, SSRs, SNPs, VNTRs to identify and locate genes or QTLs associated with desired traits in genotype-based selection fast-tracks the development of cultivars with improved yields, adaptability, qualities, abiotic and biotic stress factors resistance and resilience to climate change. Nearly all traits of economic importance in crop plants can be improved through DNA marker assisted breeding, e.g., yield and related agronomic traits, fruit or seed colour, shape, size and quality attributes, male sterility, incompatibility/compatibility traits, phenological characters, plant architecture, resistance to different races and biotypes of diseases and pests, resistance to parasitic weeds like *Striga* and *Orobanche,* resistance and tolerance to abiotic stress conditions like drought, hot and cold conditions, mineral stress and climate change resilience.

Molecular marker assisted selection

Molecular marker assisted breeding is applied for selection of superior genotypes and elimination of unnecessary materials in early generations to save resources and time that may be spent on undesirable materials. Marker assisted selection (MAS) involves planting a population that has heterogeneity, heterozygosity and is segregating for the target trait and the associated molecular markers used to tag and trace the target genes or QTLs for selecting superior genotypes and identifying rare recombinants to eliminate tightly linked undesirable genes. Plants are sampled in early stages and screened for the markers linked to the target trait and individuals or families having the markers are selected, implying indirect selection for the target trait which can also be confirmed through phenotypic expression at the right stage when the trait is expressed. In self-pollinated crops marker assisted selection can be initiated as early as F_2 and conducted up to F_5 or F_6 with due attention being paid to the expression and evaluation of target traits in each generation and subjection of lines derived from such selection to the necessary agronomic tests in subsequent generations.

With help of the markers the identified QTLs are mapped using crosses between lines that differ for the trait of interest or segregating populations, the QTLs subjected to genetic analysis and their effects on the variation of the target trait determined. Once a marker locus has been identified as linked to the QTL it is used for selection and transfer of the QTL controlling the metric trait of interest through hybridisation. Where there are several QTLs controlling a particular quantitative trait selection is done for the different respective markers linked to the QTLs in different segments of the genome controlling the trait. However, selection of several QTLs controlling a particular trait at the same time may lower the efficiency of molecular marker assisted selection for quantitative traits. The QTLs also suffer from effects of epistasis and environmental effects. Because of the action of polygenes controlling the quantitative traits the effects of various QTLs on a particular trait are verified through replicated tests across environments to ascertain the effects of QTLs and level of QTL x environment interactions so that stable QTLs with greatest contribution to the variation of a particular quantitative trait are identified (Jiang, 2013). Since molecular markers are employed to complement other breeding strategies molecular marker assisted selection and conventional selection strategies can be combined where there are several QTLs controlling a particular trait and the trait also having low heritability.

Molecular marker assisted recurrent selection

Marker assisted recurrent selection (MARS) can be employed to fast-track the improvement of polygenic characters controlled by several loci more effectively than the conventional recurrent selection approach. The improvement of polygenic traits through marker assisted recurrent selection and marker assisted backcrossing requires identification of major QTLs which account for most of the variation in the target traits to help in stacking of multiple QTLs for traits like yield and certain tolerance to biotic and abiotic stress conditions in elite breeding genotypes. In marker assisted recurrent selection the lines/genotypes are selected early before flowering using molecular markers closely linked or flanking the target genes/QTLs or located within the target genes/QTLs. Genotypic selection is done based on the molecular marker-based genotyping and the selected genotypes are used in inter-crossing to achieve recombination. Molecular markers allow for selection before pollination so that only selected plants are allowed to provide pollen, the unselected ones are either removed or their male-parts removed where possible. The selected genotypes are subjected to intermating at flowering so as to develop a population with desirable combination of favourable alleles. Similar genotypic selection and inter-crossing of selected genotypes for recombination is repeated for the required number of cycles so that superior genotypes with desired level of improvement are obtained.

Molecular marker assisted recurrent selection helps to accumulate favourable multiple genes in one population and increases genetic gain to much higher levels per unit time and resources spent as compared to the conventional recurrent selection. In this selection approach genotypic selection and inter-crossing are done in the same season, saving on time and resources. Marker assisted recurrent selection can be integrated with speed breeding to fast-track the development of the target varieties.

Molecular marker assisted backcrossing

Transfer of the identified desired genes and QTLs to the elite breeding materials can be achieved through marker assisted backcrossing (MABC) or stacking of multiple genes by gene pyramiding using molecular markers. DNA marker assisted backcrossing accelerates the process of backcrossing to transfer one or more tagged genes from a donor to a superior well adapted cultivar or to an elite breeding material for improved resistance to diseases, pests and parasitic weeds, resistance and tolerance to abiotic stress factors and crop quality aspects. Tightly linked flanking markers or markers located within the target gene(s) improve the probability and efficiency of transferring the target recessive gene(s) through selection of plants having the recessive gene in heterozygous condition without the necessity of alternate generations of backcrossing, self-pollination and progeny testing to identify the progenies carrying the target recessive gene(s), thereby saving time and resources. In conventional backcross breeding up to six generations of backcrossing with selection are needed to transfer a major gene and reinstating the traits of the recurrent parent.

Molecular markers from the donor parent are used for selection and transfer of target genes to the recurrent cultivar while the markers associated with the genes for traits in the recurrent parent but not linked to the target trait can be used for selection and faster recovery of the recurrent parent genome and elimination of undesired donor parent traits. Use of flanking markers eliminates linkage drag effects which come as undesired traits with genes/segments transferred from the donor parent by having much less of such undesired donor parent genes in the selected progenies while helping to recover rare recombinants and the genome of the recurrent parent much faster. Combination of selection of donor relevant parent molecular markers and the recurrent parent genome molecular markers improves the effectiveness of marker assisted backcrossing.

To execute marker assisted backcrossing in autogamous crops the recurrent parent is crossed to the donor parent with the target trait closely associated with the molecular marker(s) and F_1 progenies are identified in early stages for backcrossing to the recurrent parent at flowering. The backcross progenies with the molecular marker(s) are identified in early life stages for use in further backcrossing to the recurrent parent at flowering. Such backcrossing may be done until BC_4 when the progenies are screened for the marker(s) and the selected plants are allowed to self-pollinate naturally. Screening and natural self-pollination may be done for additional one or two generations before agronomic evaluations are initiated identifying promising pure lines. Molecular marker assisted backcrossing is also applicable for improvement of inbred lines in hybrid and synthetic varieties breeding programmes. Marker assisted backcrossing can be used along with speed breeding if appropriate facilities are available to fast-track gene transfer.

Molecular markers are useful in pyramiding of disease and pest resistance genes in multiple gene resistance breeding in which different genotypes have similar phenotypic expression, especially in situations where resistance is to be developed against different races of a particular pathogen or different biotypes of a particular pest. Gene pyramiding is achieved by transfer and concentration of a number of independent genes with similar effects on particular trait into one variety, e.g., disease or pest resistance controlled by more than one locus. Since the different genes have the same phenotypic effect selection for such genes and their assemblage in one cultivar is difficult but can be resolved if each of them is tagged using flanking molecular markers or molecular markers located within the genes. Without the use of molecular markers, the transfer of multiple genes is done through conventional backcrossing with difficulties of identifying plants bearing one or more genes with the same phenotypic expression. This takes several generations and is expensive in terms of time and resources. If linkage of molecular markers to different genes or different QTLs controlling the same trait is ascertained, it is much easier and faster to conduct simultaneous selection for the markers linked to different genes controlling the same character and obtain plants with different combinations of the desired genes or QTLs (Richardson et al., 2006).

In addition to marker assisted selection, DNA markers are used for better understanding of breeding materials and diversity analysis among and within populations. DNA markers help to confirm the identity of different cultivars, ascertain quality and genotypic purity during production of various categories of seed, germplasm characterisation and classification into gene pools depending on genetic diversity, taxonomy and evolution. Inbred lines for hybrid and synthetic variety development and the parental materials can be clustered into various heterotic groups based on genetic distance and hybrid performance estimates. Molecular markers are used to derive genetic distance relationships between and among different pure lines and inbred lines, populations and other germplasm with the aim of predicting performance and heterosis of hybrids that can be derived when such materials are crossed. Flanking markers linked to QTLs of traits important in general combining ability and specific combining ability analyses are identified and used to select inbred lines with high combing abilities. Markers associated with QTLs have been used to improve resistance to diseases, pests, abiotic stress factors, yield and related traits in a number of crops including maize, tomato, cotton, sunflower, sorghum, cabbage, sugar beet, rice, soybean, wheat, barley, etc., for improved hybrid and pure line varietal development.

Molecular markers are also useful in interspecific hybridisation initiatives which focus on the transfer of specific genes located on particular chromosome segments of the donor species to crops. The exact chromosome sections bearing the target genes are tagged using flanking DNA markers to facilitate interspecific transfer of the target genes from crop relatives to the crops. Genes for yield and related traits, disease and pest resistance, *Striga* and *Orobanche* resistance and tolerance to abiotic stress factors like drought, heat, cold and resistance to mineral toxicity affecting crop production can be tagged for fast and easier transfer from the crop relatives.

6.4 Genomic Selection

Genomic selection, which uses genome wide analysis data to obtain genomic merit of individuals by estimating breeding values for materials being handled, is one of the protocols for reducing the length of breeding cycle and facilitating the production of desired varieties efficiently. It helps to overcome limitations of marker assisted selection for complex quantitative traits and if it is used in conjunction with speed breeding genetic gain is much more improved, duration for varietal development is shortened and the efficiency of breeding is highly elevated (Varshney et al., 2018). Genomic selection can be used to improve complex quantitative characters and multi-trait breeding separately or in combination with other breeding protocols and does much better than phenotypic selection and marker assisted selection in terms of gain per unit time, efficiency and time required per breeding cycle (Varshney et al., 2018).

Genomic selection depends on whole genome sequencing using high-throughput next-generation sequencing technology for identification of markers spread in the whole genome and tightly linked to the target genes or QTLs to obtain the information needed for estimating required selection parameters (Bita & Gerats, 2013). It is a genome-based molecular marker selection approach which employs genotypic and phenotypic data and relies on availability of whole genome sequence data for large numbers of DNA markers, often through SNPs analysis, to construct models for evaluating individuals/lines with respect to the phenotypic manifestation. Genomic selection also relies on accurate predictions based on the estimated effects of all identified markers on the desired phenotype to obtain genomic estimated breeding values (GEBVs) proposed by Meuwissen et al. (2001). Genomic estimated breeding values are estimated as the effects of genetic markers across the whole genome to capture all the QTLs or genes that contribute towards the target trait. Through genomics the genotype is directly used to estimate the phenotypic performance. Integration of the genome sequencing data in form of molecular markers for the appropriate traits facilitates and fast-tracks genomic selection for development of desired crop varieties.

In genomic selection there is a training population for which phenotyping and genotyping are conducted and the genotypic and phenotypic data are used to estimate GEBVs for selection (Varshney et al., 2018; Meuwissen et al., 2001). The lines or individuals in the population to be subjected to selection are evaluated for the same marker parameters used in the training population to estimate model parameters which are employed to predict GEBVs for the lines or individual plants. In this way the lines or individual plants (and even populations) can be evaluated and selected for the traits which make them high yielding with appropriate qualities, resistant to various stress factors, tolerant to drought, high or low temperatures, salinity, pest and disease attack, climate change resilient or have other desirable characters. Such selection integrated with speed breeding gives the best approach for developing superior crop varieties having the desired characteristics within a short period of time. Since genomic selection and other modern molecular assisted breeding approaches use large data generated through genotyping, genomics, phenotyping, phenomics, proteomics, transcriptomics, metabolomics and other analyses it is important that application of bioinformatics tools be integrated for appropriate and professional data management and utilisation.

Problems

6.1 Why is it essential to estimate heritability and other population parameters after every cycle of selection before another cycle is initiated?

6.2 Why are morphological and physiological characters undesirable as markers in selection programmes for crop improvement?

6.3 How does phenotyping improve the breeders' efficiency in breeding programmes?

6.4 What features make DNA molecular markers superior to isozyme markers for marker assisted breeding

6.5 What are the weaknesses of QTLs use in marker assisted breeding and how can some of them be resolved?

6.6 Why is genomic selection considered the most effective selection approach for multigenic characters in crop improvement programmes and what is its main challenge?

References

Alahmad, S., Dinglasan, E., Leung, A., Derbal, N., Voss-Fels, K., Able, J. A., Bassi, F. M., Christopher, J., & Hickey, L. T. (2018). Speed breeding for multiple quantitative traits in durum wheat. *Plant Methods, 14*, 36. https://doi.org/10.1186/s13007-018-0302-y. Creative Commons Attribution International Licence (CC BY).

Alemu, K. (2015). The role and application of bioinformatics in plant disease management. *Advances in Life Sciences and Technology, 28*, 28–33. ISSN 2225-062X (online). Creative Commons Attribution International Licence (CC BY).

Amiteye, S. (2021). Basic concepts and methodologies of DNA marker systems in plant molecular breeding. *Heliyon, 7*(10), e08093. https://doi.org/10.1016/j,heliyon.2021.e0893. Creative Commons Attribution International Licence (CC BY).

Bakala, H. S., Singh, G., & Srivatava, P. (2020). Smart breeding for climate resilient agriculture. https://doi.org/10.5772/intechopen.94847. Creative Commons Attribution International Licence (CC BY).

Bita, C. E., & Gerats, T. (2013). Plant tolerance to high temperature in a changing environment: Scientific fundamentals and production of heat stress-tolerant crops. *Frontiers in Plant Science, 4*, 273. https://doi.org/10.3389/fpls.2013.00273. Creative Commons Attribution International License (CC BY).

Falconer, D. S., & Mackay, T. F. C. (1996). *Introduction to quantitative genetics* (4th ed.). Longman, London, UK. Oxford University Press, Nos. 5701810833042 & No. 5726941241802.

Fisher R. A., & Yates, F. (1963). *Statistical tables for biological, agricultural and medical research* (6th ed.). Oliver and Boyd, Edinburg and London 1963X, 146P. Preis 42 s net. John Wiley and Sons, No. 571861067293.

Gomez-Casati, D. F., Busi, M. V., Barchiesi, J., Peralta, D. A., Hedin, N., & Bhadauria, V. (2018). Application of bioinformatics to plant biotechnology. *Current Issues in Molecular Biology, 27*, 89–103. https://doi.org/10.21775/cimb.027.089. Creative Commons Attribution International Licence (CC BY).

Janni, M., Gulli, M., Maestri, E., Marmiroli, M., Valliyodan, B., Nguyen, H. T., & Marmiroli, N. (2020). Molecular and genetic bases of heat stress responses in crop plants and breeding for increased resilience and productivity. *Journal of Experimental Botany, 71*, 3780–3802. https://doi.org/10.1093/jxb/eraa034. Creative Commons Attribution International Licence (CC BY).

Jiang, G. L. (2013). Molecular markers and marker-assisted breeding in plants. In *Plant breeding from laboratories to the field chapter* (vol. 3, pp. 45–83). https://doi.org/10.5772/52583. Creative Commons Attribution International Licence (CC BY).

Kempthorne, O. (1955). The theoretical values of correlations between relatives in random mating populations. *Genetics, 40*, 153–167. https://doi.org/10.1093/genetics/40.2.153. Oxford University Press, Nos. 5701791087280 & No. 5701800342484.

Kushwaha, U. K. S., Deo, I., Jaiswal, J. P., & Prasad, B. (2017). Role of bioinformatics in crop improvement. *Global Journal of Science Frontier Research, 17*, 13–24. ISSN2249-4626. Creative Commons Attribution International Licence (CC BY).

Meuwissen, T. H. E., Hayes, B. J., & Goddard, M. E. (2001). Prediction of total genetic value using genome-wide dense marker maps. *Genetics, 157*, 1819–1829. https://doi.org/10.1093/genetics/157.4.1819. Oxford University Press, No. 5626070526316.

Razzaq, A., Kaur, P., Akhter, N., Wani, S. S., & Saleem, F. (2021). Next-generation breeding strategies for climate-ready crops. *Frontiers in Plant Science, 12*, 620420. https://doi.org/10.3389/fpls.2021.620420. Creative Commons Attribution International Licence (CC BY).

Richardson, K. L., Vales, M. I., Kling, J. G., Mundt, C. C., & Hayes, P. M. (2006). Pyramiding and dissecting disease resistance QTL to barley stripe rust. *Theoretical and Applied Genetics, 113*, 485–495. https://doi.org/10.1007/s00122-006-0314-2. Springer Nature No. 5727490532264.

Tiwari, A., Singh, P., & Kumawat, S. (2020). Application of bioinformatics in plant breeding system. *International Journal of Current Microbiology and Applied Sciences, 13*, 2825–2831. ISSN 2319-7706. Creative Commons Attribution International Licence (CC BY).

Varshney, R. K., Singh, V. K., Kumar, A., Powell, W., & Sorrells, M. E. (2018). Can genomics deliver climate change ready crops? *Current Opinion in Plant Biology, 45*, 205–211. https://doi.org/10.1016/j.pbi.2018.03.007. Creative Commons Attribution International Licence (CC BY).

Chapter 7
Introduction of Crop Cultivars

7.1 Crop Plant Introductions

Some conventional plant breeding strategies comprise of the traditional plant material introduction and exchange, simple selection and recurrent selection, or any form of selection involving recombination, to obtain pure line varieties or open pollinated varieties and hybridisation followed by pedigree, bulk, single seed descent and backcrossing breeding methods and development of inbred lines and improved populations for production of various types of hybrid varieties. The practice of crop plant introduction is as old as the beginning of crop plant cultivation when agriculture was incepted as the agricultural communities of olden times exchanged, shared and battered crop plant materials for food or planting purposes whether they were neighbours or lived far apart. The early cultivators also moved with their crops as they migrated from one place to another. These activities led to introduction and spread of crop plants across the globe.

In Africa, apart from the indigenous crops that were domesticated in the continent, the introduction and spread of various crops was enhanced over centuries by the Arab traders, European explorers, Christian missionaries, colonialists and farmers. Many of the crops currently grown in Kenya and East Africa were introduced either by the Portuguese and Arab traders or European farmers, e.g., wheat, barley, banana, rice, maize, beans, avocado, coconuts, cashew nuts, alfalfa, pineapple, strawberry, cabbages, kales, onions, sunflower, potatoes, sweet potatoes, etc. Introduction of exotic crops into Kenya has taken place within the last 400 years. However, most of the presently cultivated field crops, tree crops and vegetables were introduced within the twentieth century. Crops like cotton, wheat, barley, Brassicas, tea, potatoes, pyrethrum, sunflower, coffee and crops are among the crops that were introduced by the European farmers in East Africa or Asians during the colonial period in the twentieth century. Some other crops have also been introduced more recently by the agricultural scientists.

P. O. Ayiecho and J. O. Nyabundi, *Conventional and Contemporary Practices of Plant Breeding*, https://doi.org/10.1007/978-3-031-74998-8_7

Crop introduction can take the form of a totally new crop or cultivars and germplasm of crops already established and cultivated. The germplasm materials may include improved cultivars, landraces, weedy or wild relatives of crop plants in a region, accessions with induced mutations and even products of biotechnology and genetic engineering. Introductions are often acquired for the purpose of expanding the germplasm of already cultivated crops and addressing specific production problems. That is, if the locally cultivated and adapted varieties do not have resistance or tolerance genes to the endemic and emerging biotic production problems and abiotic stresses the breeder may look for resistance genes elsewhere and acquire them for improving the locally cultivated varieties. This has been done for various crops e.g., wheat, maize, sorghum, rice, beans, pigeon peas, coffee, barley, rice, etc. in Kenya. While the breeder will assemble the local materials and landraces for crop improvement the breeder will also rely on imported exotic germplasm acquired from other regions and institutions for crop improvement.

When crop germplasm materials are introduced into any country, they are mandatorily passed through the local quarantine station to prevent diseased and pest infested materials from being introduced into the country through the imported plant materials.

7.2 Handling of Introduced Cultivars for Crop Improvement

Once the introduced materials are confirmed to be free from diseases, pests, nematodes and weed seeds they are released to the person or institution who imported them and may be subjected to any of the established breeding strategies some of which are illustrated in Fig. 7.1.

The introduced materials can be tested for adaptability under varied conditions and released as varieties after a few seasons of evaluations if they are adequately adapted, high yielding with good quality and resistance to some local abiotic and biotic stresses. During evaluations they are compared with the locally adapted varieties that are in commercial production. This is the case with some crops and crop varieties grown Kenya, e.g., maize hybrids, rice, tomato, several vegetable crop varieties, sorghum varieties, etc., where some varieties are developed elsewhere and adopted in Kenya for commercial production after a few seasons of rigorous tests. The acquired accessions may also be subjected to selection and further improvement for adaptability, pest and disease resistance, quality and other agronomic attributes and used in breeding strategies focused on development of homogeneous pure line varieties, uniform heterozygous hybrid varieties, or highly heterogeneous heterozygous open pollinated and improved clonal varieties. The introduced accessions also serve as sources of useful genes to solve specific production problems, namely, biotic or abiotic stress factors, plant architecture, adaptability, climate change resilience, yield and quality aspects and used in crosses if they have genes for the desired traits

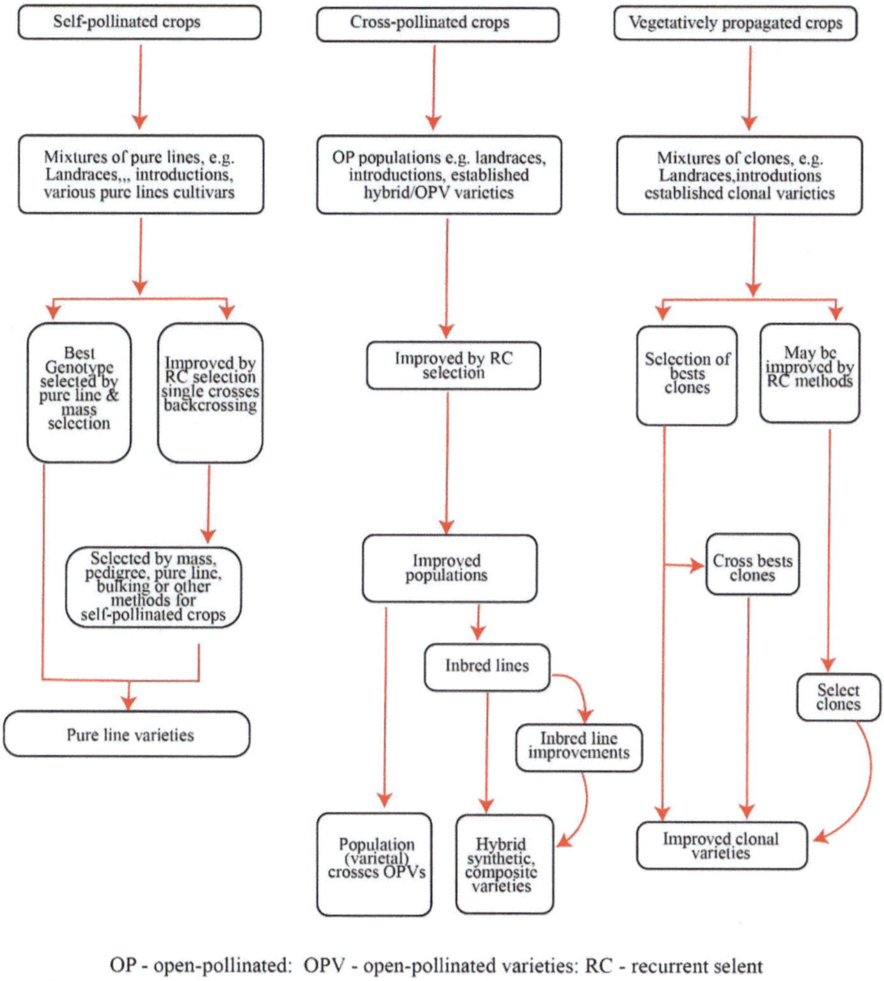

Fig. 7.1 Some plant breeding strategies

to help improve locally well adapted established varieties and landraces. They are also made available to other interested breeders and institutions.

Problems

7.1 Which are the possible ways the introduced accessions can be utilised by the plant breeders?

Chapter 8
Crop Population Improvement Through Recurrent Selection

Recurrent selection refers to population improvement selection methods involving recombination of selected genotypes applied to crop plant populations with the aim of increasing the frequency of favourable genes, desirable gene combinations and superior genotypes. Recurrent selection methods are designed to solve the limitations of traditional methods of producing inbred lines for development of hybrid varieties, synthetic varieties and open pollinated varieties in cross pollinated crops and production of pure line varieties in self-pollinated crops. Traditional methods of production of inbred lines involve selection of desirable plants from open pollinated populations and subjecting them to self-pollination and selection for a number of generations followed by sib-mating accompanied by further selection for the desired inbred lines. This is followed by making pairwise crosses or three way or double crosses to produce hybrid varieties. In autogamous crops pairwise crosses are made between selected pure lines followed by pedigree breeding or backcrossing or single seed descent or bulk breeding to develop superior pure line varieties for commercial production. Such traditional methods do not give the best possible varieties that can be derived from the breeding materials at the disposal of the breeder considering the fact that some of these methods involve selection for quantitative characters. These methods also do not provide opportunities for obtaining the best possible gene combinations and genotypes through genetic recombination to reorganise genes on linkage blocks for developing exceptionally superior varieties and materials with adequate broad genetic base. Apart from ineffectiveness of visual selection, small populations of progenies derived from the few selected plants, especially segregating populations, sometimes lead to loss of useful genes. The standard traditional methods are a hindrance to the achievement of potential ceiling imposed by genes in the original population and other germplasm available to the breeder, therefore, not allowing for the development of the best possible hybrid, open pollinated or pure line varieties. Given such situations the breeders, especially those dealing with cross-pollinated crops with particular reference to maize, developed a recurrent selection system to improve and maintain variability in breeding populations while at the same time increasing the frequencies of favourable genes and superior genotypes through

P. O. Ayiecho and J. O. Nyabundi, *Conventional and Contemporary Practices of Plant Breeding*, https://doi.org/10.1007/978-3-031-74998-8_8

recombination and selection for quantitative traits. Various schemes of recurrent selection system were developed by maize breeders as an approach of improving cross-pollinated maize with respect to quantitative traits. Apart from maize, recurrent selection methods are applicable to other cross pollinated and self-pollinated crops as well. Though there are barriers to cross-pollination in self-pollinated crops making recurrent selection difficult to apply in such crops, the presence of male sterility makes this selection system easily applicable to autogamous or often cross-pollinated crops like sorghum, finger millet, rice, wheat, barley, soybeans, beans, cowpeas, grams, peas, etc.

The most important feature of recurrent selection is that it helps in rearrangement of genes and creation of new genotypes, hence generating new variability through intermating among selected plants. Whether there is linkage or no linkage, intermating is necessary to help release genetic variability in populations on which selection can act in each generation for sustained population improvement. In the absence of linkage, genetic variability associated with the combinations of unlinked genes at different loci affecting the same trait is released fast during generations of intermating. When there is independent segregation, hybridisation of different genotypes, especially mating of heterozygotes, leads to release of the variability easily and new genotypes which were not in the parental population(s) appear. Therefore, variability due to allelic combinations involving unlinked loci for polygenic traits is more easily available for exploitation through selection and does not need genetic crossing over in the random mating populations as compared to the linked loci which need genetic crossing over to release exploitable variability. Linkage conceals variability which is released at a much slower rate during generations of intermating. If there is linkage release of variability arising from recombination is determined by extent of linkage between genes. Tight linkages can be broken by continued intermating resulting in rearrangement of genes and slow and progressive release of variability for exploitation by selection in polygenic traits though oligogenic characters also benefit from recurrent selection. Recurrent selection facilitates recombination to break and eliminate undesirable linkages while generating new and favourable gene linkages, giving the breeder opportunities to maximise selection gain from the population on which selection is conducted. Response to selection for quantitative traits in cross-pollinated crops tends to be gradual, usually starting with rapid response in the first few generations followed by protracted gradual response for several generations. In some generations there may be larger response magnitudes if recombination brings some favourable genes together in a linkage block, thereby releasing larger variability leading to corresponding higher selection gain. Progressive response to selection over several generations occurs because each cycle of selection is accompanied by a recombination generation in which intermating is carried out among the selected superior individuals supposedly having the desirable genes. This leads to formation of new gene combinations giving new opportunities for selection of better genotypes arising from fresh genetic combinations as selection is repeated in the generation derived after intermating. As the best genotypes are selected from a population whether random mated or inbred and the selected individuals are allowed or forced to undergo random mating and the progenies derived from such intermating are subjected to another round of selection a much better population is obtained because of new favourable gene combinations. If similar selection is conducted again on the improved population followed by intermating of the best genotypes/phenotypes in it the population is further improved. Through such process recurrent selection brings together favourable polygenes, raises the frequencies of the polygenes, favourable combinations of the polygenes and superior genotypes, sustains genetic variability and provides opportunities for effective selection over prolonged periods in the target population. While recurrent selection maintains high genetic variability for sustenance of stable selection progress in populations, it improves population mean phenotype as well. This selection system leads to the development of high yielding well adapted crop varieties with desired qualities. It also results in the development of breeding populations and germplasm with broad genetic base.

For good selection progress it is important that selection be initiated on the populations with good genetic variability and good performance for the character to be improved because high frequencies of favourable genes and high heritability are mandatory requirements for substantial selection response. The success with which a breeder achieves the primary breeding objectives of developing improved crop varieties depends on additive genetic variability and the selection approach since additive genetic variability is what is exploitable by selection. Appropriately designed and properly executed recurrent selection programmes will achieve the objectives of developing improved varieties and germplasm with increased frequencies of desirable genes, gene combinations, genotypes and phenotypes and an overall broad genetic base. The varieties developed from populations through well-designed selection strategies, whether they are pure lines, hybrid, synthetic or open pollinated varieties, will be much higher yielding and better adapted than those developed from similar populations not subjected to recurrent selection improvement programs. From improved populations it is easy to derive superior inbred lines with desired genes, gene combinations and genotypes that nick well to give superior hybrid and synthetic varieties.

8.1 Recurrent Selection Methods in Allogamous Crops

Recurrent selection schemes were developed in cross pollinating crops, particularly maize, to improve quantitative traits which are difficult to breed for because they are highly influenced by the environment and controlled by several polygenes with minor effects and unknown gene frequencies. The schemes improve the populations with respect to *per se* performances and additive genetic variance for the desired characters so that the improved populations can be commercialised as open pollinated varieties and also serve as sources of superior inbred lines with good combining ability. Success in recurrent selection initiatives requires the presence of high genetic variability, preponderance of additive genetic variance for characters under selection without undesirable linkages and that the population being handled should be large.

The recurrent selection methods in open-pollinated crops can be broadly categorised as intrapopulation recurrent selection and interpopulation selection schemes.

Among the intrapopulation selection methods are simple recurrent selection and genotypic (progeny testing) recurrent selection methods which include half sib selection, full sib selection, modified ear to row selection, S_1 and S_2 selection, biparental selection, etc. Interpopulation recurrent selection methods include reciprocal recurrent selection methods using various progeny structures. Though the methods were developed in maize they are applicable to other cross-pollinating crops like sunflower and some self-pollinating crops where there is male sterility as in sorghum, rice and soybeans and self-incompatibility. Other than mass selection all recurrent selection methods involve generation of progenies, evaluation of the progenies, and recombination of the selected superior individuals or relatives of the superior progenies to synthesise the new populations with the important variables involved in these operations that determine selection gain being the type of progeny structure, extent of progeny evaluation, selection intensity and method of recombination (Doggett, 1968; Doggett & Eberhart, 1968; Doggett, 1972; Moreno-Gonzalez & Cubero, 1993) , with expected selection response being

$$R = \frac{i\sigma_A^2}{\sigma_P},$$

where i, $\sigma_A^2$and σ_P are selection intensity, additive genetic variance and phenotypic standard deviation respectively for the population under selection (Doggett, 1968). If parental control factor, c, and the number of years, y, required per selection cycle (which takes into consideration the extent of progeny evaluation) are considered, the response prediction is $R = \frac{ic\sigma_A^2}{y\sigma_P}$. In terms of heritability, h^2, and selection differential, s, selection response is given as $R = i\frac{ch^2\sigma_P}{y} = \frac{ic(\sigma_A^2)\sigma_P}{y\sigma_P^2}$ and $R = s\left(\frac{\sigma_A^2}{y\sigma_P^2}\right)$ respectively. The magnitude of selection gain is determined by the variables in these prediction expressions namely, duration taken to complete one selection cycle, heritability, selection intensity, selection differential, additive genetic variance and phenotypic variance/standard deviation which can be manipulated to change selection gain magnitude.

High selection pressure leads to increased response through increased intensity of selection and high selection differential because only a small top performing proportion is selected. However, this may cause undesirable genetic drift effects if the percentage selected is too low resulting in very small population size which restricts genetic gain. Success of recurrent selection depends on the magnitude of additive genetic variance component in the population. Infusing favourable genes into the population and increasing frequencies of favourable genes and favourable gene combinations through hybridisation with other germplasm, recombination, induced mutagenesis, biotechnology methods, transgenesis and genomic editing elevate genetic variability and enhance selection gain. Inbreeding of the materials from which the parents are to be obtained for the breeding programme also raises the magnitude of additive genetic variability exploitable by selection. The magnitude of σ_A^2 exploitable by selection is influenced by the method of selection and progeny structure. In mass selection σ_A^2 is exploited in its entirety while in half sib and full sib selection the magnitudes of additive genetic variance exploitable are $\frac{1}{4}\sigma_A^2$ and $\frac{1}{2}\sigma_A^2$ respectively. As regards the effect of parental control for the materials selected there is a significant effect of the pollen source for fertilising the selected plants, progeny structure and method of recombination (Moreno-Gonzalez & Cubero, 1993). If mass selection is conducted on open-pollinating population after pollination the breeder is in control of the maternal parent only and not the source of pollen making parental control factor, c, equal to $\frac{1}{2}$. If selection is conducted before pollination and unselected plants destroyed or prevented from supplying pollen or if selected individuals are self-pollinated before selection, then c is doubled and becomes 1 because both pollen source and maternal parents have been selected (Doggett, 1968; Moreno-Gonzalez & Cubero, 1993). Similarly, in half sib selection if remnant seed of the half-sibs is planted for recombination c equals 1.0 because only selected plants provide the pollen for intermating in the recombination plot. Parental control factor, c, can be raised to 2 in situations where individual plants in a population are self-pollinated and also crossed to a tester so that the seed from self-pollination is kept as remnant seed while the test cross seed is used for evaluation and selection. The remnant seed of the selected plants is bulked and planted in the recombination plot for random mating to generate the new improved population. Improving selection gain per unit time implies reducing y per cycle of selection. Reducing the number of years per cycle can be achieved by using off-season nurseries, especially during the season for production of test progenies and recombination. In regions with bimodal rainfall at least two generations can be grown per year. The two rainy seasons can be made use of in the year and where possible another crop can be grown using irrigation facilities in the dry season. Speed breeding protocols can also be employed where facilities are available so that only the evaluation trials are grown in the main cropping season. Reduction in phenotypic standard deviation to improve selection gain is achieved by proper experimental designs, replicated tests, good field plot techniques, appropriate agronomic management practices like rotation of fields for trials and selection, optimum use of fertilisers, uniform and proper planting, weed, pest and disease control, etc., to reduce environmental components of phenotypic variation.

The improved populations derived from any cycle of recurrent selection in cross-pollinated crops can be subjected to another cycle of similar selection and/or evaluated for release as commercial open-pollinated varieties if found desirable. They are also used in production of variety cross hybrids in which open pollinated cultivars are used as parents and for development of better performing inbred lines for production of superior hybrid varieties and synthetic varieties.

8.1.1 Intrapopulation Recurrent Selection Schemes

Simple recurrent phenotypic (mass) selection

Mass selection, also referred to as simple phenotypic or visual selection, is the oldest, easiest and simplest breeding method and has been practised since the inception of cultivation of crop plants by human beings. Simple recurrent phenotypic selection requires only one generation per cycle and involves selecting individual plants from an open-pollinated population planted in isolation. Phenotypic measurements on individual plants in the open-pollinated population are used to select the best plants for the trait under selection. Open pollination gives an opportunity for random mating and recombination involving the selected plants, and therefore, the seed from the selected plants is a product of random mating. At harvesting such seed is bulked as the seed of the improved open-pollinated population which is subjected to another cycle of similar selection. This selection procedure can be repeated as many cycles as necessary depending on the target trait, resources available and breeding objectives. Selection in cross-pollinated crops necessitates that appropriate isolation measures be taken into account against undesirable pollination by pollen from other cultivars of the same crop in the vicinity.

Response for selection conducted after pollination and fertilisation when selected plants are already fertilised by random pollen from unidentified and unselected plants in the population is given by

$$R = \frac{1}{2}\frac{i\sigma_A^2}{\sigma_P}.$$

For some traits, selection can be done prior to pollination and fertilisation, hence there will be control over both parents raising the parental control factor, c, to 1. Under such circumstances, selection gain will be

$$R = i\frac{\sigma_A^2}{\sigma_P}.$$

Supply of pollen by the undesired plants can be prevented by detasselling in maize or destroying unselected plants before pollination. Alternatively, further parental control to improve efficiency of this method of selection can be achieved by self-pollinating the selected plants and the seed derived from such self-pollinated plants is bulked in equal quantities for planting. The bulked seed is planted in an isolated recombination plot where random open pollination takes place and the seed from the recombination plot is the seed of the improved open-pollinated population which is subjected to another cycle of selection. Selection involving self-pollination and a season for recombination makes this process require a total of two seasons per cycle of selection. The off-season nursery can be used for the recombination step. In biennial crops like carrot, onion, cabbage, sugar beet, selection can be done for some characters of economic importance before the plant gets into reproductive phase of its life cycle. When selection takes two years and both parents are selected selection gain prediction becomes

$$R = i\frac{\sigma_A^2}{y\sigma_P}, \text{ where } y = 2.$$

Phenotypic recurrent selection is a fast, simple and cheap method which does not require as many resources as other recurrent selection methods since it does not involve progeny testing. Through this method as many plants as possible can be handled, therefore, allowing for high selection intensity. This method of selection is also effective for characters with high h^2 like duration to flowering and maturity, adaptability to changed environment, plant height, ear height and prolificacy in maize. The effectiveness of recurrent mass selection also depends on inter-mating to generate and maintain genetic variability and gene frequencies and gives maximum response at intermediate gene frequencies of p being about 0.5 but least effective at low genetic variability when gene frequencies are near 1.0 or 0.0 which reflect low variability.

Mass selection has a number of limitations, especially in regard to quantitative traits. Environment mars the effectiveness of mass selection since the phenotype is a product of genotype and environmental effects. Because of environmental influence on the phenotype, it may not be clear whether the superiority of a particular plant is preponderantly genetic or mostly environmental. Visual selection alone does not take genotypic and environmental effects into consideration yet environmental effects are not inherited. Visual selection does not distinguish whether superiority of selected plants is due to environment or genetic effects or both. Therefore, simple phenotypic selection is not effective for traits with low genetic variability and low heritability, especially quantitative traits like yield because of the polygenic nature of such traits, environmental effects, inability to properly identify superior genotypes visually, large number of plants that are necessary to be handled and uncontrolled pollination for those traits selectable after pollination. Lack of control over the source of pollen and the need for high selection pressure also makes mass selection inefficient. Lack of control over pollen source leads to plants with unfavourable genes also supplying pollen for pollination in the open-pollinated population. High selection pressure means that only the few best individual plants will be selected, hence small populations will be selected leading to genetic drift and inbreeding, consequently resulting in restricted selection response. This selection approach is also not appropriate for improving combining ability since it does not involve progeny testing as no test crosses are used. The effectiveness of phenotypic selection can be improved by the stratified (gridded) selection (Gardner, 1961) which is a selection design that reduces environmental effects on individual plants, especially soil heterogeneity, and competition among plants because it allows for comparison of selected plants with their neighbours. Stratified or gridded mass selection method was developed by Gardner (1961) to minimise micro-environmental variation and improve efficiency of individual plant selection. The field is partitioned into small plots or cells of the same size and same number of plants to reduce plot-plot environmental variance and make individual plant selection more efficient. The plots are subjected to similar agronomic practice maintenance. Within each plot selection is done among the plants to retain the top best plants with respect to the characters under selection. The same selection approach is done in all the plots so that the same pressure is applied for all plots for the same trait and equal number of plants will be selected from each grid plot. The seed of selected plants from all grids is bulked together in equal quantities to give the seed of the improved population which can be subjected to another cycle of selection. To improve the efficiency of gridded selection (Bos, 1983) proposed use of adjusted performance by introducing a standardisation of plant yields in each cell and thereafter applying truncation selection to the adjusted performance in the cells as follows.

$$z_{ij} = \frac{(y_{ij} - y_i)}{s_i},$$

where y_i is mean yield of all plants in i^{th} grid, z_{ij} and y_{ij} are adjusted and unadjusted yields of plant j in grid i and s_i is the phenotypic standard deviation in grid i.

Recurrent genotypic (progeny testing) selection

In genotypic selection the plant genotype is evaluated before selection is done so that selection is based on the genotypic value rather than on the phenotypic performance. The genotypic value of an individual is measured through its progenies which received its genes. This is a measure of the breeding value or genetic worth of a parent which is also the value of the genes an individual passes to its progenies as judged by the mean performance of its progenies. Therefore, genotypic selection necessitates generation of progenies or relatives which are used to evaluate the genotypic values of the individuals or families to be selected. Progeny testing evaluates the genetic value, that is, average gene effects of the selected plants. In family selection the breeder is engaged in creating a family structure through generation of progenies which are evaluated to help select the best families or parents followed by recombination of selected families or selected parents. The steps involved in this selection scheme are self-pollination or other form of inbreeding and/or crossing the plants to be selected to a tester to generate progeny families for evaluation. The results from evaluations are used to select the superior genotypes or families which are subjected to random mating in a recombination plot to generate seed for the improved population to be subjected to another cycle of similar selection. At least three seasons are needed for each cycle of selection involving progeny testing which include season for generation of the progenies, season for evaluation of the progenies and season for recombination of the selected plants. Progeny testing may be replicated over years and locations to obtain adequate and reliable information. In some cases, recombination may be done for two or three seasons before another selection cycle is initiated Therefore, in the genotypic selection, often referred to as family selection, discrimination among plants/families is based on family means obtained from replicated progeny evaluations. Family means are more reliable and less affected by environmental variation than for single plant selection. There are various family selection methods which include ear-to-row selection and its modifications, half sib selection, full sib selection, half sib progeny test selection, selfed progeny testing selection involving S_1 and S_2 and various modifications of some these methods.The differences among the methods involving progeny testing also lie on the family structure and kind of tester used.

Half sib selection methods

Ear-to-row selection

Ear-to-row selection method is a modification of mass selection to involve progeny testing and was developed for maize breeding by Hopkins in 1897 with a modification proposed by Lonnquist (1964) to improve its efficiency. In this selection method, visually superior ears are picked from individual plants in a field of an open-pollinated maize population to be improved. The seed from each of such visually selected superior ears is used to grow progeny rows for evaluation such that seed from each ear is grown in its own row of half sibs. Based on such half sib family row evaluation the superior progeny rows are selected and most of good-looking plants within the superior progeny rows are selected at maturity for harvesting. The seeds from harvested plants are bulked together to give seed of the improved open-pollinated population which can be subjected to another round of such selection. In this procedure parental control factor, c, is $\frac{1}{2}$ for half sib family selection, the magnitude of additive genetic variance exploitable by half sib family selection is $\frac{1}{4}\sigma_A^2$ and the expected gain from selection among families is

$$R_{HS} = \frac{1}{8}\frac{(i\sigma_A^2)}{y\sigma_{PHS,}}$$

where i is selection intensity, σ_{PHS} is the phenotypic standard deviation of ear row (half sib families) and y is the number of years taken to finish the cycle of selection.

A modification of ear-to-row method called ear-to-row-to-ear or modified ear-to-row selection was developed by Lonnquist (1964) to improve the efficiency of ear-to-row method. The modified method involves harvesting an ear from each plant visually identified as superior and seed from each ear is grown for progeny evaluations replicated over locations. That is, for each selected ear there is a row of or a plot of half sib progenies grown at each of the test sites. At one of the sites in the same year the evaluation plot serves as the recombination plot as well since in each plot at this site, each of the ear progenies being evaluated is planted in four rows or convenient number of rows (treated as female parent) and alternated with four rows or convenient number of rows of pollen source. This plot is isolated from other maize plots. The pollen source rows comprise of a mixture of equal number of plants of each progeny (half sib) family in the evaluations. That is, equal amounts of seed from each ear being evaluated are bulked and planted as the pollen source. For each entry all the plants in the rows to be treated as the female parents in the recombination/evaluation plot are detasselled to ensure that all the plants in such a plot cross to the bulk pollen source. At harvesting best detasselled plants in the recombination/evaluation plot are selected from the superior half sib family plots and harvested. The superiority of the half sib families is determined by the mean performances across all the evaluation sites. That is, selection among the ear progenies (half sib families) is based on their mean performances across the evaluation sites. Selection within the half sib families at the evaluation/recombination site is based on individual plant phenotypic performance. Therefore, this is a mixture of progeny testing selection and simple phenotypic selection. The numbers of superior half sib families and individual plants selected depend on the selection intensity, i_1, among the half sib families and selection intensity, i_2, within the half sib families. The seed obtained from the best plants harvested from the recombination/evaluation plot is bulked together in equal proportions to give the seed of the improved cultivar which can be subjected to another cycle of similar selection. Expected selection gain for modified ear-to-row selection has two components arising from selection among half sib families and the component attributed to selection within families.

Since parental control factor, c, is $\frac{1}{2}$ the expected gain from selection is

$$R = \frac{1}{8}\frac{(i_1 \frac{1}{8}\sigma_A^2)}{\sigma_{PHS}} + \frac{3}{8}\frac{(i_2\sigma_A^2)}{\sigma_{wHS}}$$

where σ_{wHS} is the phenotypic standard deviation within half sib families (Moreno-Gonzalez & Cubero, 1993).

Modified ear-to-row procedure has an advantage of being completed in one year because it combines progeny evaluation and recombination in one season and also combines two selection methods, i.e., family selection and individual plant (phenotypic) selection. However, if there is sufficient seed from the ears, progeny evaluation can be replicated over years and across location, with recombination being done in the last season of evaluation. The selection gain can also be improved by having the recombination plot grown with seeds from best plants selected from the superior half sib families only. Though this doubles parental control factor, making c become 1, it has the disadvantage of postponing the recombination activity to another season, prolonging the selection cycle by at least one season. Modified ear-to-row method was developed specifically for maize but it can be applied to other open pollinated crops like sunflower while for crops like sorghum it works well if there is male sterility.

Recurrent half sib family selection

Apart from ear-to-row or modified ear-to-row selection other half sib recurrent selection approach entails obtaining half sibs seed from several plants in an open-pollinated crop by harvesting each individual plant separately and the seeds of each plant is divided into two portions. This method which was proposed by Compton and Comstock (1976) involves keeping one portion of the seed of each harvested plant as half sib remnant seed while the other portion is planted in the field with seeds of other plants for half sib evaluation such that seeds of each plant are planted in a row or plot as a family. The evaluation field will have several half sib families being compared to one another. The tests can be replicated in locations and/or over seasons as determined by the amount of available seed and other necessary resources. Based on such tests the best performing half sib families are selected and the remnant half sib seeds of the selected best performing half sib families are bulked in equal quantities and planted in an isolated recombination plot for random pollination. The seed obtained from the recombination plot makes the seed of the improved open pollinated population which can be subjected to another cycle of similar selection.

Apart from maize this selection procedure is applicable to all other cross-pollinated crops.

Parental control factor, c, for this selection method is 1 while the time needed for one cycle of the process is at least three seasons, namely, the season for production of open pollinated half sib seed, season for half sib family evaluation tests which may be replicated over seasons and the season(s) for recombination. Production of open pollinated half sib seed and recombination activity can be done in the off-season nurseries to reduce the number of years. The expected selection

$$R = \frac{1}{4}\frac{i\sigma_A^2}{y\sigma_{PHS}}.$$

Use of half-sibs in selection can be further improved by combining half-sib family and generation of self-seed in multi-flowered crops. Several good-looking plants in an open-pollinated population to be improved are selected and each plant is self-pollinated and at the same time allowed to undergo normal open pollination or crossed to a broad based open-pollinated tester to generate half sib seed. The half sib seed is used for evaluation while seed derived through self-pollination is used for recombination (Moreno-Gonzalez & Cubero, 1993). When the population itself is the tester, as in cases of the different ear-to-row selection schemes and half sib family selection described above, the seed from open pollination in the population itself is used as the half sib family seeds. For each plant the seed derived from self-pollination is kept as the remnant seed while the half sib seed derived from open pollination in the population or cross to a broad based open pollinated tester is grown in an evaluation field. Several half sib families derived from the population are grown, each in its own row or plot for comparison to select the best half sib families. The remnant selfed seeds of the plants whose half sib families performed best in the evaluation are taken in equal quantities, bulked and planted in a recombination block where natural random pollination for recombination takes place. The seed derived from random mating in the recombination block makes the seed for the improved population which can be subjected to further cycles of such selection. In this selection scheme parental control factor, c, is 2 since selection is for both sexes with the crop for recombination being derived from selfed seed. In this selection procedure the tester may not be the population itself but an inbred line or an open pollinated cultivar, hybrid like a single cross, a double cross, a three-way cross, a composite or a synthetic cultivar, giving a top cross test. This selection scheme can be used to improve a population with respect to general combining ability (GCA) if the tester is of broad genetic base or specific combining ability (SCA) if the tester is of narrow genetic base like an inbred line. General combining ability is dependent mainly on the additive effects of genes carried by the parents and is therefore a measure of the parents breeding values. Specific combining ability arises from non-additive gene effects as it is attributable to dominance and epistasis arising from specific gene interactions in specific cross combinations.

Full sib test selection

In an open pollinated population to be improved by full sib test recurrent selection several good-looking plants are subjected to pairwise crossing to generate full sibs. The full sib seed for each pairwise cross is divided into two portions; one portion is kept as the remnant full sib seed while the other portion is grown out in the field for evaluation of the full sib families. The seed of each cross is planted in its own full sib family row or plot in the evaluation field. The full sib family tests are used to decide on which ones are superior enough to be selected. The full sib remnant seed for each selected family is taken in equal proportions, bulked and planted in a recombination plot for random open pollination in an isolated plot for recombination. The seed harvested from the recombination plot is the new improved open-pollinated population seed which is planted for another cycle of such selection.

This selection method needs at least three seasons per cycle, one season for generation of full sib seed, season for evaluation of full sib families and the third season for recombination. The full sib tests can be replicated in locations and seasons. The recombination process can be done during the off-season period so as to reduce the numbers of years, y, in the selection formula. In this scheme, selection is for both sexes because only best families are inter-crossed in the recombination block and parental control factor, c, is 1. The expected selection gain is given by

$$R = \frac{1}{2}\frac{i\sigma_A^2}{y\sigma_{FS}},$$

where σ_{FS} is phenotypic standard deviation for full sib families (Moreno-Gonzalez & Cubero, 1993).

The S_1 family selection

In S_1 family selection self-pollinating several vigorous S_0 plants in a genetically broad-based open-pollinated population with high mean for yield and other traits is undertaken to generate S_1 seed. The S_1 seed is divided into two portions so that one portion is kept as the remnant seed while the other portion is planted for S_1 family evaluation. The S_1 seed for each plant is planted in its own row or plot as an S_1 family and compared with several other S_1 families in the evaluation tests which may be replicated across locations and seasons as determined by the amount of S_1 seed and resources available. The families identified as superior in the evaluation tests are selected and their remnant S_1 seed is bulked in equal quantities and planted in a random pollination field to enhance recombination. The seed derived from the recombination block makes the improved population seed which is subjected to another cycle of similar selection. In this method both sexes for the improved population are selected and parental control factor, c, is 2. The expected selection gain is

$$R = \left(i \frac{\sigma^2_{A'S1}}{y\sigma_{PS1}} \right),$$

where $\sigma^2_{A'S1}$ is the additive genetic variance among the S_1 families which has some dominance bias, σ_{PS1} is the phenotypic standard deviation for the S_1 families (Moreno-Gonzalez & Cubero, 1993).

Three generations are needed per cycle, two of which, production of S_1 seed and the recombination generation, can be done during the off-season. If selection can be done before pollination, evaluation and recombination can be combined so that rejected S_1 families are detasselled or destroyed and only selected families get involved in pollination so as to reduce the selection period.The S_1 selection is particularly useful in improving populations for development of superior inbred lines for generation of high yielding hybrid and synthetic varieties. This method is also applicable to self-pollinated crops like soybean, wheat, and often cross-pollinated crops like sorghum and grain amaranths and is made easier for such crops if there is male sterility.

The S_2 family selection

In S_2 family selection procedure several good-looking plants in an open pollinated population are identified and subjected to self-pollination to get the S_1 seed. The S_1 seeds are planted as families and superior looking plants within each family are self-pollinated to obtain S_2 seed. The S_2 seed is divided into two portions whereby one portion is kept as the remnant seed while the other portion is planted in the evaluation experiments to test and identify the best S_2 families. The evaluation tests can be replicated over sites and seasons. The remnant S_2 seeds of the selected S_2 families are taken in equal quantities and bulked for planting in an isolated recombination plot for open random pollination. The seed harvested from random mating block is the seed of the improved population which is subjected to another cycle of similar selection process. However, for effective recombination another recombination season can be undertaken before initiating the next cycle of selection. Parental control factor, c, is 2 while the number of seasons is at least four. Production of S_1 and S_2 seeds and recombination can be done using off-season facilities to shorten the duration needed for each cycle of selection. The expected selection gain in one generation is

$$R = \frac{(3i\sigma^2_{A's2})}{2y\sigma_{PS2}},$$

where $\sigma^2_{A'2}$ is additive genetic variance which has some aspect of dominance bias, σ_{PS2} is phenotypic standard deviation for the S_2 families (Moreno-Gonzalez & Cubero, 1993).

8.1.2 Interpopulation Recurrent Selection Methods

Recurrent reciprocal selection

Inter-population selection schemes are aimed at improvement of hybrid varieties using selected genetically divergent pairs of populations. Selection of divergent populations, each with broad genetic base, for interpopulation improvement increases the opportunities of obtaining improved populations and elite inbred lines for development of superior hybrid and synthetic varieties. In addition, the populations will be improved with respect to *per se* yield, combining abilities and other important traits. Interpopulation improvement schemes are reciprocal selection strategies which borrow from selection methods proposed by Comstock et al. (1949) and are often designed to improve hybrids between the target populations. Reciprocal recurrent selection helps in improving two unrelated open pollinated populations with respect to each other by changing population genetic structures through altering gene frequencies in a complimentary manner such that additive and dominance gene actions and epistasis are exploited for higher heterosis in the hybrids developed between the populations. Hybrids developed using such reciprocally improved populations also maximise the exploitation of general and specific combining abilities. Reciprocal selection methods are applicable to multiflowered crop species like maize with at least two ears, Brassicas, sunflower, rye, castor, onion, tillered sorghum cultivars, etc. Some of reciprocal recurrent selection methods are half sib reciprocal recurrent selection, full sib reciprocal recurrent selection and their modifications.

Half sib reciprocal recurrent selection

In half sib reciprocal recurrent selection (Comstock et al., 1949) two genetically diverse and broad-based populations 1 and 2 are selected based on their adaptability, yielding ability, disease and pest resistance and superiority in other traits. The populations should be prolific with at least two ears in case of maize or multi-flowered in which some flowers can be selfed while the others are used in a cross. In Population 1 good looking plants are selected and self-pollinated at the same time crossed with bulk pollen from Population 2 which acts as the tester for Population 1. The progenies of each plant in Population 1 fertilised with bulk pollen of Population 2 is a half sib family. The selfed seed is kept as remnant seed while the seeds derived from the test cross are planted for evaluation of the half sib families in the next season. The test cross progeny evaluation results are used to select Population 1 plants whose test cross half sib progenies performed best. The selfed seed of the superior selected plants are bulked in equal proportions and planted in an isolated recombination plot where the plants undergo normal open pollination in the season following half sib evaluation. The seed harvested from the recombination plot serves as the seed of improved Population 1. Similarly, a simultaneous selection exercise is done in which Population 1 is used as the tester for Population 2. Good-looking Population 2 plants are pollinated with bulk pollen from Population 1 and at the same time self-pollinated. The selfed seeds of the various Population 2 plants are kept separately as remnant seed while the half sib progenies of the test cross are evaluated to help select superior Population 2 plants. The selfed seed of Population 2 plants whose test cross progenies excelled in the evaluations are bulked in equal proportions and planted in an isolated recombination plot. The seed harvested from the recombination plot is the seed of the Population 2 which has been improved using Population 1 as the tester. Each of the improved populations is subjected to another cycle and subsequent cycles of reciprocal selection.

The expected gain from half sib reciprocal selection is the sum of gain from the two Populations 1 and 2 and is given by

$$R = \frac{1}{y}\left[\left(\frac{1}{4}i_1\frac{\sigma_{A1}^2}{\sigma_{P1}}\right) + \left(\frac{1}{4}i_2\frac{\sigma_{A2}^2}{\sigma_{P2}}\right)\right],$$

where i_1 and i_2 are selection intensities for population 1 and 2, σ_{A1}^2 and σ_{A2}^2 are additive genetic variances for populations 1 and 2 and σ_{P1} and σ_{P2} are the phenotypic standard deviations for Populations 1 and 2 based on family means (Moreno-Gonzales & Cubero, 1993).

Because Populations 1 and 2 are open pollinated the test cross progenies evaluate GCA. However, if Populations 1 and 2 are of narrow genetic bases which is always the case for inbred lines, the tests and improvement will be for SCA. Tests for SCA are usually done in later stages of selection when inbred lines are developed from the two genetically diverse improved populations. Therefore, this method provides opportunities for improving the populations with respect to both GCA and SCA.

This method needs at least three seasons to complete one cycle of selection, namely, the season for generation of selfed seed and test cross seed, season for test cross progenies evaluation and the season for recombination. All the operations for both Populations 1 and 2 are done simultaneously, that is, production of S_1 seed and test cross seed for the two populations is done in the same season and evaluation of the half sib test progenies for the two populations is done in the same season. Recombination of the selected genotypes for the two populations is also done in the same season with each of them grown in isolation. The method has the problem of large amounts of materials to be handled which limits the number of test progeny families tested and therefore the number of plants that can be selected from each population. Limitation in the number to be selected leads to genetic drift and constraints to selection response and early selection plateaus.

Modified reciprocal recurrent half sib selection

Paterniani (1967) proposed a modification for half sib reciprocal recurrent selection of Comstock et al. (1949). Open-pollinated populations, 1 and 2, from which half sib seeds are obtained separately through open pollination, are carefully selected for the process. In population 1 half sib seeds from several good-looking plants are divided into two portions for each plant so that one portion is kept as remnant seed while the other portion is planted out in individual ear-to-rows in an isolated plot where such rows are alternated with the rows of population 2 plants. The plants in ear-rows of population 1 are detasseled and therefore made females so that population 2 plants provide bulk pollen to the female ear-rows for generation of test cross seed for population 1. Since population 2 supplies pollen to the ear-rows of population 1 it serves as the tester for population 1. In another isolated plot similar ear-row half sib families of population 2 are planted, detasselled, and are alternated with rows of population 1 plants to provide bulk pollen for the detasselled ear-rows of population 2. That is, population 1 is also used as the tester for population 2. For population 1, each ear-row is harvested separately so that the top cross test seeds of plants in the ear-row are bulked and planted out for evaluation in replicated trials in the next season with other ear-row test cross seeds to identify and select the best half sib families of population 1. Similar evaluation of top cross test progenies obtained from ear-rows of population 2 is done to identify and select the superior families of population 2. Remnant seed of the selected half sib families are bulked for each population separately and used for recombination in isolated plots to form improved 1 and 2 populations.

Full sib reciprocal recurrent selection

Full sib reciprocal recurrent selection scheme (Comstock et al., 1949) requires two genetically broad based and prolific maize Populations 1 and 2. In each population one ear is self-pollinated and the other one is used in a pairwise cross to generate full sibs between plants of Populations 1 and 2 so that the populations act as testers for each other. For each of the two diverse Populations, 1 and 2, good looking plants are selected and crossed in specifically arranged pairings. At maturity the S_1 seed for each plant is saved as remnant seed while the pairwise cross seed which gives the respective full sib seed for each cross is planted for full sib family evaluation. Each full sib family is planted in its own row/plot for comparison with several other full sib families in the same experiment. For each population the full sib evaluation helps to identify plants which gave the best performing full sib families. The remnant S_1 seed for the respective best performing full sib families in each population are bulked in equal amounts and grown separately for recombination in isolated random mating block. Seed harvested from the two separate recombination blocks gives seed of each of the improved population (1 and 2 respectively). In this method twice as many families can be handled as compared to half sib reciprocal recurrent selection. The two improved open pollinated populations or inbred lines obtained from them are used to develop hybrid and synthetic varieties at any stage as the populations are progressively improved. The advantages of full sib reciprocal recurrent selection over reciprocal half sib selection are that only one half of families are evaluated in each cycle to provide the needed information, full sib families' additive genetic variance exploitable is double that of half sib families. Full sib reciprocal recurrent selection helps in improvement of *per se* performance of the two populations, their combining abilities, especially SCA and heterosis in hybrids between the two populations or lines derived from them.

Expected genetic gain for full sib reciprocal recurrent selection, ΔG_{FSRRS}, is given as

$$R = \frac{1}{y}\left[i\frac{1}{4}\frac{(\sigma_{A1}^2 + \sigma_{A2}^2)}{\sigma_{PFS}}\right]$$

where i is the selection intensity, σ^2_{A1} and σ^2_{A2} are additive variances for Populations 1 and 2 respectively and σ_{PFS} is the phenotypic standard deviation for the full sib families (Hallauer & Eberhart, 1970).

Combined reciprocal S_2 and Full sib recurrent selection (CRRS)

Moreno-Gonzalez and Hallauer (1982) designed combined S_2 and full sib reciprocal recurrent selection (CRRS) to improve the efficiency of full sib reciprocal recurrent selection. The method needs four seasons and involves generation of S_1 seed for each population, 1 and 2. For each S_1 seedlot a portion is planted in the next season such that for population 1 S_1 rows are paired with population 2 S_1 rows. A portion of S_1 seedlot is kept as remnant seed for each plant in each population. For a pair of population 1 S_1 and population 2 S_1 rows good looking plants are selfed and at the same time used in pairwise crosses, i.e., population 1 S_1 x population 2 S_1. Therefore, in the second season S_2 and full sib of population 1 S_1 x population 2 S_1 progeny seeds are generated. In the third season the full sib families and S_2 families are evaluated to help select the best families. Based on such evaluations the S_1 remnant seed of the corresponding good performing full sib and S_2 families in each population are selected for recombination in an open pollination isolated field in the fourth season. This is done for each population separately. Seed harvested from the recombination plots give the seed of the improved population 1 and improved population 2.

Recurrent genotypic selection for combining ability

In cross-pollinated crop breeding progeny evaluations and selection of populations and inbred lines for combining ability is always part of the hybrid and synthetic cultivar development. The test crosses which may be top crosses or poly-crosses are used to improve populations or inbred lines performances *per se* and combining abilities through assessing the genetic worth of the genotypes under evaluation. Evaluation and selection for combining ability necessitates generation of progenies between the genotype to be evaluated for combining ability and a tester. The choice of the type of tester is an important aspect of this selection process since the tester used will determine whether the progenies obtained in the test cross are for GCA evaluation or SCA evaluation or both. If the aim is to improve GCA, broad genetic base testers like open pollinated populations, synthetic varieties and composites are used as testers while narrow genetic base testers like inbred lines are employed to improve SCA. General combining ability helps in exploitation of additive gene effects while SCA is useful where dominance, overdominance and epistasis are of significance especially for hybrid development.

Evaluation and selection for combining ability follow similar methods above where half sibs and full sibs are generated and are most conveniently applied to crop species with multiple flowers like brassicas, sunflower, maize with at least two ears, tillering crops like sorghum. In a genotype to be evaluated and selected for GCA superior looking plants are self-pollinated at the same time pollinated using bulk pollen from a broad genetic base tester. Seeds from self-pollination are kept as remnant seed while the seeds from the test cross are planted out to evaluate the half sib progenies with respect to GCA and other necessary parameters. The use of test cross progenies derived using a genetically broad-based tester makes this method appropriate for testing for additive gene effects since variation among the test cross progenies is attributable to half sib covariance which is interpreted in terms of additive gene effects as being responsible for GCA variation among the progenies tested.

Selection procedure for specific combining ability uses half sib progeny test selection and is similar to selection for GCA except that the tester used is of narrow genetic base like an inbred line or full sib progenies may be produced between genotypes under test and the tester. In the genotypes under selection for SCA, which may be an inbred line or an open-pollinated population, superior looking plants are self-pollinated and at the same time crossed to the tester. Harvested selfed seed will be kept as the remnant seed while the test cross seed is planted to evaluate the material for SCA and *per se* yield performance and other parameters. Through this selection scheme a cultivar or an inbred line is improved with respect to its specific combination in a cross with a specific inbred line for improved performance of the hybrid to be developed between them. The evaluations assess the non-additive effects as a result of specific gene combinations and interactions in a cross in addition to additive gene effects. The products of this selection are mainly for specific hybrid variety development and sometimes the specific tester is used as one of the parents in the hybrid population.

8.2 Population Improvement in Autogamous Crops

Population improvement methods used for cross-pollinated crops are not easily applied to self-pollinated, especially where there is no male sterility, because of problems associated with difficulties of making crosses and limited amounts of seed obtained from hybridisations. Difficulties of application of recurrent selection in many autogamous crops are due to practicability and costs of making crosses to produce sufficient amounts of cross seed for evaluations and inter-crossing to effect recombination. Most self-pollinated crops also have problems of inadequate pollen supply. However, the principles of recurrent selection used for allogamous crops equally apply to autogamous crops employing various methods to conduct population improvement with accompanying genetic recombination and increased genetic variability, particularly in the presence of male sterility. Population improvement schemes resolve the limitations of pedigree, bulk, single seed descent and backcross breeding methods and their modifications for developing superior pure line varieties. These breeding methods have the weakness of leading to products with limited and narrow genetic bases since they simply involve crossing of pairs of selected parents followed by self-fertilisation or backcrossing giving no opportunities for infusing new genes into the breeding programme and recombination to generate and sustain genetic variability in breeding materials. Self-pollination or backcrossing that follows pairwise hybridisation keeps linkage blocks intact. Other shortcomings of such methods are limitations to the number of crosses and cultivars that can be handled at any given time, fixation of undesired genes, limited genetic variability in the cultivars leading to restricted selection response, early selection plateaus and the need for large populations when selection is for quantitative traits.

As is the case with allogamous crops the development of improved populations in autogamous crops starts with selection and hybridisation of selected genotypes to effect recombination for creation, maintenance and exploitation of additive genetic variability in the breeding materials. Through recombination potential variability, especially additive genetic variability, becomes available for selection in the population and eliminates early selection plateaus which restrict selection responses due to undesirable tight linkages. Because they lead to increased additive genetic variability, schemes incorporating intermating and recombination in self-pollinating crops result in improved heritability magnitudes and increased selection response for metric traits. The improved populations derived through recurrent selection methods are used to develop pure line varieties for commercial production and parents for hybrid varieties in some crops like sorghum and rice where hybrid varieties are also commercialised.

8.2.1 Intermating Methods for Population Improvement in Autogamous Crops

Population improvement through recombination schemes for autogamous crops can be done in the absence of male sterility by making limited number of hand pollinations or facilitated through the use of male sterility to make pairwise and multiple crosses employing selected genotypes followed by segregation and selection of superior pure lines from such crosses to be used in further hybridisations and evaluations. The F_1 or segregating generations like F_2, F_3 or subsequent generations of the initial crosses are used for intermating to effect genetic recombination before pure lines are generated (Fig. 8.1).

Superior pure lines developed through mass selection, pure line selection, pedigree breeding, single seed descent, bulk breeding and backcross breeding which are selected for some target attributes to be combined are used in pairwise crosses to obtain F_1's. Superior individual segregant plants in the F_2 or F_3 or F_4 generations of each cross are selected and used in hybridisations by enforcing intermating (*inter se* mating) among the selected plants within a cross to effect recombination. Seeds of such intermating within the crosses are harvested separately and planted for another round of selection of superior segregant plants followed by enforced *inter se* mating of the selected plants within a cross. By this process of repeated selection and intermating among the selected individual plants belonging to segregating generations of various crosses, recombination, breakage and reorganisation of gene linkages is achieved. From F_3 or F_4 or other subsequent generations arising from such intermating, the segregating populations of superior crosses are subjected to

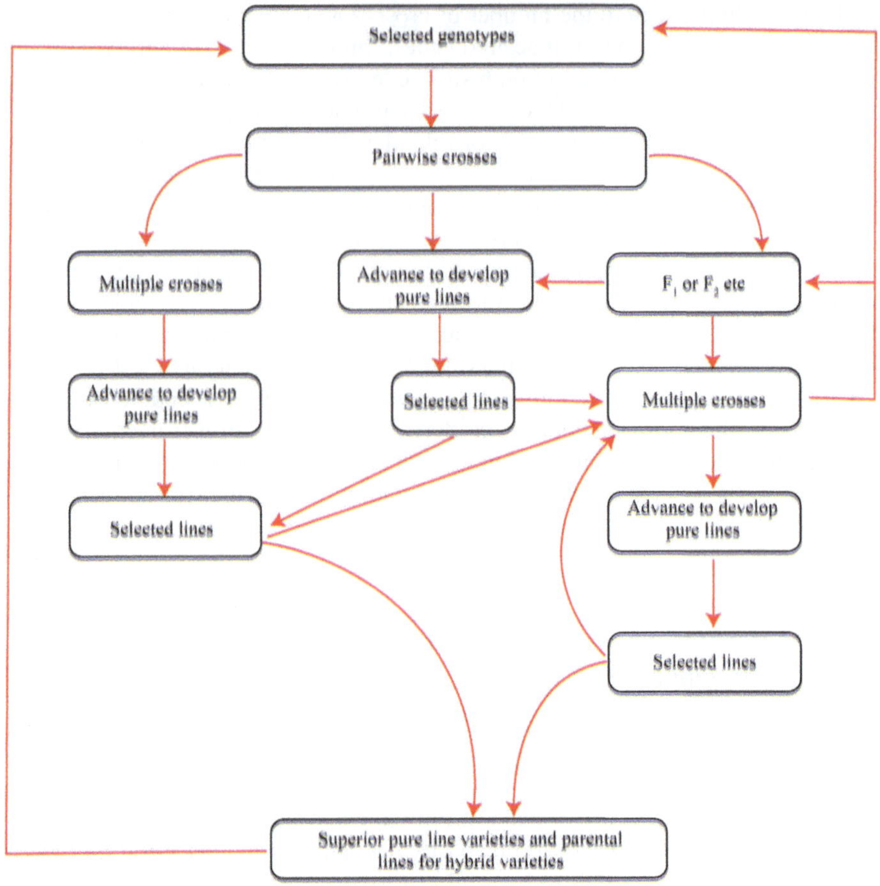

Fig. 8.1 Possible routes of deriving improved pure line cultivars through population improvement in autogamous crops

pedigree breeding, single seed descent, bulk breeding or other conventional methods to extract superior pure lines.

Superior pure lines obtained by pure line breeding and those developed from crosses between selected genotypes through conventional methods and having desired level of homozygousity, e.g., F_5 or F_6 or F_7, can also be subjected to crossing in all possible combinations through intermating or diallel mating to effect recombination. If the pure lines are to be obtained from crosses between selected genotypes it is preferable to use single seed descent because of the speed with which pure lines are developed. The period needed to achieve the desired level of homozygousity before evaluations are conducted can be shortened by using off-season nurseries for single seed descent advancement and speed breeding where appropriate facilities are available. The crosses obtained by intermating or diallel mating systems among the selected pure lines are advanced by single seed descent, starting from F_2 until a desired level of homozygousity is achieved (e.g., in or F_5 or F_6 or F_7.) or by other breeding method until pure lines are obtained. At this point, when single seed descent advancement is terminated, the seed of each plant is multiplied to give a line for use in performance trials to select the top lines. The selected best lines are again crossed in all possible combinations and the crosses are then advanced to subsequent generations by single seed descent, starting from F_2 followed by evaluation and similar crossing of the selected lines in all possible combinations.

Progenies of pairwise crosses involving selected pure lines can also be used to accomplish recombination by way of multiple crosses where the F_1 progenies or selected F_2 plants are used in double crosses of the type (AxB) x (CxD), (E x F) x (G x H), etc. The progenies of such multiple crosses are advanced to later generations to obtain pure lines which are employed in another round of pairwise or diallel crossing systems and multiple crosses for another round of generation advancement and selection.

In the above schemes superior pure lines are extracted periodically after adequate cycles of selection and appropriate level of homozygosity is reached. The derived pure lines are evaluated in replicated yield trials to identify superior lines for commercial production and use as parents for hybrid varieties in crops like sorghum. Products of these schemes can be improved at any stage by further hybridisation with other germplasm and backcrossing if they are lacking in certain desired attributes found in other germplasm.

Diallel selective mating was developed by Jensen (1970) for autogamous crops to be used along with the conventional methods. The method is designed to create and maintain genetic variability and broaden the genetic base of breeding populations through intermating of superior genotypes while at same time allowing for screening of the germplasm at various stages, incorporating and infusing new germplasm into the program. Diallel selective mating (DSM) addresses the limitations of conventional breeding methods like pure line, pedigree, single seed descent, bulk and backcross breeding methods and helps to improve their efficiencies. It incorporates the use of biparental crosses along with conventional breeding methods in segregating populations of various cross series derivatives at different stages to develop pure line genotypes.

Diallel selective mating (Jensen, 1970) series involves the following operations.

1. It begins by establishing several breeding populations starting with a set of parental lines selected from the population, referred to as basic parental series (P_1), which are crossed in a diallel series or partial diallel depending on the number of parents.
2. The F_1 progenies from the diallel series (P_1) cross are used to produce F_2 population which are advanced by conventional breeding to develop pure line varieties for commercialisation.
3. The F_1 progenies from the diallel series (P_1) cross are also used as parents in another diallel cross called F_1 diallel series cross or P_2 series. The F_1 diallel series crosses are multiple crosses of the type $(A \times B) \times (C \times D)$ since the F_1's themselves are directly used as parents in the diallel cross.

4. The F_1 progenies derived from the multiple crosses of the F_1 diallel (P_2) series are allowed to self-pollinate naturally to produce an F_2 population which is then subjected to mass selection and then advanced through other generations up to F_5 where pure line selection using conventional methods are applied to produce commercial pure line varieties.
5. Some of the selected F_2 or selected F_1 plants from the multiple F_1 diallel (P_2) series cross are used in another diallel mating series, called first diallel selective mating series or P_3 series, producing progenies of higher-level multiple crosses.
6. The F_1 from the first diallel selective mating (P_3) series are allowed natural self-pollination to produce F_2 and advanced to subsequent generations for extraction of superior pure lines to be evaluated and released for commercial production.
7. The F_1 plants of the first diallel selective mating (P_3) series are also selected and used in another diallel mating series called second selective diallel mating (P_4) series.
8. The F_1 obtained from the second selective diallel mating (P_4) series are allowed natural self-pollination for generation of pure line varieties through conventional breeding methods.
9. The F_1 of the first diallel selective mating (P_3) series and progenies of any cross-series level can also be used in backcrossing to other parents in the program. Any germplasm can be infused into the program at any stage.

All crosses involving P_2 series, P_3 series and P_4 are multiple crosses. From the populations developed using diallel selective mating method a number of selection programs can be initiated concurrently and improved populations derived periodically.

8.2.2 *Use of Male Sterility for Population Improvement in Autogamous Crops*

The presence of male sterility overcomes limitations and difficulties for recurrent selection in self-pollinated crops because it makes it extremely easy and cheap to conduct hybridisation without the need for hand emasculation and manual crossing to obtain adequate test cross seed and intermating. The only handicap in some crops is limited pollen production. However, it is easier to transfer pollen by hand to male sterile plants where necessary than hand or chemical emasculation. Male sterility controlled by recessive and dominant major genes is known in several self-pollinated crops like wheat, sorghum, soybeans, barley, rice and many other crops. It is extremely essential that commercial pure line varieties derived from products of intermating and selection regimes employing male sterility have no traces of male sterility and must be homozygous for the fertility gene. Like the case of cross-pollinating crops, easy and cheap production of hybrid varieties in self-pollinating crops requires the use of cytoplasmic male sterility factors along with recessive genic male sterility and the dominant male fertility allele acting as the restorer gene. The cytoplasmic-genic male sterility system is available and being use in a number of crops like sorghum, rice, wheat, pearl millet, soybean, barley and some other crop self-pollinated crops

Introduction of nuclear male sterility allele into a population is achieved through crossing followed by backcrossing to the recurrent population to which male sterility gene is to be incorporated. The transfer of recessive male sterility gene, rf, entails crossing plants from the target male fertile cultivar (RfRf) to the plants of the male sterile cultivar (rfrf). The F_1 will all be fertile while the F_2 will segregate for the dominant fertility and sterility alleles, Rf and rf. Male sterile plants from F_2 generation are backcrossed to the male fertile parents. This is followed by self-pollination of the progenies of the first backcross to obtain male sterile segregants which are in turn backcrossed to the male fertile cultivar. After each generation of backcrossing to the male fertile recurrent parent followed by self-fertilisation of the backcross progenies, only male sterile plants having intense expression of the recurrent parent traits are selected for further backcrossing. Backcrossing of male sterile progenies alternated with self-pollination is done until all the traits of the recurrent male fertile parent are incorporated back with sterility. The number of backcrosses of male sterile progenies to the recurrent male fertile parent depends on the characters of the male fertile cultivar to be reincorporated. The process of male sterility transfer can be made faster and easier through molecular marker assisted backcrossing, employing markers for male sterility and markers for the genome of the recurrent parent. Once male sterility transfer is accomplished random mating is achieved in the population by harvesting seed from male sterile plants only in each generation. The seed is bulked and planted again for such random mating to maintain a population having both male fertile and male sterile plants because of segregation. If a population has recessive male sterility allele along with dominant male fertility allele and male sterile and male fertile plants are grown together harvesting of seed from male sterile plants ensures that the seed obtained arises from cross pollination. In a population having recessive male sterility allele and dominant male fertility allele with sterility maintained through random mating there are male sterile plants having genotype rfrf and male fertile plants with genotypes RfRf and Rfrf. Male sterile plants are cross-pollinated with random pollen from male fertile plants in the population and each will give half-sib progenies when its seeds are planted. If rfrf male sterile plants in the population are fertilised by Rfrf plants the derived progenies have 1:1 ratio for male fertile: male sterile plants while the male sterile plants pollinated by male fertile RfRf plants give only male fertile progenies. The male fertile plants in such a population will undergo natural self-fertilisation, producing S_1 seed. If most of the male fertile plants in the population are harvested and their seed planted the progenies will be comprised of male sterile and male fertile progenies in varying proportions as determined by the frequencies of male sterility allele.

Apart from selection and intermating methods described above, autogamous crop population segregation for male sterility and fertility alleles can also be subjected to recurrent selection methods similar to those employed for allogamous crop populations. Mass selection can be conducted on autogamous crop populations having male sterility and male fertility genes. If mass selection is done for male sterile plants only the desired superior male sterile plants are identified in the field and harvested at maturity. Seeds from the selected male sterile plants are bulked in equal amounts to give the seed of the improved population. The expected gain from such selection which takes into account that parental control factor, c, is $\frac{1}{2}$, is given by

$$R = \frac{1}{2}\frac{i\sigma_A^2}{\sigma_P}.$$

When the seed of the improved population is grown in the field it gives both male fertile and male sterile plants. Mass selection can also be conducted by identifying superior looking male fertile plants only and compositing seed harvested from the selected plants to give the seed of the improved population. Due to self-pollination in the male fertile plants parental control factor, c, is 1, making selection gain from selection of the male fertile plants be

$$R = i\frac{\sigma_A^2}{\sigma_P}.$$

Planting of such bulked seed gives a population having a mixture of male fertile and male sterile plants. Selection on such population can be continued by selecting the majority of superior male sterile plants only, or superior male fertile plants only, or alternating such selection of superior male sterile plants and superior male fertile plants in successive generations.

Half sib selection can also be conducted on autogamous crop populations segregating for recessive male sterility by identifying and tagging good looking superior male sterile plants in the field. Poor-looking and undesired inferior male fertile plants can be removed before pollination. Seed obtained from each of selected superior male sterile plants are half sib seeds. For each plant the seed is divided into two portions so that one portion is kept as remnant seed while the other portion is used for half sib family replicated evaluations. In the evaluations seed from each plant is grown in half sib family row (head-to-row for case of sorghum) in a field where several half sib families are grown for comparison. Based on the evaluations the best half sib families are identified and their remnant seed is bulked in equal quantities to give the seed of the improved population. When the bulked seed is planted, the arising population is comprised of both male fertile and male sterile plants. In the recombination field undesirable poor looking male fertile plants are removed before pollination so that they don't supply pollen. The male sterile plants can be subjected to another round of similar selection. The genetic gain for this selection is similar to half sib selection in allogamous crops.

$$R = \frac{1}{4}\frac{i\sigma_A^2}{y\sigma_{P_{HS}}}.$$

The population segregating for male sterility can also be subjected to full sib selection by making several pairwise crosses between selected good looking male sterile and male fertile plants. Seeds for each pairwise cross is harvested from male sterile parent to give full sib family seed. The full sib family seeds are grown the next season in full sib family rows for comparative evaluation in replicated tests while the remnant seed of each full sib family is reserved. Full sibs are selected based on such evaluations and the remnant seed of each selected full sib family is bulked with the others in equal proportions and grown out in a recombination block for intermating. Such a plot will be comprised of both male fertile and male sterile plants. In the recombination field undesirable poor looking male fertile plants are removed so that they don't supply pollen while superior male sterile plants are identified so that seed can be harvested from them at maturity. Seed harvested from such plants is bulked to give the seed of the improved population which is grown and another set of pairwise crosses are made between selected good looking male sterile and male fertile plants for a second cycle of similar selection. Selection gain for this selection scheme is similar to full sib selection gain given under full sib selection for allogamous crops.

$$R = \frac{1}{2} \frac{i\sigma_A^2}{y\sigma_{FS}}.$$

For S_1 family selection desirable male fertile plants are identified for harvesting at maturity. In case of sorghum such plants can also be bagged to ensure complete self-fertilisation. For other self-pollinating crops there is no need for bagging. The S_1 seed for each selected plant is divided into two so that a portion is reserved as remnant seed and the other portion is planted in S_1 family evaluation field where all the S_1 families are compared for selection of the best performing families. For the selected S_1 families the remnant seed is bulked and planted for recombination. In the recombination field undesirable male fertile plants are removed so that they don't supply pollen while good looking male sterile plants are identified and tagged for harvesting. Seed is harvested from the tagged male sterile plants and planted for repeat of the selection process in which good looking male fertile plants are identified and their S_1 seed harvested at maturity for another round of evaluation and selection. Selection gain prediction for this selection procedure is the same as the prediction given for S_1 selection for allogamous crops.

$$R = \frac{i\sigma_{A'S1}^2}{y\sigma_{PS1}}.$$

Using the same procedure S_2 selection can be conducted in self-pollinated crops in which recessive male sterility is maintained. Reciprocal recurrent selection can be employed to improve autogamous crops using recessive male sterility gene or cytoplasmic-genic male sterility where the focus is to produce superior lines for making commercial hybrid varieties or superior commercial pure line varieties.

Unlike recessive male sterility, dominant male sterility does not require a generation of self-pollination to obtain segregants that are male sterile. Also in case of dominant male sterility the progenies of male sterile plants always appear in ratio of 1:1 for dominant ♂ sterile plants: ♂ fertile (recessive) plants. Phenotypic mass selection is conducted on a cultivar having dominant male sterility by identifying and tagging superior male sterile plants in the population. At maturity seed is harvested and bulked in equal proportions from the selected plants to form the improved population which can be subjected to another cycle of similar selection. The selected male sterile plants undergo recombination due to out-pollination. Selection response is

$$R = \frac{1}{2}\frac{i\sigma_A^2}{\sigma_P}.$$

Half sib selection is conducted in populations having dominant male sterility by identifying superior looking male sterile plants. After harvesting the selected plants, a portion of the seed of each plant is kept as remnant seed while the other portion is used for half sib family evaluations where the seeds are planted in separate progeny rows as half sib families. The best half sib families are selected and their remnant seed is bulked to give the seed of the improved population which can be subjected to similar selection process. Selection gain prediction is similar to the half sib selection response prediction above.

If pure lines are to be developed from any cultivar subjected to population improvement and is segregating for dominant male sterility, the male sterile superior plants are selected and seed obtained is grown out to identify male fertile segregant plants among their progenies. The male fertile progenies from male sterile plants are homozygous for the recessive fertility allele and are harvested to give the new population from which pure lines are obtained using conventional breeding methods for commercial production. Only those lines not segregating for male sterility, i.e., having only male fertile plants, are selected for further evaluation to identify best pure lines with complete male fertility.

Problems

8.1 Describe any possible improvement of mass selection for cross pollinated crops.

8.2 Which are the weaknesses of ear-to-row selection and how are such problems resolved?

8.3 What are the advantages of full sib reciprocal recurrent selection?

8.4 What limitations of traditional breeding methods makes it necessary to carry out recurrent selection in self-pollinated crops?

8.5 Describe a procedure of achieving recombination in recurrent selection designed for self-pollinating crops in presence of male sterility.

8.6 How do you incorporate a recessive male sterility gene into a cultivar having dominant male fertility?

References

Bos, I. (1983). About the efficiency of grid selection. *Euphytica, 32*, 885–893. https://doi.org/10.1007/BF00042170. Springer Nature No. 5732640297153.

Compton, W. A., & Comstock, R. E. (1976). More on modified ear-to-row selection in corn. *Crop Science, 16*, 122. https://doi.org/10.2135/cropsci1976.0011183X001600010034x. John Wiley and Sons No. 5733081135038.

Comstock, R. E., Robinson, H. F., & Harvey, P. H. (1949). A breeding procedure designed to make maximum use of both general and specific combining ability. *Agronic Journal, 41*, 360–367. https://doi.org/10.2134/agronj1949.00021962004100080006x. John Wiley and Sons, No. 5625521183588.

Doggett, H. 1968). Mass selection systems for sorghum. *Crop Science, 8*, 391–392. https://doi.org/10.2135/cropsci1968.0011183X000800030038x. John Wiley Sons No. 571980239101.

Doggett, H. (1972). Recurrent selection in sorghum. *Heredity, 28*, 9–29. https://doi.org/10.1038/hdy.1972.2. Springer Nature No. 5731971217247.

Doggett, H., & Eberhart, S. A. (1968). Recurrent selection in sorghum. *Crop Science, 8*, 119–121. https://doi.org/10.2135/cropsci1968.0011183X000800010038x. John Wiley and Sons No. 5731980774577.

Hallauer, A. R., & Eberhart, R. E. (1970). Reciprocal full-sib selection. *Crop Science, 10*, 315–316. https://doi.org/10.2135/cropsci1970.0011183X000030033x. John Wiley and Sons No. 5733510274694.

Gardner, C. O. (1961). An evaluation of mass selection and seed irradiation with thermal neutrons on yield of corn. *Crop Science, 1*, 241–245. https://doi.org/10.2135/cropsci1961.0011183X000100040004x. John Wiley and Sons, No. 5625530525202.

Jensen, N. F. (1970). A diallel selective mating for cereal breeding. *Crop Science, 10*, 629-635. https://doi.org/10.2135/cropsci1970.0011183X001000060006x. John Wiley and Sons, No. 5625540463713.

Lonnquist, J. H. (1964). A modification of ear-to-row procedures for the improvement of maize populations 1. *Crop Science, 4*, 227–228. https://doi.org/10.2135/cropsci.1964.0011183X000400020033x. John Wiley and Sons, No. 5625530151438.

Moreno-Gonzalez, J., & Hallauer, A. R. (1982). Combined S_2 and crossbred family selection in full-sib reciprocal selection. *Theoretical Application Genetics, 61*, 353–358. https://doi.org/10.1007/BF00272856. Springer Nature No. 5733531122736.

Moreno-Gonzalez, J., & Cubero, J. I. (1993). Selection strategies and choice of breeding methods. In M. D. Hayward, N. O. Bosemark & I. Ramagosa (Eds.), *Plant Breeding: Principles and Prospects*. Springer Dordrecht. https://doi.org/10.1007/978-94-011-1524-7_19. Springer Nature No. 5731970465420.

Paterniani, E. (1967). Selection among and within families in a Brazilian population of maize (*Zea mays* L.). *Crop Science, 7*, 212–216. https://doi.org/10.2135/cropsci1967.0011183X000700030012x. John Wiley and Sons, No. 5626060623708.

Chapter 9
Development of Pure Line Varieties

Pure line varieties, whose development and commercialisation are a preserve for autogamous crops and often cross-pollinated crops, are almost 100% homozygous at nearly all loci. They are the most common varieties in self-pollinated and often self-pollinated crops except for a few cases where hybrids are also developed and commercialised like sorghum and rice, tomatoes and some high valued vegetables. All conventional breeding methods for self-pollinated crops follow more or less the same breeding pattern involving three major operations, namely, creation of genetic variability, selection and varietal maintenance. Creation of genetic variability can be achieved through hybridisation (single or multiple crosses or backcrosses) and recombination, induced mutations and biotechnology methods. The selection method that breeder chooses will depend on the breeding objectives, how fast the breeder has to develop a pure line variety, the characters to be improved and the available resources.

In autogamous crops various conventional breeding methods which are used to exploit available genetic variability for development of adapted high yielding pure line varieties include mass selection, pedigree, bulk, single seed descent and backcrossing methods and their modifications. If the objective is narrow like the transfer of one or a few major gene traits to an established variety simple crossing followed by a few generations of backcrossing may suffice to give the desired pure line in a short period. If the objectives are wider and related to traits like yield, adaptability and other traits influenced by polygenes more elaborate methods like pedigree breeding will be employed in the development of the desired pure line varieties. The conventional breeding methods can be used along with selection methods involving intermating and recombination, induced mutagenesis, molecular marker assisted breeding, biotechnology, genetic modification and genome editing techniques to concentrate favourable genes and gene combinations into desirable genotypes for development of superior pure line varieties.

P. O. Ayiecho and J. O. Nyabundi, *Conventional and Contemporary Practices of Plant Breeding*, https://doi.org/10.1007/978-3-031-74998-8_9

9.1 Selection Without Hybridisation: Mass and Pure Line Selection

Landraces of various autogamous crops comprise of mixtures of pure line genotypes which are almost 100% homozygous at nearly all loci. A landrace is a cultivar that is endemic to an area in which it has been subjected to traditional agricultural practices for several decades and is therefore a product of human and natural selection operating in the area to which it is indigenous. Landraces are adapted to respective regions around the world to which they belong, especially in developing countries. Since landraces of autogamous cultivars comprise of mixtures of true breeding lines, they are highly heterogeneous and homozygous and easy to improve by simple conventional breeding methods like mass selection and pure line selection, leading to development of pure line varieties.

Selection methods for self-pollinated crops are designed along the pure line theory developed by W. Johannsen in 1903 with the assumption that a population (variety or landrace) of a self-pollinating crop is comprised of true breeding distinct homozygous lines. Simple mass selection and pure line selection conducted on cultivar mixtures are the simplest breeding methods used to improve self-pollinating crops.

9.1.1 Mass Selection

Mass selection is a simple method of improving landraces and can be conducted on populations derived from crosses which have been allowed to undergo normal self-fertilisation and segregation for a number of generations like products of bulk advancement of segregating populations. Mass selection can be applied on products of recombination selection schemes, newly introduced or assembled cultivars and commercial varieties which might have become mixtures of genotypes. Established commercial pure line varieties which have been cultivated for several years lose their purity due to mutation, accidental mixing and occasional outcrossing due to pollen from neighbouring variety fields and therefore such varieties may be made pure again by mass selection.

To conduct mass selection the target cultivar is planted in the field and several superior looking plants in regard to the target trait are selected and their seed bulked. Individual plant selection is based on visual assessment of the phenotype only without progeny testing and majority of plants in the population are selected leaving out only a few bad looking ones.

After selection in the first season the bulked seed forms the seed of improved population. Selection may be repeated on the improved population as dictated by the character under selection, degree of heterogeneity, genetic variability, selection pressure, breeding objectives and purity of the desired end product. Such selection will continue so long us there is variability and response. Mass selection on a population or a cultivar will lead to response only if there is genetic variability and high heritability for the trait under selection. Because of self-fertilisation parental control factor, c, is 1. The response prediction equation is given by

$$R = \frac{i\sigma_A^2}{\sigma_P}.$$

In autogamous crops mass selection exhausts genetic variability in one or a few generations of selection as determined by the thoroughness of the selection process, trait under selection and the environmental influence on the trait. If genetic variability is exhausted there will be selection plateau since there will be no more response, an observation achieved faster in self-pollinated crops than in cross-pollinated crops where it takes several generations for attainment of the selection plateau.When the breeder is satisfied with the selection process the population/cultivar obtained through selection is subjected to evaluation in replicated comparative tests where it is compared to the parental cultivar and commercial varieties. Such evaluations are replicated over seasons and sites before decision is reached about the superiority of the product(s) of selection which may consequently be a single superior pure line or a set of superior pure lines or a cultivar that may still be comprised of pure lines but much less heterogeneous than the original cultivar.

The merits of mass selection are that it is simple, cheap and fast as regards the resources and time needed to accomplish one cycle of selection. Mass selection is effective for traits with high heritability and least affected by the environment and those controlled by a few major genes. On the other hand, it is not effective for characters with low heritability and its effectiveness is marred by environmental effects because of the influence of the environment on the phenotype. Cultivars developed by mass selection do not have high level of genotypic uniformity and have some level heterogeneity.

9.1.2 Pure Line Selection

Pure lines, which are the foundation of autogamous crop breeding, can be developed from mixed populations, segregating populations following hybridisation and populations arising from selection schemes involving intermating of selected genotypes for direct release as commercial varieties, use in development of multiline and hybrid varieties. Pure lines can be used as sources of specific genes for resistance to biotic and abiotic stress factors, quality aspects, plant architecture, male sterility and other desired traits. Pure lines are also used in crossing and intermating for autogamous crop improvement applying methods like diallel selective mating and other selection schemes incorporating recombination.

Pure line selection involves growing and evaluating progenies of the selected plants and is applied on landraces, commercial cultivars, introduced accessions and segregating populations. Since it involves progeny testing it is suitable for traits with low heritability, traits under the control of polygenes and influenced by the

Table 9.1 Possible outline for pure line selection

Year	Activities
1	Select superior looking 200 to 1000 plants from a mixed cultivar population
2	Grow progenies of selected plants in individual progeny rows. Harvest superior progeny rows and composite seed from plants within each row. Each progeny (from a given row) is then treated as a line/strain
3	Grow the selected lines in replicated observation plots. If seed is sufficient the strains can be grown in preliminary yield tests to assess their performances. Superior strains are selected and their seeds separately composited. Quality assessments are done
4 to 7	Continue with yield tests and quality assessment to compare the performances of the selected strains. Observations are made for disease and pest resistance and tolerance to other forms of stress are done. This helps to eliminate undesirable lines and identify the most promising line
8	Best line identified for release as a new pure line variety

environment. A pure line is a highly homozygous individual progeny derived solely by continued self-pollination or some form of inbreeding and such pure lines are true-breeding. That is, all plants arising from a pure line are genetically alike and more often resemble one another phenotypically and that any differences among the plants derived from a pure line are purely due to environmental effects and not genetic. Such plants are also genetically similar to their parental plants. Differences among different pure lines are attributable to both genetic and environmental effects. A pure line cultivar is expected to be more uniform than a cultivar developed by mass selection.

In pure line selection the cultivar to be improved is planted in the field and a number of plants which look superior are selected for target agronomic traits to be improved and the selected plants must be disease and pest free, strong and vigorous with signs of good performance. The progenies of selected plants are planted out in the field for evaluation in the next season such that the progenies of each selected plant are grown in a progeny family row/plot for evaluation and elimination of poor performing progeny families (Table 9.1). Mild phenotypic selection may also be done among individual plants within a superior progeny family if there are some rare segregants in some progeny lines. Seeds of plants within a selected progeny line are bulked and advanced to the next season of progeny comparative evaluation among the selected progeny lines.

Preliminary yield tests may be done for a few seasons for further elimination of undesirable lines/families as testing progresses. Once the superior lines have been identified replicated yield tests are conducted over locations and years until only one best line or only a few superior ones are left to be compared with the commercial varieties in further replicated yield trials and ultimate release for commercial production.

9.2 Hybridisation and Methods of Handling Segregating Populations

Most conventional breeding methods also take advantage of inherent variability created through hybridisation or mutation followed by selection for pure line varietal development. Hybridisation is done to bring desired traits into one variety with the aim of developing a genotype that is superior to the parents involved in generating it, and possibly superior to the varieties in production with respect to adaptability, yield performance, resistance to the prevailing biotic and abiotic stresses, quality aspects and other desirable characters. When selecting prospective parents for hybridisation to develop a superior variety it is ensured that one of the parents selected is of proven performance in the intended area of production, well adapted and high yielding and that the second parent or the other parents are selected for superiority in some specific characters in which the adapted genotype is deficient like resistance to the prevailing or new diseases and pests, quality attributes, specific adaptation parameters, flowering and maturity periods, short stature and other forms of plant architecture, etc. Therefore, the other parent(s) is/are used to correct and complement specific limitations of the adapted parent through single crosses, backcrosses, three way or double crosses or various forms of multiple parent crosses for gene pyramiding. The type of crosses is determined by the number traits to be combined and their distribution among the parents, the pattern of inheritance, the number of loci and alleles involved and the method of handling the segregating generations. Once the crosses are obtained the segregating generations are handled using various methods, namely, pedigree selection, bulk breeding, backcrossing, single seed descent, mass selection and their modifications, as determined by the resources available to the breeder, the inheritance patterns of the target traits and how quickly the breeder has to release the new variety. The crosses can also be used in recurrent selection schemes to develop populations or lines to improve breeding materials at hand and expand the working stocks of the breeders' germplasm. If male sterility is available it helps to facilitate the hybridisation process and intermating among selected genotypes for recombination followed by selection of the desirable genotypes with the best gene combinations.

9.2.1 Pedigree Breeding

Pedigree breeding method which leads to development desirable genotypes from segregating populations through progeny testing is commonly and widely used to develop superior pure line varieties and pure line parents for hybrid varieties development. A key step to achieving the intended objectives in pedigree breeding is selection of parents which must be done in such a way that best adapted line is crossed with a good line having the desired characteristics lacking in the adapted line. Once the cross has been made between or among the selected parents the crosses are advanced

Table 9.2 Possible pedigree breeding plan

Year/season	Generation	Activities
1	P_1 x P_2	Cross the selected parents P_1 and P_2 under appropriate conditions for success in making the cross. Parents to be crossed are selected for specific traits to be combined
2	F_1	F_1 plants space planted in appropriate conditions. Grow enough to produce F_2 seed of the desired size. Discard those that appear as off-types and non-hybrids. Brief notes to recorded
3	F_2	F_2 generation plants are space planted in the field for adequate expression of traits of interest. About 2000 to 6000 F_2 plants may be planted. Selection of single plants that appear early and having desirable qualitative traits of interest. Artificial disease and pest epidemic or abiotic stress factors may be applied to assist in selection. Records are taken for the plants of this generation. Only mild selection is done, particularly for simply inherited traits
4	F_3	Space planting for rows of selected plants (i.e., family progeny rows) so that row selection can be initiated. Checks can be planted in every 5 to six rows. Artificial disease and pest epidemics and conditions for abiotic stress factors of interest can be created. Selection is done for simply inherited traits and certain quality attributes. Selection is done among the rows so that whole rows that are undesirable can be discarded. Single plant selection within superior rows is also done because of segregation within the families. If enough seed is obtained in F_2 preliminary yield tests (in form of early generation yield tests) can be initiated as early as this generation for F_2-derived F_3 lines or yield tests can be postponed to later generation. Records are taken
5 to 7	$F_4$to F_6	Continue with between family selection. Some selection for superior segregants within families in F_4. Within family selection reduced significantly in F_5. Within family selection avoided by F_6 except for outstanding and significant segregant genotypes from superior families. Larger plots per family (at least two rows per family) by F_5. Preliminary yield tests initiated in F_3 continued or yield tests are initiated by F_4 or F_5 if they had been postponed. Selection for abiotic and biotic stress resistance. Only best appearing families/lines selected by 7th year. Lines left should be much less than 50. Checks are used for comparison. Keep taking records. The pedigree information helps in selecting the best families in related groups
8 to 12	$F_7$to F_{11}	Yield tests continued where comparison with commercial varieties is done. Quality tests and selection for abiotic and biotic stress resistance. Records are taken. Regional/locational trials can be done. Best pure line identified
13 to 14	$F_{12}$to F_{14}	Multiply seed of identified pure line and release the new variety for commercial production

to F_2 and subsequent generations with selection for the superior plants and lines (Table 9.2). The scheme starts with space planted F_2 as separate family rows derived from individual F_1 plants. The F_2 plants are subjected to mild selection based on desirability, vigour, presence of target traits and appropriate combination of such traits. Progenies from F_2 plants are grown in space planted individual progeny rows in the F_3 generation in which mild selection is done on the basis of desired characteristics with proper record about performance, various characteristics and family ancestry. Selection for high heritability traits like plant type, earliness, some disease and pest resistance, etc, may be done on individual plants within the selected F_3 progenies as dictated by breeding objectives. That is, in F_2 and F_3 generations attention should be on simply inherited Mendelian traits like plant architecture, plant height and phenology characters and stress conditions like disease infection and pest epidemics can be created to enhance selection. However, if sufficient seed was produced in F_2 early preliminary yield tests can be initiated in F_3 for the F_2-derived F_3 lines or such tests can be postponed to F_4 generation. Selection for resistance to other forms of stress are also initiated in F_3 generation.

The selected F_3 plants are used to derive F_4 progenies in family plots and, depending on materials at hand, resources, experimental design and management, strong selection between the families is done in F_4 and continued through F_5 up to F_7 when uniformity is achieved in progenies within individual families. The progenies derived from selected plants are subjected to reselection as generations advance until genetic purity is reached because of high level of homozygosity within families. Selected families which are treated as lines due to high homozygosity levels are subjected to replicated tests at two or more sites and over seasons. Some selection is done for superior segregants within families in F_4. Within family selection is reduced significantly by F_5. Selected desirable lines derived from this process will each have its seed harvested in bulk so that there is enough seed for each line for replicated evaluations. Replicated yield tests are done from F_7 or F_8 or F_9 to identify the best yielding lines with desirable traits. Only one line should be left by F_{11}.

Early Generation Selection

Early generation yield testing and selection in segregating populations helps to eliminate undesirable genotypes as early as F_3 and F_4 generations, speeds up the breeding process and saves on resources in self-pollinated field crops like wheat, barley, rice, soybean, beans, cowpea, pea and others. The need for early generation testing is determined by the breeding objectives, available resources and pressure to release varieties as quickly as possible. Though space planting, small number of plants, heterosis, heterozygosity and accompanying segregation make selection for quantitative traits difficult in early generations it has been established that it is possible to identify promising segregant genotypes as early as F_2 that may give superior lines in subsequent generations (Sneep, 1977). Early generation selection requires minimal selection in F_2 generation so that adequate seed can be produced to plant sufficiently large F_3 lines for selection under normal crop stand (Cooper, 1990). Effective early generation testing and selection for yield also requires that the F_3 generation consist of a certain minimum number of plants per family with dense planting and high plant populations of at least 228 plants per F_3 family (Sneep, 1977). Early generation selection is initiated in F_3 by conducting yield testing and evaluation of F_2-derived F_3 lines to select the best segregant lines for further tests in F_4 and F_5 so as to identify superior lines for comprehensive replicated yield evaluations in F_6 and F_7 generations (Cooper, 1990). In addition to yield tests the families are also evaluated for earliness, height, quality attributes, resistance to diseases and pests and abiotic resistance attributes touching on adaptability. Selection for stress resistance is achieved through creation of abiotic stresses where necessary and artificial disease or pest epidemics under appropriate conditions.

9.2.2 Bulk Breeding

Bulk breeding involves advancing F_2 and subsequent generations without artificial selection by harvesting nearly all the plants and bulking the seed to plant the next generation until high level of homozygousity in F_6 or F_7 is apparent. Thereafter line selection is initiated and seed is increased for promising lines for further testing.

Bulk breeding was proposed by Nilsson-Ehle in 1908 using a cross of winter wheat. It is a method which can be used to obtain pure lines from a cross after segregating populations have been grown in bulk for a number of generations so that natural selection assists the efforts of the breeder by eliminating weaker genotypes in the segregating populations. Bulk breeding is based on the fact that natural selection influences the composition of the population leading to alteration of gene frequencies as the segregating population advances towards homozygousity.

In bulk breeding the early generations derived from a cross or materials obtained from any of the population improvement schemes are grown as bulk populations in successive generations under normal cultural conditions with regular commercial planting rate. The number of generations of bulking will depend on the resources available to the breeder and breeding objectives. In this scheme the seed of a given cross is harvested from the F_1 plants and bulked for planting to get the F_2 generation whose seed is also bulked to grow the F_3 generation. The F_4 population is similarly obtained from bulked F_3 seed. Such bulking may stop in F_6 or F_7 generation since by F_6 and F_7 the majority of the plants will be highly homozygous with very little further segregation and progenies of plants from such generations will start breeding true. During the bulking generations no artificial selection is done apart from elimination of obviously undesirable plants like lodged or badly diseased ones and those with serious pest attack. Selection is initiated in F_7 or F_8 when individual desirable and good-looking plants are selected and their seeds planted in individual progeny rows or plots as lines for selection among the lines. The superior lines will be selected based on yield and other traits and advanced to subsequent generations for further selection so that only a few pure lines will have been left for further yield trials.

This method is simple, easy and has low work load in the bulking generations with no pedigree record keeping making it feasible for the breeder to handle several crosses and large populations in each generation. It is more efficient than pedigree method in terms of resource utilisation and effort. Bulk breeding method is fast-tracked by using off-season nurseries for bulking generations so that more than one generation is grown in a year and where facilities are available speed breeding procedure can be used for bulking generations. Disadvantages of the method are that it may not lead the desired changes in genetic constitution of plant populations as fast as expected, natural selection may not lead to adequate yield improvement and it may take long to produce an acceptable variety. Where the segregating populations being bulked are heterogeneous for height natural selection may favour the tall genotypes which may not necessarily be the highest yielding genotypes. In crops like barley, wheat, rye, oats, rice and sorghum short genotypes have higher yield potential than tall genotypes under favourable productive environments but they don't compete well with tall plants and are eliminated in dense mixed populations. Therefore, natural selection can be contrary to the plant breeding objectives by eliminating some plants with desirable genes, making it necessary to conduct mild artificial selection during bulking generations so as to retain such genes. Large populations at the end of bulking generations also create challenges for artificial selection, especially if the resources are limiting. Since no pedigree records are kept during bulking generations it is not possible to keep track of the inheritance of various important traits over generations. However, the products of bulking are reservoirs of genetic variability and are useful sources of populations with increased yield and favourable adaptability for development of superior pure lines.

9.2.3 Single Seed Descent

Single seed descent (SSD) breeding procedure which is a modification of bulk breeding was described by Brim (1966). The intention of single seed descent is to advance the generations of crosses quickly and obtain pure lines as fast as possible. It entails advancing each plant in F_2 population from a cross or any population improvement scheme by one seed without selection to F_5 or F_6 generation such that each F_6 or F_7 family traces back to a single F_2 plant. In F_5 or F_6 there will be a high level of homozygousity due to continued natural self-fertilisation and a very high proportion of plants will be almost true breeding.

In single seed descent breeding method once a cross is made, whether single cross or multiple cross and F_2 generation is obtained by growing F_1 seed, only a single seed from each of the F_2 plants will be used to raise the F_3 generation (Fig. 9.1). Similarly, only one seed is taken from each individual F_3 plant to get F_4 generation so that each F_2 plant is represented by one plant in F_4 generation. The F_5 generation is similarly derived from F_4 generation through single seed propagation per plant. Single seed propagation may be stopped in F_5 or F_6 when there is no more significant segregation and there is attainment of high level of homozygousity. After single seed propagation is stopped seed is increased by harvesting all the seed from each F_5 plant separately and the seed is planted in the next generation such that the progenies of each plant form a separate line (family) in F_6 for evaluation and selection of superior lines. Some mild selection for individual plants within superior families may be

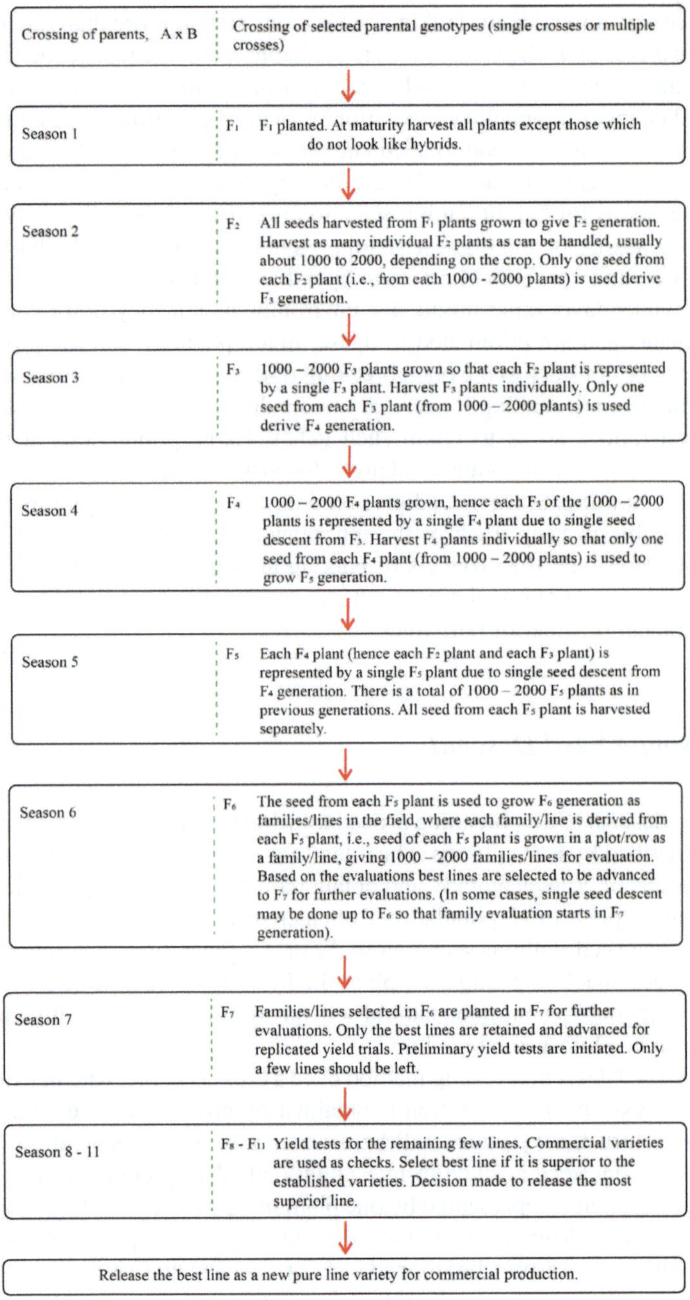

Fig. 9.1 Outline of single seed descent breeding method

done where there is some segregation, but much greater emphasis is put on selection among families. Preliminary yield trials are started in F_7 or F_8. If there is adequate seed replicated tests of the lines are done for two or more generations to select the highest yielding pure lines for further comparison with established varieties.

The SSD breeding approach is simpler, quicker and cheaper than the pedigree method and helps to overcome the limitations of bulk breeding which arise due to natural selection. It minimises natural selection, helps to maintain broad representation of genotypes in the breeding material and also increases and maintains additive genetic variability in breeding populations (Brim, 1966). Single seed descent method saves on workload for record taking and works well for traits with high heritability. Since there is no selection to eliminate undesired plants during generations of advancement by single seed not much effort is needed in the early generations until after multiplication of lines is done for selection among lines. When the breeder is dealing simultaneously with a large number of crosses in the program SSD may be the preferred breeding scheme. Because of small populations for each cross, advancement of generations by single seed descent can be conducted within small nursery space using off-season facilities, thereby shortening the duration for pure line development. For crops maturing in three months three generations can be done in one year using the off-season nurseries and irrigation. Speed breeding protocols can be applied to both single seed descent and single pod descent procedure by growing F_1 to F_5 generation under speed breeding conditions for rapid generation advancement to fast-track pure line variety development. Single seed descent can also be incorporated in a number of population improvement selection methods like diallel selective mating and other methods involving intermating and selection in self-pollinated crops.

One of the major limitations of single seed descent is drift because single plant approach used to advance segregating generations is likely to lead to random genetic drift. This may lead to some genes being fixed and others getting lost and the lost alleles will not be recovered unless saved materials are analysed and those with required genes are put into the breeding programme again. This makes it an undesirable breeding scheme for metric traits controlled by several minor genes and influenced by the environment.

A modification of single seed descent is employed for annual grain legumes where single pod descent (SPD) is applied for generation advancement to obtain homozygous lines for evaluation and pure line variety development. This procedure ensures that each F_2 plant is represented in F_5 generation. Single pod selection is done by selecting one healthy pod from each plant in F_2 generation so that seeds from the selected pods are used to derive F_3 (Saxena et al., 2019). The seeds from the single pod obtained from an F_2 plant are grown in a pot or a hill to obtain the F_3 plants, hence each F_2 plant is represented by a pot or a hill of two to four plants (depending on the number of seeds from the pod). Some days after emergence when the seedlings are well established and vigorous all the plants except the most vigorous and heathy one, are removed, leaving only one plant per pot/hill. At maturity only one healthy pod is harvested from each F_3 plant and the seeds from the harvested pod are similarly planted in a pot or hill to give the F_4 generation, with two to four plants representing an F_3 plant. These plants are thinned to one plant per hill as was done in F_3 generation. The same procedure is used to derive the F_5 generation. At maturity all the seeds from each F_5 plant are harvested together with seeds of each plant kept separate for growing the F_6generation. The F_6 generation is handled as F_5-derived families by growing the seeds from each F_5 in a row/plot as a progeny family for seed multiplication and preliminary screening. The selected F_6 families/lines are also grown as lines in F_7 for further seed multiplication and selection of the most promising lines. Further evaluation is done in F_8 to identify the most desirable pure line(s).

9.2.4 Backcross Breeding

Backcrossing is used to transfer one or a few useful genes from a donor cultivar, often referred to as a donor parent, which may be a landrace, pure line cultivar, wild or weedy relatives, or some other identified source of the target gene(s) to a cultivar to be improved which may be a pure line or an established commercial variety. The cultivar to be improved, referred to as the recurrent parent in the backcross, is usually the adapted desired commercial or a promising pure line being developed for release as a commercial variety and pure line to be used as an isoline in a multiline variety. Therefore, backcross breeding method is used to improve an otherwise well adapted superior variety with respect to most attributes but lacks in one or a few characters for which it is to be improved by sourcing the target trait(s) from the donor parent(s). Through backcrossing single dominant or single recessive gene traits like disease and pest resistance, phenology characteristics, plant height, pigmentation, seed size and shape and even polygene traits can be transferred from one cultivar to another. The trait or traits to be transferred must be clearly, distinctly and intensely expressed, independent of the environment in both the donor and recurrent parents, with high heritability and preferably of major gene inheritance and easy to select for in all stages of backcrossing.

In backcross breeding a cross is made between the donor parent and the recurrent parent to produce F_1 generation, more often using the recurrent parent as the female parent so as to retain the favourable cytoplasmic background and cytoplasmic genes of the recurrent parent. However, in some rare situations where the cytoplasmic background of the donor parent is better than that of the recurrent parent the donor parent is used as the female parent. Once the cross is made, backcrossing is done to the recurrent parent so as to reconstitute all the characters of the recurrent parent except for the character(s) being transferred from the donor parent. During backcrossing selection is done for the trait to be transferred so that it is maintained in the backcross progenies otherwise it may be lost. Selection is also done for the traits of the recurrent parent in each backcross generation so as to fast-track the recovery of the genome of the recurrent parent.The most easily transferred are major genes with distinct and intense expression as in cases of dominant or recessive genes for which the genotypes carrying target major genes are easily identified in the segregating populations of backcrosses. In this breeding approach single recessive genes are not exposed in the backcross generations unless there is a generation of selfing. Self-pollination of the progenies after every two backcrosses is necessary for unmasking the homozygous recessive genotypes which are backcrossed to the recurrent parent. The transfer process is made easier, faster and more efficient if the genes to be transferred are tagged using isozyme or DNA molecular markers, especially for characters expressed in late stages of the plant life cycle or those not clearly distinct, making molecular marker use an essential requirement in backross breading exercise. Backcrossing to improve quantitative traits is more challenging than for improvement of qualitative traits unless molecular markers are used to tag the target QTLs. If heritability is high for the polygenic character, it may be slightly easier to improve such a quantitative character using backcrossing.

Genetic Changes Due to Backcrossing

During the process of backcrossing the genomic constitution of the backcross progenies changes progressively towards that of the recurrent parent and in each generation of backcrossing the genome of the donor parent in the progenies is reduced by half. The progressive change is similar to inbreeding through self-fertilisation because it leads to homozygousity at the same rate as that of self-fertilisation. For example, self-fertilisation of a genotype Aa for one generation leads to 50% Aa, 25%AA and 25%aa progenies, that is 50% heterozygotes and 50% homozygotes. Similarly, for a cross AA x aa all the F_1 progenies will be Aa. If no selection is enforced the progenies and their proportions derived from backcrossing to AA in various backcross generations will be as illustrated in Table 9.3. By BC_6 an extremely high proportion (99.609%) of the recurrent parent genome will have been recovered. In BC_8, BC_9 and BC_{10} generations the recurrent genome, A, in the progenies will be 99.8055%, 99.902 and 99.951 respectively while the donor parent genome will be 0.195%, 0.098% and 0049% respectively. This illustrates that under backcrossing in self-pollinating crops homozygousity is attained at the same rate as in self-pollination so that after t generations of backcrossing the amount of heterozygousity among the progenies at a given locus will be $H = (\frac{1}{2})^t$. This expression also gives the amount of the donor parent genome in the backcross progenies. On the other hand, homozygousity or the amount of recurrent parent genome in the backcross progenies after t generations is given by $1 - (\frac{1}{2})^t$ for a one locus situation. These derivations assume no selection during backcrossing. However, when backcrossing is done to transfer a target trait selection is usually done after every generation of backcrossing to maintain the trait being transferred and also for the traits of the recurrent parent. As selection is done for both traits of the recurrent and the target trait only five or six backcross generations are needed to accomplish the objectives of a backcrossing programme. In addition to transfer of desired genes backcrossing has been employed in cytoplasmic conversion, production of isogenic lines for construction of multi-line varieties and germplasm conversion.

Table 9.3 Genotypic constitution of progenies in progressive backcrossing

Cross or backcross parents		Cross/backcross	Progeny generation and genotype proportions derived from the cross/backcross	Average % of recurrent parent (A) genome among the progenies
RP,	DP	$RP \times DP$, i.e., $AA \times aa$	F_1; all Aa	50%
F_1		$RP \times Aa$	BC_1; $\frac{1}{2}AA$, $\frac{1}{2}Aa$	75%
BC_1,	RP	$RP \times (\frac{1}{2}AA, \frac{1}{2}Aa)$	BC_2; $\frac{3}{4}AA$, $\frac{1}{4}Aa$	87.5%
BC_2,	RP	$RP \times (\frac{3}{4}AA, \frac{1}{4}Aa)$	BC_3; $\frac{7}{8}AA$, $\frac{1}{8}Aa$	93.75%
BC_3,	RP	$RP \times (\frac{7}{8}AA, \frac{1}{8}Aa)$	BC_4; $\frac{15}{16}AA$, $\frac{1}{16}Aa$	96.875%
BC_4,	RP	$RP \times (\frac{15}{16}AA, \frac{1}{16}Aa)$	BC_5; $\frac{31}{32}AA$, $\frac{1}{32}Aa$	98.438%
BC_5,	RP	$RP \times (\frac{31}{32}AA, \frac{1}{32}Aa)$	BC_6; $\frac{63}{64}AA$, $\frac{1}{64}Aa$	99.218%
BC_6,	RP	$RP \times (\frac{63}{64}AA, \frac{1}{64}Aa)$	BC_7; $\frac{127}{128}AA$, $\frac{1}{128}Aa$	99.609%

RP = Recurrent parent (AA), DP = Donor parent (aa), BC = Backcross generation.

Transfer of a Recessive Gene Through Backcrossing

In breeding programmes backcrossing may be used to transfer a recessive gene or dominant gene from the donor parent to recurrent parent. In Fig. 9.2 a procedure that can be followed to transfer a recessive gene from the donor parent, aa, to the recurrent parent, AA illustrated. The allele to be transferred, a, is recessive to the undesired allele, A.

If most of the traits of the recurrent parent have been recovered backcrossing is terminated in BC_4 generation or BC_5 generation or it can be continued as deemed necessary. When backcrossing has been terminated superior plants are selected and their progenies planted in family rows/plots for selection among the families. Family selection with some mild selection within the selected families can be done for a few generations until the best lines are obtained for yield and adaptability evaluations.

Transfer of a Dominant Gene Through Backcrossing

If the allele to be transferred is R, and is dominant to the undesired allele r, with donor parent having the genotype RR and the recurrent parent having genotype rr at this locus the procedure for transfer of the desired dominant allele follows the outline given in Fig. 9.3.

Once backcrossing has been terminated the selected Rr progenies are allowed to self-pollinate naturally and the seed harvested from them will comprise of RR, Rr and rr genotypes. When rr seed is grown the obtained rr plants will show the recessive phenotype and will be discarded easily. If the RR and Rr seeds are planted the derived dominant phenotype plants with RR and Rr genotypes are distinguished through progeny testing whereby the seeds derived from them through self-pollination will be

Original cross:	♀ AA x ♂ aa
F_1; backcross to AA: (First backross)	AA x Aa
BC_1 progenies:	AA, Aa; Allow all the progenies to self-pollinate naturally and plant the derived seed to identify aa
Products of selfing:	AA, Aa, aa; Select the plants that show the recessive phenotype, i.e., retain aa plants only that express most of recurrent parent traits and use in back cross
Second backcross (BC_2) :	AA x aa; Backcross selected aa plants to recurrent parent AA
BC_2 progenies; backcross to AA: (Third backross)	AA x Aa
BC_3 progenies:	AA, Aa; Allow all progenies to self-pollinate naturally and plant the derived seed to identify aa
Products of selfing:	AA, Aa, aa; Select the plants that show the recessive phenotype, retain aa plants only that express most of recurrent parent traits and use in back cross
Fourth backcross:(BC_4)	AA x aa; Backcross selected aa plants to recurrent parent AA
BC_4 progenies; backcross AA (Fifth backcross)	AA x Aa
BC_5 progenies;	AA, Aa Allow all progenies to self-pollinate naturally and plant the derived seed to identify aa
Products of selfing:	AA, Aa, aa; Select the plants that show the recessive phenotype, retain aa plants only that express most of recurrent parent traits and express the transfered gene
The selected plants (having aa genotype) used in evaluations and identifications of desired pure pure lines cultivar	

Fig. 9.2 Recessive gene transfer using backcrossing

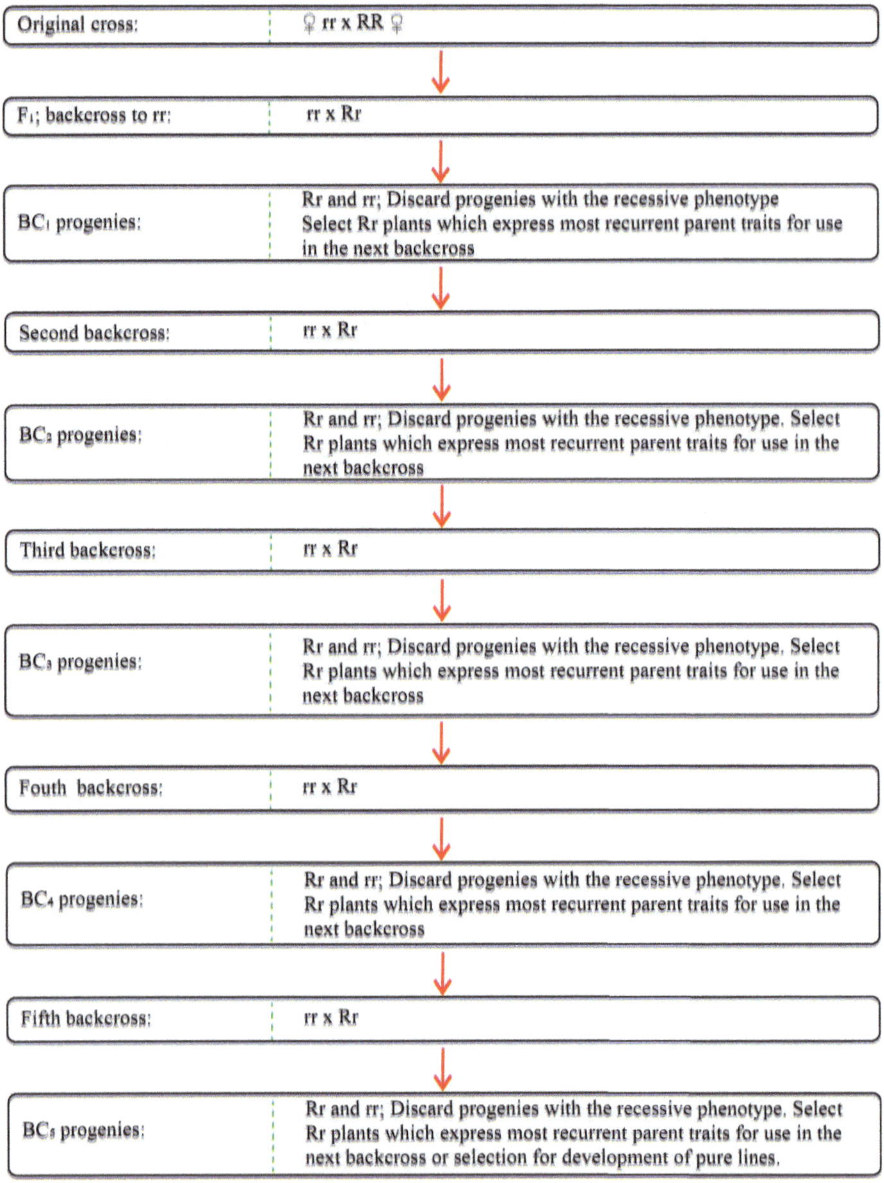

Fig. 9.3 Dominant gene transfer using backcrossing

planted out to differentiate between the segregating and the non-segregating plants. The non-segregating plants, RR, are selected for evaluation and identification of superior pure lines.

Problems

9.1 Give an account of superior pure line developed from double cross involving wheat inbred lines.

9.2 Under what conditions does a plant breeder find it necessary to use backcross breeding method?

9.3 A plant breeder conducted pure line selection procedure on two landraces of cowpea, namely, Rakido and Mweni.

The pure line obtained from Rakido was high yielding and well adapted but susceptible to a bacterial disease that causes leaf spot. The pure line from Mweni was well adapted and resistant to the bacterial disease but lower yielding.

The breeder decided to combine high yield and disease resistance in the two pure lines using backcross breeding method. Disease resistance allele, Y, is dominant to disease susceptibility allele, y.

Give an outline of the process the breeder used to develop a good cowpea variety that is high yielding, well adapted and disease resistant by combining the characters in the two pure lines.

9.4 Which features make single seed descent a desirable breeding approach?

9.5 Mention the problems associated with application of bulk breeding procedure in self-pollinated finger millet.

References

Brim, C. A. (1966). A modified pedigree method of selection in soybeans. *Crop Science, 6*, 220. https://doi.org/10.2135/cropsci1966.0011183X000600020041x. John Wiley and Sons, No. 5625531261530.

Cooper R. L. (1990). Modified early generation testing procedure for yield selection in soybean. *Crop Science, 30*, 417–419. https://doi.org/10.2135/cropsci1990.0011183X003000020037x. John Wiley and Sons No. 5737020257702.

Saxena, K. B., Saxena, R. K., Hickey, L. T., & Varshney, R. K. (2019). Can speed breeding approach accelerate genetic gain in pigeon pea? *Euphytica, 215*, 202. Springer Nature, No., 5702640029122. https://doi.org/10.1007/s10681-09-2520-4

Sneep, J. (1977). Selection for yield in early generations of self-fertilising crops. *Euphytica, 26*, 27–30. https://doi.org/10.1007/BF0032064. Springer Nature No. 5737010251280.

Chapter 10
Hybrid Varieties

10.1 Hybrid Variety Performance and Diversity of Parents

A hybrid variety is comprised of F_1 population from a cross and is used as the commercial variety cultivated by farmers. While the idea of use of hybrids was first suggested by W. J. Beal in 1878 hybrid variety technology originated from the discovery of hybrid vigour in maize hybrids by maize breeders in early 20th century leading to the development of maize variety crosses. The use of inbred lines to develop hybrid varieties using single cross hybrids as commercial varieties was proposed by Shull (1909). However, cultivation of F_1 hybrids as commercial varieties was not economically viable because the hybrid seed was expensive and difficult to produce and handle at planting. The problem was resolved by the development of double cross hybrids which turned out to be more economically viable, and making hybrid seed cheaper to produce and hybrid varieties cultivation more economical. The invention of double cross hybrid technology in maize made the development and commercial production of hybrid maize shoot up in 1930's and 1940's and hybrid varieties became highly popular globally and to date hybrid varieties are the basic pillars of commercial crop production of cross-pollinated crops.

The first commercial production of hybrids was in maize but this technology is now applied in several other open pollinated crops and a few self-pollinated crops as well. The floral morphology of maize made it possible to develop hybrid seed production system and commercialisation of hybrid varieties in early 20th century. Successful commercialisation of hybrid varieties also depends on the presence of male sterility with restorer genes or self-incompatibility making production of large amounts of commercial hybrid seed feasible as in maize, sorghum, cotton, onions, sugar beets, Brassicas, tomato and other crops. Sustained commercial production of hybrid varieties requires a practical and economical approach of producing the hybrid seed and maintaining the parents. Economic viability of hybrid variety production is conditioned on the value of the hybrid crop being much higher than the value of the conventional crop and the sales of the hybrid seed by the seed producer must offset

P. O. Ayiecho and J. O. Nyabundi, *Conventional and Contemporary Practices of Plant Breeding*, https://doi.org/10.1007/978-3-031-74998-8_10

the costs of hybrid seed production and distribution and leave an appreciable profit margin for the producer. Heavy costs of hybrid varieties production are due to the maintenance of the parents and seed production costs which are normally passed to the farmers who buy and grow the hybrid seed. The farmers must be able to meet the costs of the seed and other inputs and at same time realise profits from the hybrid crop.

The performance of hybrids is related to heterosis which is a function of gene frequencies and genetic diversity of the parents used. The hybrid should not only have heterobeltiosis (performance of a hybrid above the performance of the better parent) but should also perform better than the established cultivars in production for it to be recommended for commercial production in a given target area. Heterosis expressed as deviation from the mid-parent performance is given as

$$H_{F_1} = M_{\dot{F}_1} - M_{\dot{P}},$$

and as percentage of mid-parent performance

$$\%H_{F_1} = \left[\frac{(M_{F_1} - M_{\dot{P}})}{M_{\dot{P}}}\right] \times 100,$$

where $M_{\dot{P}}$ is mean of the two parents (mid-parent performance), $M_{\dot{F}_1}$ is mean performance of the F_1 hybrid.

Heterosis, expressed as the difference between F_1 hybrid and the better parent performance is

$$H_{F_{1b}} = M_{\dot{F}_1} - \dot{P}_b,$$

and expressed as percentage of the better parent performance,

$$\%H_{F_{1b}} = \left[\frac{(M_{\dot{F}_1} - \dot{P}_b)}{\dot{P}_b}\right] \times 100,$$

where $\dot{P}_b$ is mean performance of the better parent.

The performance of a hybrid is pegged on diversity, performance and combining ability of its parents which must be subjected to tests to find out which inbred lines/varieties/populations combine favourably to give the best performing hybrids. The more diverse the parents are the better is the hybrid performance and use of exotic cultivars in hybrids enhance heterosis and *per se* performance of the hybrid. Heterosis correlates with parental diversity and that the superiority of hybrid performances over their mid-parent values is proportional to genetic distance between the parents involved in crosses (Ghaderi et al., 1984). Diversity of origin or geographical diversity of different parents in a cross can be used as a measure of distance between the parents in the absence of information regarding other criteria and employed to evaluate distance-heterosis correlation. If a double cross or a three-way cross is to be used as the commercial hybrid the diversity among the parents is best exploited for heterosis by bringing together the genomes of the diverse parents at the final stage of making the hybrid (Eckhard & Bryan, 1940). For example, if inbred line D and E are related by virtue of being derived from the same population or closely related populations and F and M are similarly related the best and most heterotic double cross hybrid is $(D \times E) \times (F \times M)$ instead of $(D \times M) \times (E \times F)$ or $(D \times F) \times (E \times M)$. Similarly, the three-way cross hybrids that will maximise heterosis most are $(D \times E) \times F$, $(D \times E) \times M$, $(F \times M) \times D$ and $(F \times M) \times E$.

10.2 Types of Hybrid Varieties

The different types of commercial hybrids made in various crops are single cross hybrids, double cross hybrids, top cross hybrids, three-way cross hybrids, double top cross hybrids, advanced generation cross hybrids and synthetic or composite varieties. A single cross hybrid, which is a cross between two inbred lines, namely, $D \times E,\ F \times M,\ D \times F,\ E \times F$, etc involving parents $D,\ E,\ F$ and M, is highly heterozygous and homogeneous while the other types of hybrids are both heterozygous and heterogenous. The single cross hybrids between inbred lines are commercialised under situations where high level of uniformity and quality is required and the hybrids give high value crops like sweet corn, popcorn, some vegetable and other horticultural crops. Single cross hybrid varieties are also commercialised for grain maize and sorghum. Single crosses are used to develop three-way, double crosses and other multiple crosses. A double cross hybrid has four parents and is of the type $(D \times E) \times (F \times M)$ where the hybrid is derived between two single cross hybrids and is highly heterozygous and heterogenous because of segregation resulting from the use of the hybrids as parents. Because the double cross hybrids are more economical in terms of seed production they are extensively produced and cultivated in maize, sorghum, sunflower and sugar beet. Three-way cross hybrid has three lines in its parentage, e.g. $(D \times E) \times F$, $(D \times F) \times M$ where a single cross hybrid is used as the seed parent, making commercialisation of such hybrids comparable to the double cross and top cross hybrid. Three-way cross hybrids are also common among grain maize, sorghum and sunflower as commercial hybrid varieties. Three-way crosses are highly heterozygous and heterogeneous because of segregation arising from the single cross hybrid used as one of its parents. A top cross is a hybrid between an inbred line and an open-pollinated variety and takes the form variety $\times D$, variety $\times E$ and variety $\times F$ crosses. Apart from being used in combining ability evaluation of inbred lines, top cross hybrids are commercialised in maize and pasture crops because of ease of production of their seed since the variety is used as the seed parent. Advanced generation cross hybrids are produced using advanced generations of crosses like F_2, F_3 instead of F_1, e.g., F_2 of $(D \times M) \times F_2$ of $(M \times E)$. Synthetic and composites are multiple crosses where at least five parents, usually inbred lines, are used by crossing double crosses with single crosses or with other double crosses.

Such a multiple cross is thereafter maintained by open pollination allowing the genotypes therein to inter-cross freely in all possible combinations. A synthetic variety has fewer parents than a composite, usually five or six, while a composite has several parents. Synthetic varieties are used as commercial varieties for maize, sunflower, sorghum and other allogamous crops where the farming systems cannot support the use of the commercial hybrid seed production. Double top cross hybrid is a cross between a single cross and a variety where the single cross is used as female parent. Commercialisation of double top cross hybrids is done in some places for crops like maize because of ease of production of the hybrid seed.

10.3 Hybrid Varieties Development Process in Allogamous Crops

10.3.1 Development of Parental Inbred Lines

Key activities in hybrid breeding programs are development of high yielding inbred lines, identification of superior inbred lines with high combining ability and *per se* performance to produce superior hybrids and sustained improvement and maintenance of the inbred lines. Apart from production and maintenance of superior inbred lines, the development of improved open-pollinated populations with desirable traits and high combing ability for use as sources of inbred lines or parents is also an important component of hybrid varieties programme. Selected open-pollinated base populations to be used as sources of inbred lines should be of broad genetic base, unless it is highly necessary and in unique situations, populations with narrow genetic base and limited variability must be avoided as sources of the inbred lines. Preferably, the selected populations should be products of recurrent selection and have the characters targeted by breeding objectives. Inbreeding process for development of inbred lines is usually initiated in a number of genetically diverse high yielding base populations. Open pollinated populations with desirable genes and good performance are selected since performances of the hybrid varieties are determined by the frequency of favourable alleles, desirable gene combinations and genotypes present in the base populations. The process of developing inbred lines starts with planting the selected open pollinated populations in the field and selecting several good-looking vigorous plants which are disease-free, pest-free and have good features for self-pollination. At maturity the self-pollinated plants are harvested and their seed grown to give progeny rows/plots such that seeds from each self-pollinated plant are grown in a row or plot as a family. The original self-pollinated plants in the base population comprise the S_0 generation while their progenies obtained by first generation of self-pollination make S_1 generation, hence referred to as S_0 and S_1 plants respectively. Good looking, vigorous desirable plants within the S_1 generation family plots are selected and self-pollinated to give S_2 seed. The derived S_2 seeds are in turn planted in progeny rows/plots as families for further selection and self-pollination. Self-pollination of selected plants is continued for a few more generations until S_5, S_6 or S_7 after which further inbreeding is done through sib-pollination by conducting random pollination among closely related plants in a family instead of self-pollination to avoid severe and excessive level of inbreeding depression which may lead to loss of lines. During inbreeding many undesirable and deleterious recessive genes that are usually hidden or masked by their dominant counterparts in the heterozygous populations are exposed due to homozygousity and eliminated and plants having such genes in homozygous condition are discarded as undesirable. Therefore, in the first few generations of inbreeding many derived inbred lines are eliminated due to poor performance, lack of vigour, disease and pest susceptibility, poor adaptability and other undesirable traits and only superior lines are retained. Inbreeding through self-pollination and sib-pollination helps to fix desirable characters in homozygous condition so that the developed lines may be maintained without genetic changes. After five to six generations of self-pollination the remaining selected inbred lines are distinctly different from one another and variation within the lines is extremely low because plants within a line are highly homozygous and genetically identical. Variation among the lines is more evident than within line variation and selection emphasis is among the lines instead of within lines.

10.3.2 Evaluation of Inbred Lines

Production of good hybrid varieties is determined by the yield potential and combing ability of the parental genotypes. The inbred lines and other potential parents are subjected to rigorous evaluations for combining ability, yield *per se,* resistance to biotic and abiotic stress factors and performance in other agronomic traits so as to select the best genotypes. Usually, such evaluations start from S_4 and continue into subsequent generations when the inbred lines are already highly homozygous. However, evaluations of the inbred lines can also be done starting as early as S_1 for *per se* yield performance and early elimination of poor genotypes.

Evaluation of GCA, SCA and per se performance

Appropriately designed and properly executed evaluations identify the best inbred lines and open pollinated cultivars which can be used as parents to give the most superior hybrid combinations that perform better than their parental inbred lines and the open pollinated populations from which the inbred lines are derived. If there are n inbred lines the possible number of single crosses among them without reciprocals is given by a diallel cross prediction as $\frac{n(n-1)}{2}$. With seven parents this is 21 crosses, for ten parents there are 45 single crosses, for 50 lines the number of possible single crosses is 1225 and for 100 parents there are 4950 single crosses. The numbers of possible three-way crosses and double crosses among n parents are given as $\frac{n(n-1)(n-2)}{2}$ and $\frac{n(n-1)(n-2)(n-3)}{8}$ respectively. If there are four inbred lines D, E, F and M there will be six possible single crosses without reciprocals, namely, $D \times E, D \times F, D \times M, E \times F, E \times M$ and $F \times M$ and twelve three way crosses among them, namely, $(D \times E) \times F, (D \times F) \times E, (D \times M) \times E, (E \times F) \times D, (E \times M) \times F, (F \times M) \times D, (D \times E) \times M, (D \times F) \times M, (D \times M) \times F, (E \times F) \times M, (E \times M) \times D$ and $(F \times M) \times E$. Double crosses for the four lines are $(D \times E) \times (F \times M)$, $(D \times F) \times (E \times M)$ and $(D \times M) \times (E \times F)$. The large numbers of possible crosses among a large number of available potential parents make evaluation of all possible crosses a difficult task. However, top cross method makes evaluation and identification of superior parents with good *per se* performance, high GCA and categorisation of potential parents with respect to GCA much easier. The kind of tester to be used in the top cross is determined by whether the evaluation is for GCA or SCA or both. If the focus is to estimate and evaluate GCA of the inbred lines and other potential parents an open-pollinated broad genetic base tester like open pollinated variety, composite variety, synthetic variety or advanced generation of a hybrid is preferred. However, if the evaluation is for SCA narrow genetic base inbred lines are the preferred testers.

In the top cross the inbred lines are compared to identify those that give superior top cross progenies. Through top cross method the number of potential parents is reduced to manageable numbers very fast by eliminating inferior inbred lines and leaving only those with superior GCA estimates. In top cross the number of crosses is equal to the number of parents to be evaluated only. If there are 50 or 100 parents there will be only 50 or 100 top crosses instead of 1225 or 4950 different crosses for a diallel cross. The top cross method gives a small number of crosses that is easy to manage using the regular plant breeding field designs like lattice square designs. For example, if the lines to be evaluated for selection are D, E, F, M, G, H, I and J using a common tester K there are only eight different crosses instead of 28 single crosses among the parents, namely,

$D \times K \rightarrow DK$ cross
$E \times K \rightarrow EK$ cross
$F \times K \rightarrow FK$ cross
$M \times K \rightarrow MK$ cross
$G \times K \rightarrow GK$ cross
$H \times K \rightarrow HK$ cross
$I \times K \rightarrow IK$ cross
$J \times K \rightarrow JK$ cross

The test cross seed is harvested from the test lines and grown in half sib family rows/plots for evaluations and the inbred lines whose progenies reflect superior performance and high GCA are identified as superior and selected. The top cross progeny evaluations in replicated trials eliminate inferior inbred lines and other possible parents leaving only a few superior ones which are used in all possible crosses to identify the genotypes which give hybrid combinations with high SCA and *per se* performance for effective exploitation of both additive and non-additive gene effects. The lines are clustered for diallel crossing purposes and heterotic groups, depending on different characteristics, relationships and diversity among themselves, geographical areas of origin, flowering and maturity dates, adaptability, resistance to diseases and pests and other traits of concern to the breeders, farmers and consumers.

Table 10.1 Array for possible single crosses involving four inbred lines

Inbred lines	D	E	F	M
D		$D \times E$	$D \times F$	$D \times M$
E			$E \times F$	$E \times M$
F				$F \times M$
M				

Prediction of hybrid performances

Once the inbred lines with high combining abilities are selected the next exercise is to identify the heterotic specific crosses with highest yields that perform better than the varieties in production under the prevailing conditions. The predictions for double cross and three-way cross hybrid performances are based on the performances of their non-parental single crosses. The identification of the best performing double cross among four inbred lines D, E, F and M is done by predicting and comparing the performances of all the possible double crosses among the inbred lines (Anderson, 1938). The array of possible single crosses between these lines are given in the array in Table 10.1.

While the respective performances of the single crosses in the cells above are represented by Y_{DE}, Y_{DF}, Y_{DM}, Y_{EF}, Y_{EM}, and Y_{FM} the possible double crosses are $(D \times E) \times (F \times M)$, $(D \times F) \times (E \times M)$ and $(D \times M) \times (E \times F)$. The non-parental single crosses for double cross $(DXE) \times (F \times M)$ are $D \times F, D \times M, E \times F$ and $E \times M$. Similarly, the non-parental single crosses for the double cross $(D \times F) \times (E \times M)$ are $D \times E, D \times M, E \times F$ and $F \times M$ and those for the double cross $(D \times M) \times (E \times F)$ are $D \times E, D \times F, E \times M$ and $F \times M$. In simple terms, predicted performance for each of these double crosses is based on the mean performances of its non-parental single crosses (Jenkins, 1934) (Eberhart et al., 1964) follows:

For $(D \times E) \times (F \times M)$ the predicted performance $= \frac{1}{4}(Y_{DF} + Y_{DM} + Y_{EF} + Y_{EM})$.

For $(D \times F) \times (E \times M)$ the predicted performance $= \frac{1}{4}(Y_{DE} + Y_{DM} + Y_{EF} + Y_{FM})$.

For $(D \times M) \times (E \times F)$ the predicted performance $= \frac{1}{4}(Y_{DE} + Y_{DF} + Y_{EM} + Y_{FM})$.

Predictions for the performances of three-way crosses to identify best three-way hybrid are also based on performances of the respective non-parental single crosses in each three-way cross (Skaracis & Smith, 1984). For the four inbred lines D, E, F and M there are twelve possible three-way crosses, e.g., $(D \times E) \times F$, $(D \times F) \times E$ and $(E \times F) \times D$, etc. The performance of each of them is predicted by mean of the non-parental single crosses as given for three-way cross $(D \times E) \times F$,

$$(D \times E) \times F = \frac{1}{2}(Y_{DF} + Y_{EF})$$

10.3.3 Improvement of Inbred Lines and Hybrid Varieties

Progressive improvement of the base populations using recurrent selection schemes is one of the best approaches for improving hybrid varieties since it brings about corresponding improvement of inbred lines derived from the populations. If a population has been improved the inbred lines derived from it will equally be superior as compared to those derived from the original unimproved population because the improved population will have higher frequency of favourable genes, better combinations of favourable genes and high frequency of superior genotypes enhanced through recurrent selection.

Though genetically broad-based populations are important in the sustainability of hybrid development, continuous improvement of existing inbred lines is one of the main activities in hybrid breeding programs. The prime objective of continuous improvement of existent inbred lines is to enhance the inbred lines with respect to specific traits like lodging, pest and disease resistance, tolerance to abiotic stress factors, climate change resilience and quality parameters and *per se* performance of inbred lines. Inbred line improvement also elevates combining ability of specific inbred lines and yielding ability of their hybrids.

Inbred lines existent in breeding programmes are improved by crossing them to donor parents possessing specific attributes and backcrossing derived progenies to the respective inbred lines to be improved as recurrent parents. The source of the gene can be another inbred line or some other material. Backcrossing to the recurrent inbred line is done for enough generations to incorporate back the traits of the inbred line being improved while at same time maintaining the character being transferred through selection. About five to six backcrosses are usually sufficient. Products of such backcrossing are used to derive improved inbred lines. Backcrossing is also applied in reciprocal improvement of the inbred lines through convergent improvement (Murphy, 1942) if the lines are to be crossed later to obtain single cross hybrids between themselves. Two inbred lines, A and B, targeted for reciprocal improvement are crossed and the progenies derived are divided into two lots. One lot is backcrossed to one of the parental inbred lines, A, considered as the recurrent parent while the other inbred parent, B, is considered as the donor for the traits which inbred line A is lacking. As backcrossing progresses selection is done for the traits of B to be transferred to A. Similarly, the other lot of progenies is backcrossed to B while selecting for the traits to be transferred to B from donor inbred parent A. In each backcross plants with desirable characters are selected for further backcrossing. Such backcrossing is continued for each inbred line for about four to five generations then improved inbred lines are developed through self-fertilisation or sib-mating. Molecular marker assisted backcrossing for the gene(s) to transferred from the donor and the genome of the recurrent parent makes the process of inbred line improvement more efficient and faster. A modified form of convergent improvement called multiple convergent was proposed by Richey (1946) in which the inbred lines to be improved are crossed to a number of populations having the desired traits. The crosses made between various sources of target traits and inbred lines are backcrossed to each inbred line as the recurrent parent in each cross for about five generations to reinstate traits of the respective inbred lines while retaining the traits being transferred. When sufficient backcrossing is done the progenies of separate backcrosses are inter-crossed. The products of such inter-crossing with improved genetic base and target traits are used to develop improved inbred lines through inbreeding, evaluation and selection of progenies with the desired combination of characters.

In addition to the improvement of individual inbred lines established double or three-way cross hybrids are also improved by substituting poor inbred parents with newly identified or newly developed good inbred lines as parents in such hybrids. To identify better combining inbred lines the superior single cross parent for the established double cross or three-way cross is used in test crosses as a common tester to identify better inbred lines as potential parents in the other parental single cross for the double cross hybrid. Once new inbred lines have been identified through the top cross test, they are used as parents for the single cross to be used in the double cross hybrid or used as a new parent in a three-way cross hybrid. Such replacement of the inbred line parents is done continuously in breeding programmes as need arises, always using the better parental single cross hybrid as the top cross tester to identify new superior parents.

10.4 Hybrids Varieties in Autogamous Crops

Since the discovery of hybrid varieties technology, it remained a reserve of cross-pollinating crops like maize, sunflower, Brassicas, onions, sugar-beets, carrots for several decades because their reproductive system makes it easy to produce hybrid seed in a large scale. However, the technology is now applied in several autogamous crops across the world, especially where there is usable self-incompatibility system or male sterility as in sorghum, cotton, rice, wheat, barley, soybean, and horticultural crops of high value. In self-pollinated crops presence of male sterility is an essential requirement for commercialisation of hybrid varieties. Because of the presence of male sterility commercial production of hybrid sorghum is well established in many countries while hybrid rice is widely grown in China, Japan and Philippines. In sorghum there is cytoplasmic male sterility with recessive nuclear sterility genes and their respective dominant restorer genes. In rice the discovery of cytoplasmic sterility and nuclear sterility with restorer genes revolutionised hybrid rice production as from 1970's. In bread wheat cytoplasmic male sterility and a dominant restorer exist for use in hybrid production.

Despite the presence of other limitations like low pollen production and presence of mechanisms that hinder natural pollen dispersal and flower morphology in the autogamous crops, easy and natural production of pure lines which can be crossed to exploit heterosis makes hybrid production very appealing. For the sake of hybrid production seed production in self-pollinated crops it is desirable to have parents that open their flowers adequately, female parents having prolonged stigma receptivity and male parents with abundant light pollen production and prolonged pollen viability.

The traditional breeding methods for autogamous crops can be used along with recurrent selection schemes designed to concentrate favourable genes and gene combinations into desirable genotypes for development of superior pure lines to be used as parents for hybrid varieties. Once adequate selection and recombination has been achieved in an autogamous crop population superior pure line with favourable and desired gene combinations to be used as parents for the hybrids are easily produced through natural self-pollination and conventional breeding methods. Development of pure lines as parents for hybrids is done through pure line selection, pedigree breeding, backcross breeding and single seed descent followed by testing of the lines for *per se* performance and combining ability to identify lines that give superior combinations for production of heterotic high yielding hybrid varieties. Commercialisation of hybrid varieties in self-pollinating crops requires high levels of heterosis and the yield of the hybrid varieties must be much higher than the best local pure line varieties and production of the hybrid varieties must offset the cost of development and distribution of such varieties.

Problems

10.1 Predict the best performing double cross and 3-way cross hybrids that can be obtained by use of maize inbred lines A, B, C, D and E given in the following table. Data given are mean cob weight in g per plant.

	A	B	C	D
A	450	490	480	520
B		400	600	550
C			430	580
D				380

10.2 Give an outline of key steps followed in the development of a single cross variety in sunflower, starting with open pollinated parental populations.

10.3 Give any three reasons for improving established inbred lines in a maize hybrid breeding programme.

10.4 Which methods are used to improve inbred lines?

References

Anderson, D. C. (1938). The relation between single and double cross yields of corn. *Journal of American Society Agronomy, 30*, 209–211. https://doi.org/10.2134/agronj1938.00021962003000030005x. John Wiley and Sons, No. 5700901103609.

Eberhart, S. A., Russell, W. A. & Penny, L. H. (1964). Double cross hybrid prediction in maize when epistasis is present. *Crop Science, 4*, 365–366. https://doi.org/10.2135/cropsci1964.0011183X000400040008x. John Wiley and Sons, No. 573748126262464.

Eckhard R. C. & Bryan A. A. (1940). Effect of method of combining the four inbred lines of a double cross of maize upon the yield and variability of the resulting hybrid. *Journal of American Society of Agronomy, 32*, 347–353. https://doi.org/10.2134/agronj19400002196003200050005x. John Wiley and Sons, No. 5700910423472.

Ghaderi, A., Adams, M. W. & Nassib, A. M. (1984). Relationship between genetic distance and heterosis for yield and morphological traits in dry edible bean and Faba bean. *Crop Science, 24*, 37–42. https://doi.org/10.2135/cropsci1984.0011183X002400010009x. John Wiley and Sons, No. 5737140260972.

Jenkins, M. T. (1934). Methods of estimating the performance of double crosses in corn. *Journal of the American Society of Agronomy, 26*, 199. 1.2134/AGRONJ1934.00021962002600030004X. John Wiley and Sons, No. 5737471367718.

Murphy, R. P. (1942). Convergent improvement with four inbred lines of corn. *Journal of American Society of Agronomy, 34*, 138–150. https://doi.org/10.2134/agronj1942.00021962003400020005x. John Wiley and Sons, No. 5737540371412.

Richey, F. D. (1946). Multiple convergence as a means of augmenting the vigour and yield of inbred lines of corn. *Journal of American Society of Agronomy, 38*, 936–940. https://doi.org/10.2134/agronj1946.00021962003800100010x. John Wiley and Sons, No. 5737560171554.

Shull, G. H. (1909). A pure line method of corn breeding. *The American Breeders Association, 5*, 51–59. https://doi.org/10.1013/jhered/os.5.1.51. Oxford University Press, No.5701260066926.

Skaracis G. N. & Smith, G. A. (1984). Prediction of three-way top cross sugar beet hybrid performance. *Crop Science, 24*, 55–60. https://doi.org/10.2135/cropsci1984.0011183X002400010013x. John Wiley and Sons, No. 5737491469675.

Chapter 11
Heterogeneous Varieties: Landraces, Synthetic, Composite and Multi-line Varieties

11.1 Landraces and Products of Selection

In cross-pollinated crops a number of open-pollinated populations are in cultivation either as endemic cultivars or improved varieties. The major categories of such open-pollinated cultivars, sometimes referred to as open-pollinated varieties (OPVs), are endemic landraces, products of selection, synthetic varieties and composites. Because random mating is sustained in these population through open pollination the genetic structures of these populations are similar or close to Hardy-Weinberg populations and are, therefore, highly heterozygous and heterogeneous.

Landraces are cultivars of broad genetic structure and locally adapted to the areas they evolved in. They are traditional varieties for the indigenous farming communities who own them and have cultivated them under traditional agricultural systems with continuous on-farm mild selection for traits of interest to the communities over many decades. They are highly variable for most of the agronomic traits, ear size for maize or head size for other cereals, fruit size, pod size, seed or kernel colour and size, plant height, phenology, adaptability to the various biotic and abiotic conditions and other characters and are excellent raw materials for the different methods of selection. Because of random mating and population equilibrium the cultivars are stabilised in agronomic performances and often give fairly modest yields that are satisfying to the communities growing them.

There are open pollinated varieties developed for cultivation under more modern but less sophisticated agricultural production system in which farmers can't meet costs of modern hybrid seed. These are products of modern breeding activities by subjecting landraces or other populations to simple mass selection and recurrent selection and thereafter released as fairly high yielding superior open-pollinated varieties. While such open pollinated varieties can be cultivated their main use is as parents in variety crosses or as sources of inbred lines for hybrid variety production.

In self-pollinated crops the most well-known pure line mixtures are landraces which are highly variable for agronomic traits, head size, tillering capacity, maturity

P. O. Ayiecho and J. O. Nyabundi, *Conventional and Contemporary Practices of Plant Breeding*, https://doi.org/10.1007/978-3-031-74998-8_11

dates, seed colour and size, plant size and adaptability to the various biotic and abiotic conditions. Natural self-pollination has ensured that landraces in various crops are comprised of different homozygous genotypes, making them mixtures of pure lines. Landraces are excellent raw materials for selection and if a landrace is subjected to mass selection the products may still be a mixture of pure lines though it may have much fewer pure lines as determined by intensity of selection and the character selected. Pure line selection on landraces leads to production of pure line cultivars. Other populations of autogamous crops that are composed of mixtures of homozygous lines are products of bulk breeding before selection is initiated. Such populations are also highly heterogeneous, may be highly homozygous, depending on the number of generations of self-pollination incurred and are good raw materials for pure line selection and mass selection. Multiline cultivars are also comprised of pure lines having been intentionally synthesised by mixing appropriately developed isolines.

11.2 Synthetic Varieties

A synthetic variety is produced by crossing selected genotypes in all possible combinations and thereafter maintaining the obtained population by open pollination. That is, a synthetic variety is derived by *inter se* crossing among inbred lines, clones, and open pollinated cultivars and is highly heterozygous and heterogeneous. Synthetic varieties help in the exploitation of both additive and non-additive gene effects, can be used as sources of inbred lines and as testers in top crosses. Synthetic varieties production is a prerogative of open pollinated crops and are particularly useful in marginal production conditions where crop yields are not adequate to support the production of hybrid varieties and that the farmers cannot sufficiently pay for the cost of the hybrid seed. In some regions breeding programmes develop both hybrid varieties and synthetic varieties for commercial production as done for maize in Kenya. Production of synthetic variety seed is simpler and cheaper than hybrid variety seed and farmers can save the seed they harvest from their fields for growing the next crop.

11.2.1 Production of Synthetic Varieties

Synthetic varieties development comprises two important steps, namely, identification and selection of superior parents with the potential to give high yielding and stable synthetic varieties and mating the selected parents *inter se* to obtain the best combinations for the synthetic varieties.The parents for synthetic varieties may include inbred lines, open pollinated cultivars and clones as determined by the crop under improvement, available breeding materials and resources. Inbred lines are developed in a similar way as described under production of hybrid varieties. Diverse open-pollinated populations are identified and improved through recurrent selection for use as sources of inbred lines. The diversity of populations from which the parents for constructing the synthetic varieties are derived is important since the performance of a synthetic variety is also determined by heterosis which is a function of diversity among the parents. The potential parents are subjected to evaluations employing top cross testers to identify the most promising parents with good combining abilities and high yields and also decision has to be made on the appropriate number of parents and the best parental combination to produce the desired performing synthetic variety.

Decision on the number of parents in construction of a synthetic variety is linked to heterosis and its sustainability in the synthetic variety despite segregation. The F_1 generation has the highest level of heterosis expression for metric traits in any cross and heterosis is reduced by $\frac{1}{2}$ in F_2 for a cross between two parents, especially for inbred lines. However, if several parents are crossed in all possible combinations to produce a synthetic F_1 the fraction by which heterosis is reduced in F_2 is a function of the reciprocal of the number of parents used to derive the cross. In crosses where 3, 4, 5, 6, n parents are mated *inter se* to derive the synthetic F_1 heterosis in the corresponding F_2 generation will be reduced by a factor of $\frac{1}{3}$, $\frac{1}{4}$, $\frac{1}{5}$, $\frac{1}{6}$ and $\frac{1}{n}$ respectively. The higher the number of parents mated *inter se* to generate a multiple parent cross the lower is the proportion by which hybrid vigour is reduced in F_2 generation, especially if parents are inbred lines (Busbice, 1970). Though high number of parents lowers the reduction in heterosis of the synthetic variety, having too many parents may lead to the inclusion of some poor performing parents with undesirable genes that reduce heterosis and the performance of the synthetic hybrid. Generally, about five or six parents with highest combining abilities give a good performing synthetic variety. Therefore, the performance of a synthetic variety depends on the diversity and number of parents, the mean performance and combining abilities of the parents and the mean performance of all the possible crosses among the parents.

As the synthetic cross is sustained by open pollination which is equivalent to random mating, heterosis is reduced in F_2 generation only and the level and performance of the synthetic cultivar remains the same in subsequent generations. This is because the open pollinating population generated by *inter se* mating of a number of parents comes to the state of equilibrium as regards its genetic structure after one generation of random mating. This is in accordance with the Hardy-Weinberg Law for large populations in relation to the population genotypic and phenotypic frequencies so long as there is normal diploid segregation, no mutation, no pollen flow from another population and no selection. A simplified prediction for the second-generation synthetic variety performance in a metric trait, based on the predicted mean of the advanced generation of hybrids, which helps to identify the best and highest yielding synthetic varieties is

$$Syn_2 = Syn_1 - \frac{1}{n}[(Syn_1 - \tilde{Y})],$$

where Syn_2 is the predicted yield of the synthetic cross in F_2 generation; Syn_1 is yield of the first-generation synthetic population which is estimated as the mean yield of all possible F_1 crosses among the n parents and $\tilde{Y}$ is the mean performance of the n parents used in the synthetic (Busbice, 1970).

Improving the performance of synthetic varieties

Since yield of a synthetic variety is dependent on the genetic worth of its component parents the performance of a synthetic variety can be improved by elevating the mean performances of both F_1 and parental inbred lines. This achievable be by improving the source populations for parental inbred lines and improving the performance of inbred lines as regards their GCA, SCA and *per se* yield performances.

Inbred lines can be improved through the methods described under hybrid varieties while open pollinated populations used as sources of inbred lines are improved through recurrent selection methods to increase the frequencies of favourable alleles, favourable gene combinations and infusing desirable genes from other sources into such populations. Synthetic varieties derived through *inter se* mating of improved high yielding parents with better combining abilities will be superior and much better yielding than those derived from poor performing parents.

11.3 Composites

Composite varieties are produced by random mating among selected genotypes but unlike synthetic varieties a large number of parents is used to construct a composite variety. A composite is developed by mixing diverse component genotypes and maintaining the mixture by random open pollination making a composite highly heterozygous and heterogeneous. Composite varieties are synthesised in maize, sunflower, sorghum, pearl millet and other allogamous crops using elite and genetically diverse materials for commercial production mainly in regions with marginal production for the crop in question and where seed production system is not well established. The development and modification of composite varieties is easy, fast, cheap and simple and farmers can easily use seed from their harvests for planting next season. Because of diverse nature of the components, composite varieties have good combinations of desired traits like yield, adaptability, resistance/tolerance to biotic and abiotic stress factors, plant type, good quality traits, other desirable traits and also act as germplasm reservoirs and sources for inbred lines.

The source parental populations to be used to form a composite should be diverse with respect to geographical origin, various agronomic characters and must have the desired traits like disease, pest resistance and adaptability to various abiotic conditions and, where necessary the needed genes can be infused into potential parental populations. The composites can be constructed using advanced generations of single crosses, double and three-way crosses, variety hybrids, synthetic varieties, open pollinated varieties and other composites.

Selection of parents to use in composites is based on evaluations and screening under appropriate conditions where potential parental materials are assessed for adaptability, yield performance and resistance to stress factors of concern. The genotypes selected as potential parents are further assessed for the performance in hybrid combinations and combining abilities using testers with broad genetic base. Though the number of composite components may vary with objectives and the crop, the number of component parents should be kept around 10 and not exceed 20. It is important that the components be balanced as regards their contribution to the composite, i.e., equal amounts of seed from each component are taken to synthesise the composite. Once parental materials are identified the composites are constructed by subjecting the selected component parents to a diallel cross without reciprocals and bulking the seeds of the crosses to get the composite. After the seeds of the crosses are mixed the composite is allowed at least three generations of random mating to enhance recombination.

11.4 Multiline Varieties and Cultivar Mixtures

In self-pollinated crops heterogeneous cultivars also take the form of multi-line varieties and cultivar mixtures, especially for use in breeding for resistance to diseases and pests. Use of multiline varieties and cultivar mixtures is a way of exploiting multigenic vertical resistance in self-pollinating crop species where use of pure line varieties with single major gene resistance often leads to disease outbreaks. The idea of multiline varieties was first presented by Jensen (1952) to develop major gene resistance in self-pollinated crops with the intention of preventing occurrence of epidemics. Multiline varieties are mixtures of lines, referred to as isogenic lines or isolines, that are genetically and phenotypically similar in all agronomic aspects and other traits except for resistance genes and reactions to different races of a particular parasite. In a multiline variety each component genotype carries a single gene for resistance to a specific pathogen race or pest biotype but may not be effective against other races of the pathogen or biotypes of the pest.

Since the different components of a multiline cultivar have resistance genes against different races of a given pathogen a multiline cultivar drastically lowers the disease occurrence chances by reducing the amount of disease inoculum and impedes disease spread thereby ruling out any disease or pest epidemic outbreak. The multiline composition also reduces the chances of a particular race inoculum falling on a susceptible plant in the field and makes the resistance more durable and stable. In a multiline cultivar a high proportion of pathogen spores produced on susceptible plants will fall on resistant plants and will not multiply, thereby retarding growth and spread of the pathogen in the cultivar. Some of the susceptible plants may even escape attack and reach maturity and be harvested before being attacked by the disease. The genetic diversity for resistance in the multiline cultivar provides buffering system in the crop, protects the crop population by preventing disease epidemics, prolongs durability of resistance and improves crop yield. On the other hand, cultivation of pure line varieties with single gene resistance tends to remove competition among pathogen races and any new virulent races will have selection advantage and spread fast on a susceptible line.

The process of creating a multiline variety entails making separate crosses between the established variety with various cultivars identified as having different resistance genes to different races of the pathogen or different biotypes of a particular pest. Backcrossing is done for each cross to the established variety as the common recurrent parent for each of the crosses. In each generation of backcrossing for each cross selection is done for resistance to respective disease race/insect pest biotype and all the good traits of the recurrent parent, a process that can be fast-tracked and made more efficient by use of DNA molecular markers associated with R genes for marker assisted backcrossing. The cytoplasm of the recurrent parent is also retained by using the recurrent parent as the female parent.

At the end of backcrossing there will be several lines which will be the same as the number of available different resistance genes for various races of the pathogen/biotypes of the pest. Such lines or isogenic lines will be similar to the recurrent parent and among themselves in agronomic traits and all respects like height, maturity, plant type, grain quality and other characteristics except for resistance genes to the different pathogen races/insect pest biotypes. The seed of such isogenic lines are blended together in equal proportions to create a mixture of isogenic lines or simply a multiline. The multiline thus created is heterogeneous with regard to resistance and has a more stable and durable resistance. The multiline cultivars in which each of the component isolines has more than one non-allelic resistance genes are superior to the multiline cultivars in which component lines have a single resistance gene each, especially if they have the non-allelic resistance genes in disjointed sets (Marshall & Pryor, 1979).

When developing multiline varieties outstanding high yielding recurrent parents that are well adapted and popular varieties are selected. A variety with a broad genetic base that has been derived through multiple crossing and intermating methods of improving autogamous crops is preferable as a common recurrent parent to develop a multiline variety.The composition of a multiline is adjusted according to the emerging pathogen races/insect pest biotypes and as need arises and any component that becomes susceptible to any race/biotype is replaced by a more resistant isoline.

Though they give better yields than the respective parental pure line varieties because of less disease/pest damage they have not gained popularity in agricultural production and are not commercially cultivated because of poor agronomic attributes, time needed to develop them, failure to respond to changing market demands and frequency of virulence alleles in the pathogen population (Pink & Hand, (Pink and Hand, 2002)). Various resistance genes often have tight linkages to unfavourable genes that are difficult to break leading to undesirable and non-uniform agronomic and quality attributes. Undesirable genes linked to resistance genes sourced from different cultivars lower the quality of agronomic traits in the component isolines in different ways leading to overall poor agronomic acceptability of the multiline varieties. Too much attention accorded to development of different isolines with varied resistance genes derails focus and resources away from other agronomic characters. The process of incorporation of individual race-specific genes to different specific isolines through crossing and backcrossing requires a lot of work, time and is expensive in terms of resources. However, such a problem can be solved by thorough and proper selection, application of molecular marker assisted backcrossing and transgenesis.

11.5 Participatory Breeding

Apart from the traditional on-farm selection which the farmers have conducted since the inception of agriculture for developing their landrace cultivars, farmers are sometimes engaged by the professional plant breeders in the formal plant breeding exercise through participatory methods along with other stakeholders. They are often engaged in evaluation and selection of potential varieties developed by the breeders and participatory plant breeding procedure. In evaluation and selection of potential varieties developed by the breeders only a few potential varieties that have been developed by the breeder are grown for evaluations and selection by the farmers and other stakeholders who include other relevant scientists, facilitators, consumers, processors and value addition experts. This process, also referred to as participatory varietal selection, is conducted in the farmers' fields in consultation with the farmers to test the developed varieties under the farming conditions so that the farmers and other stakeholders can apply their knowledge to select the preferred genotypes having the desired characteristics that are adapted to varied agro-ecological conditions (Merga, 2017; Begna, 2022). The farmers assess the genotypes under test based on the traits that are of interest to them. Even though the farmers are only involved at the tail-end of varietal development, participatory varietal selection is done with farmers' needs and consumer preference in mind, is aimed at identifying varieties that are acceptable to the farmers, convincing the farmers to abandon their low yielding cultivars and adopt the newly developed higher yielding superior genotypes and reduces the cost of varietal evaluation and identification of cultivars that fit the varied farming conditions (Begna, 2022).

In participatory plant breeding farmers are collaboratively engaged in breeding activities together with professional plant breeders and other stakeholders and researchers like agronomists, health and nutrition experts, facilitators, processors and value chain addition experts, sales promotion experts, consumers and policy makers to make decisions at various stages of varietal development. Participatory plant breeding is an on-farm approach in which farmers participate in defining key breeding goals and priorities, provision of germplasm, hosting the various stages of the process on their farms and evaluation and selection of superior genotypes at various stages of breeding process. The breeding strategy is complimentary to and compatible with regular breeding strategies, promotes the development of agro-ecology and agrobiodiversity, helps to generate and sustain germplasm diversity through development and maintenance of varieties and germplasm with broad genetic base and stable good local adaptation (Begna, (Begna, 2022); Ceccarelli & Grandos, 2022). The approach also allows the breeder to understand local farming conditions, challenges and management of plant diversity by farmers.

Participatory plant breeding process starts with engaging the farmers and other stakeholders to define the goals and objectives, the type of variety to be developed, making the choice of the genetic materials and populations to be used in hybridisation and subsequent selection and the environments and locations/sites for the experiments (Merga, 2017). Some of the materials to be selected as parents or populations to be subjected to selection and improvement are locally adapted farmers' varieties/landraces. The locations selected for breeding exercise have to be representative and typical of the agroecological conditions in which the farmers operate and in which the varieties to be developed are expected to be grown, taking into consideration the diversity of environmental conditions. While the breeder can conduct the hybridisation of the parents to generate genetic variability, all other activities are done on farmers' fields and require the full participation of the farmers and other relevant stakeholders. Involvement of farmers in selection ensures that the farmer-preferred traits that are important to the farmers and farmers' needs are included in the selection and varietal development process (Begna, 2022; Ceccarelli & Grandos, 2022). Resistance and tolerance to various forms of biotic and abiotic stress factors are also addressed, taking into account the challenges posed by climate change. Based on the environmental diversity and farmers' needs, several locations/areas are used for the tests and selection and different location/areas receive different packages of the populations of the same crop to be tested for selection so that there are different entries in different locations and at the end of the process several different varieties are developed, each for a different specific area (Merga, 2017), leading to the development of several different varieties. Selection is done independently in different locations.

In self-pollinating crops the participatory plant breeding incorporates breeding methods like bulk breeding, pedigree and pure line selection and even backcrossing along with molecular marker assisted selection and genomic selection where necessary. Selection is usually initiated in F_3 and proceeds to subsequent generations to obtain homozygous lines and the number of lines is progressively reduced by selection to a few lines to be evaluated for identification of a suitably adapted superior variety for the area where the tests and selection have been conducted. For cross-pollinated crops heterogeneous populations obtained through intermating are distributed to farmers in different areas for population improvement selection. Recurrent selection is incorporated in participatory plant breeding involving various cycles of selection and recombination to obtain improved populations which are used to derive improved open pollinated varieties, parental inbred lines for synthetic and hybrid varieties. Farmers also have other roles to play like leadership for the project in the community, information dissemination, bringing on-board other community members to participate, breeding material supply, some input supply and seed distribution once a superior variety has been identified (Begna, 2022).

Problems

11.1 Under what circumstances are production of synthetic and composite varieties recommended?

11.2 What features make composite varieties appealing to farmers?

11.3 Why have multi-lines not been popularised as commercial crop varieties?

11.4 What is the role of participatory plant breeding on crop improvement?

References

Begna, T. (2022). Importance of participatory variety selectin and plant breeding in variety development and adoption. *Advances in Crop Science Technology, 10*, 497. https://doi.org/10.4172/2329-8863.1000497. Creative Commons Attribution International Licence (CC BY).

Busbice, T. H. (1970). Predicting yield of synthetic varieties. *Crop Science, 10*, 265—269. https://doi.org/10.2135/cropsci1970.0011183X001000030017x. John Wiley and Sons, No. 5625530955683.

Ceccarelli, S., & Grandos, S. (2022). Return to agrobiodiversity: participatory plant breeding. *Diversity, 14*(2), 126. https://doi.org/10.3390/d14020126. Creative Commons Attribution International Licence (CC BY)

Jensen, N. F. (1952). Intra-varietal diversification in oat breeding. *Agronomy Journal, 44*, 3034. https://doi.org/10.2134/agronj1952.00021962004400010009x. John Wiley and Sons, No. 5738300361250.

Marshal, D. R. & Pryor, A. J. (1979). Multiline varieties and disease control. II. The dirty crop approach with components carrying two or more genes for resistance. *Euphytica, 28*, 145–159. https://doi.org/10.1007/BF00029185. Springer Nature, No. 5738340632322.

Merga, W. (2017). Review on participatory plant breeding. *International Journal of Research Studies in Agricultural Sciences, 3*(9), 7. https://doi.org/10.20431/2454-6224.0309002. Creative Commons Attribution International Licence (CC BY).

Pink, D. A. C. & Hand, P. (2002). Plant resistance and strategies for breeding resistant varieties. *Plant Protection Science, 38*(Special Issue), 9–13. https://doi.org/10.17221/10310-PPS. Creative Commons Attribution International Licence (CC BY).

Chapter 12
Varietal Development in Clonally Propagated Crops

A clone is a group of plants derived from a single plant through vegetative asexual propagation. By nature, all vegetatively propagated crops are highly heterozygous and plants belonging to a single clone are homogeneous. Most asexually propagated crops have reduced flowering with low seed production while some of them like bananas, yams and sisal do not produce seeds at all due to autopolyploidy causing sterility. There are some vegetatively propagated crops which reproduce sexually as well and are highly cross-pollinated and show severe inbreeding depression if subjected to any form of inbreeding. Also, in such crops unmanageable high level of variability arises due to segregation when attempts are made to enforce sexual reproduction in them. Many vegetatively propagated crops are polyploids and perennial or biennial like olives, sugarcane, potato, sweet potato, bananas, sisal, yams, strawberry. Some herbaceous, shrubby and biennial species are treated as annuals for agronomic production and breeding purposes unless there is reason for them to go through the full biennial life cycle like the need for botanical seed production so that the seed is harvested at the end of biennial life cycle (e.g., onions, carrots, sugar beet, potatoes, sweet potatoes, yams, cassava). Tree crops like rubber live for several years and take some years before flowering while herbaceous or shrubby crops like sugarcane, bananas, cassava, pineapples, sweet potatoes and strawberries take a few years before they are removed for fresh planting. Clones can survive for several years, even hundreds of years for some species but can also be short lived especially if they are disease infested.

12.1 Vegetative Propagation Methods

Clonal crops are propagated vegetatively by various natural and artificial methods using different tissues and organs which are products of mitosis. Natural methods include runners, stolons, offsets, suckers, crowns, bulbs, corms, bulbils, tubers, rhizomes and tuberous roots. Artificial methods are stem and root cuttings, budding,

P. O. Ayiecho and J. O. Nyabundi, *Conventional and Contemporary Practices of Plant Breeding*, https://doi.org/10.1007/978-3-031-74998-8_12

grafting and layering. Runners are above ground stems used for propagation of strawberry and buffalo grass while rhizomes are lateral underground stems used for propagation of crops like ginger, orchid and turmeric. Stolons are also horizontally growing stems which produce roots at points of contact with soil as in *Cynodon*. In some monocotyledons specialised underground organs form bulbs as in garlic and onions. Stem tubers are underground swollen storage organs used for propagation of potatoes and Jerusalem artichoke while tuberous roots are swollen thick modified roots which also function as storage organs and are used to propagate crops like cassava, sweet potato and dahlia. Sucker division is shoot arising from a plant below the ground and commonly used in propagation of banana, sisal and pineapple. Corms which are also underground organs are used in *Crocus* and *Colocasia*. In sisal and onions plantlets arise due to vivipary in the inflorescence axils forming bulbils and are commonly used for propagation. In pineapple, the offsets are lateral shoots taking rosette shape, developing from main stem and are used for propagation. Crowns which form on top of pineapple fruit are used for propagation while leafy shoots, referred to as slip, growing at the base of the pineapple fruit is another propagation organ.

Artificial vegetative propagation also employs vegetative plant tissues and organs arising from mitotic division giving plants that are true-to type as their parental plants. In cuttings a piece of leaf, stem or root provided with favourable conditions develop into a new plant. Cassava and sugarcane stem cuttings are buried horizontally in the soil for rooting and formation of new shoots from buds on the stems. The stem cuttings may be of hardwood (mulberry, apple), semi-hardwood (cassava, coffee, macadamia, mangoes), softwood (*Chrysanthemum*) and herbaceous cuttings as in sweet potato. Leaf cuttings or whole leaf propagation is known in African violet and Begonia. Grafting and budding are used to combine different cultivars for propagation purposes. Grafting connects two pieces of plant tissues from different plants, one piece being the rootstock and the top piece being the scion, to unite to grow as one plant. In budding the scion is only a small piece with one bud and no leaf while in grafting the scion has a few leaves. Another artificial propagation method is layering where adventitious roots are induced to form from a stem while that portion of the stem is still attached to the mother plant. The rooted stem is then detached to become a new plant growing independently.

12.2 Selection of Clones

Within a clone, plants are highly heterozygous but genetically homogeneous and identical to one another because they are products of mitotic division of the same genotype. Any phenotypic variation among plants within a clone is purely environmental or may have genetic variation due to somatic mutations which are quite rare. Mechanical mixing of propagules and occasional sexual reproduction in some species may cause variation in clones, creating selection opportunities for improvement of clones. Variation within a clone arising from somatic mutations is responsible for chimeras which are mutated sectors on plants in whose tissues the cells might have undergone such changes. Chimera is a section of an individual or organ having tissues or cells that are genetically different from the rest of the organism/organ. If chimeras arising from mutation are used for propagation new genetic variant plants arise in the clone. Bud selection may be initiated in a clone if a bud derived from a cell with a mutation gives rise to a sector that can be selected and used for propagation of a new variant clone. Somatic mutations and disease infection lead to reduction of vigour and lowered yields of clonal cultivars, a process referred to as clonal degeneration. Apart from viruses, there are other pathogens, particularly bacteria, which cause clonal degeneration through planting material-transmitted diseases, interference with reproductivity and maintenance of clonal varieties. A good example is sugarcane ratoon stunting disease caused by a small bacterium transmitted in planting material which causes serious clonal degeneration in sugarcane. Diseases that are propagule-borne keep multiplying in the clones through propagation material.

An advantage of vegetatively propagated crops is that they can be maintained indefinitely by asexual reproduction and once a genotype has been selected for its superiority it forms a new variety. Clonal selection can be done on landraces, established varieties, introduced clones, natural mutants propagated by bud selection, hybrids derived by crossing of selected clones, materials subjected to induced mutagenesis and products of biotechnology procedures or any other form of breeding material. Induced mutagenesis, hybridisation and biotechnology methods like callus culture, cell suspension culture and genetic transformation are often used to create variability in clones for selection. If variability has been created or if there is already inherent variability in a cultivar, clonal selection is conducted to identify superior plants to be used to produce new clonal varieties. Clonal selection helps to maintain genetic integrity of clones, eliminate disease infection from clones, create new clonal varieties and can also be used along with hybridisation. Clonal selection only utilises variability already present among the clones or within a clone and gives no opportunity for creation of new variability through recombination.

Clonal selection is similar to mass selection since it is based on the phenotypic observations on individual single plants and is usually unreliable for quantitative traits with low heritability like yield. On the other hand, selection for qualitative traits based on single plant observations is more reliable for characters like disease and pest resistance, maturity dates, height and other characters under oligogenic control. In early stages of clonal selection when single plant selection is done focus is to retain individual plants with desirable traits that are of qualitative nature and high broad sense heritability. Selection for high yielding ability is done in later stages when there is enough planting material for replicated trials.

In the first year of clonal selection several superior plants are selected based on target traits with high broad sense heritability like certain quality aspects, disease and pest resistance and general vigour (Table 12.1). In the second year, vegetative planting materials are obtained from the selected plants so that each selected plant is grown as a clone in individual clone plots for selection of clones with desirable traits. Special disease and pest resistance selection nurseries may be used and established varieties are included in the experiment as checks for comparison. Adequate planting material is obtained from the selected clones in the second year so that preliminary

Table 12.1 A possible clonal selection plan

Year/Season	Operations
1	Several hundred superior plants with desirable traits selected with focus on traits of high heritability like quality aspects, disease and pest resistance, etc.
2	Materials from selected plants grown as individual clone plots for selection of superior clones with desirable traits selected. Standard checks may be used for comparison
3	Selected clones are subjected to preliminary yield trials along with continued selection of the clones for quality, disease and pest resistance. If possible, plants can be grown in special disease or pest resistance screening nurseries. An established commercial variety or mother clones is used as a standard check. Based on the evaluations only a few outstanding clones are selected
4 to 8	Further replicated yield evaluations are conducted to identify the best clone as potential new variety. Best clone to be released for commercial production

yield trials for the selected clones is initiated in the third year in addition to selection for characters done in the previous seasons. Based on such selection in the first three years only a few clones are retained for planting in the fourth year. In the fourth year, there is likely to be enough planting material from each of the selected clones for multi-location replicated yield trials when the selected clones are compared with established commercial varieties as standard checks. The replicated trials are done for about four years to identify the best clone to be recommended for release as the new variety and its planting material is multiplied for distribution to farmers.

12.3 Clonal Hybridisation

Hybridisation is only possible for those species which can be propagated both asexually and sexually. If hybridisation is to be initiated between different clonal genotypes parents are carefully selected depending on specific traits to be combined and none of the parents should have undesirable traits that will be expressed in the hybrid. Hybridisation targets bringing together favourable traits found in different clones into one clone like good yield performance, adaptability to the prevailing abiotic conditions, vigour, resistance to diseases and pests, quality aspects and any other trait of concern. In certain species mild inbreeding can be done for one or two generations before superior genotypes are identified as parents for hybridisation. Sometimes selection of parents may be based on GCA and SCA evaluations using top cross or crosses involving the specific genotypes targeted as potential parents.

Following hybridisation of selected clones several F_1 seedlings are obtained by growing the F_1 seed in space planted plots for maximum expression of characters and avoidance of interplant competition. Once the cross is made the genotype is fixed in the F_1 hybrid and selection is thereafter done among the F_1 plants of different crosses without concern for any segregation in the crosses since progenies are only propagated vegetatively from F_1 and subsequent seasons. For each cross every plant in F_1 is a different genotype because of segregation in the genomes of individual heterozygous parents which are used to produce hybrids, hence the F_1 plants and subsequent vegetative generations are basically segregants which are genetically different from one another, displaying a lot of variability among themselves and can be used to exploit additive, dominance, epistasis and heterosis through selection.

Selection among the F_1 plants is based on vigour and targets simply inherited traits only and only those plants that are genuine hybrids combining characters of their parents are selected. Selected superior F_1 plants are propagated vegetatively and plants from each selected F_1 plant is treated as a different clonal family genotype from others and is planted in a plot as a family. Each F_1 plant family also belongs to the larger family of the cross from which it is derived. Therefore, vegetative propagules from superior F_1 plants are grown in blocks, each block comprising of plots of various clonal families derived from a particular cross. Selection is done for desired characters like vigour, plant types, resistance to diseases and pests among the clonal families, applying moderate selection pressure (Table 12.2). Selected clones are propagated vegetatively for further selection between clones and can be subjected to preliminary yield tests since there will be adequate planting material. Standard check varieties and/or parental clones are used for comparison and high selection pressure

Table 12.2 Clonal hybridisation and selection

Year/Season	Operations
1	Selected clones, e.g., A and B are crossed. The F_1 seeds are harvested at maturity
2	Grow the F_1 seed under space planting. Plant as many as obtained so as to have several hundreds of F_1 plants. Plants that are not genuine hybrids are removed. Select several superior F_1 plants basing selection on vigour, expression of traits to be combined, especially traits of high broad sense heritability. Even in a hybridisation involving two parents it is possible to get wide variation in F_1 because of segregation within the parental genomes, hence different plants obtained from a cross are genetically different
3	Individual clones from selected F_1 plants are grown with suitable checks. Vegetative propagules obtained from each individual F_1 plant are treated as propagules of an individual clone and planted in a plot for comparison with other clones. Superior clones are selected based on traits of high broad sense heritability. Selection is done for resistance to stress factors, especially resistance to diseases and pests. There should be moderately high selection pressure so that a manageable number of clones for yield tests are retained
4	There is enough planting material from each selected clone to allow for preliminary yield tests with appropriate checks for comparison. A few superior clones are selected for further testing and selection
5 to 9	Replicated multilocation yield tests using standard commercial varieties for comparison are continued. An outstanding superior clone identified and recommended for release as a new variety

is applied to reduce the number of clones to be advanced to subsequent evaluations and selection. By the fourth season there is likely to be sufficient vegetative planting material and only a small number of hybrid clones left. At this stage the selected superior hybrid clones are subjected to replicated yield trials and compared with the standard commercial varieties as checks. Multilocation replicated yield tests for the remaining few superior selected clones are conducted for about four seasons before decision is made on the clone to be released as a new variety.

Problems

12.1 How are the challenges of clonal degeneration addressed to maintain yields of vegetatively propagated crops?

12.2 How are better performing clonal varieties derived through hybridisation?

Chapter 13
Mutation Breeding

Changes within the gene involving alteration in the bases are often referred to as point mutations since they affect only small segments comprising a base or a few bases in the genetic material. Such mutations may be base substitutions where a base pair is replaced by another without change in total number of base pairs in DNA molecule or frame shift mutations which arise due to addition or deletion of one or a few base pairs. There are transition mutations in which one base is replaced by another of the same type, i.e., a purine is replaced with another purine or a pyrimidine is replaced with another pyrimidine while in transversion mutation purine is replaced by a pyrimidine and vice versa. Point mutations are reflected in the changed phenotypes because mutations alter the alleles giving rise to changes in the genotypes which lead to phenotypic variation. Mutations may be backward or forward, i.e. $A_1 \leftrightharpoons A_2$ and may be spontaneous or induced. Spontaneous mutations occur naturally in populations at very low rates in the order of 1×10^{-8} to 1×10^{-5}, confirming the stability of the genetic material. The gene may be considered unstable if it has high spontaneous mutation rates in the order of 1×10^{-3}. Mutations are induced using chemical or physical mutagens or biological mutagens for genetic studies or breeding purposes. Mutagens are agents that are capable of causing heritable changes in the genetic material, and are of three categories, physical mutagens, chemical mutagens and biological mutagens. While recognition of induced mutations as a source of variation was observed as early as 1908 by Gager it was not until towards the end of 1920's when genetic basis of induced mutations was verified by H. J. Muller and l. J. Stadler who used X-rays to induce mutations in various organisms. The application of induced mutagenesis for crop improvement was initiated in 1940's when the use of induced mutations for crop plant breeding was developed and various useful induced mutations observed in wheat, barley, oats, pea, flax, mustard and rape.

Since natural mutations occur at extremely low frequencies of 10^{-5} or less per generation breeders apply artificial methods to accelerate mutation rate to create variability in breeding materials that usually take much longer to obtain. Induced mutagenesis helps the breeder in creating variability which may be in form of gene

P. O. Ayiecho and J. O. Nyabundi, *Conventional and Contemporary Practices of Plant Breeding*, https://doi.org/10.1007/978-3-031-74998-8_13

point mutations, chromosome translocations, deletions, inversions, duplications and other chromosome aberrations that are expressed phenotypically, giving opportunities to select mutants with desired effects on the plant. Though most mutations are lethal or deleterious to the organism, interfering with balance of gene effects, several favourable induced changes which occur are recoverable through selection and can be incorporated into other cultivars or parental cultivars through hybridisation and selection to give improved varieties.

Induced mutations are important where point mutations are required for single gene or oligogenic traits and are used for cultivar mutational rectification by altering specific traits like susceptibility to diseases, pests and abiotic stress factors, plant architecture, days to flowering and maturity through creation of mutations followed by selection and hybridisation where necessary. For annual crops, rectification of specific traits in established cultivars usually takes a shorter period of about 1–2 years as compared to crossing and backcrossing methods which can take up to 5–6 years to transfer a specific trait. Apart from point mutations for major genes, induced mutagenesis can be used to improve multigenic traits like yield though major gene-controlled traits are easier to deal with than quantitative traits. Induced mutagenesis is also applied in conjunction with tissue culture and other biotechnology methods for crop improvement and alteration of ploidy levels of crop plants to facilitate interspecific or inter-genera hybridisation for transfer of identified traits from crop relatives to cultivated crops.

Autogamous crops suffer restriction to genetic variability generation due to lack of recombination because of their self-pollination nature as compared to the allogamous crops in which cross-pollination leads to recombination and natural generation of genetic variability every generation, making induced mutation breeding more appealing in the autogamous crops for creation of genetic variability. After induction of mutations, it is easier and much faster to recover recessive mutations in self-pollinated crops through natural self-fertilisation without enforced inbreeding and loss of heterosis as in cross-pollinated crops. However, induced mutagenesis has been used for improvement of several of cross-pollinated crops like sunflower, Brassicas and others. Induced mutagenesis is extremely useful for mobilising genetic variability in clonally propagated crops like sisal, pineapples, sweet potatoes, bananas, cassava, sugarcane, potato, etc., due to difficulties associated with crossing and breeding of most vegetatively propagated crops. Through induced mutagenesis several hundreds of varieties have been developed in various crops with improved characteristics and enhanced performances either by direct selection of induced mutants as improved varieties or use of mutants in hybridisation to improve other cultivars with respect to various traits. Towards the end of the twentieth century there were at least 2250 cultivars in various crops derived from induced mutations and by 2015 this number had climbed to over 3200 mutant varieties belonging to some 225 different crop species e.g., cereals, fibres, ornamentals, oilseeds, pulses, vegetables, vegetatively propagated crops, fruits, root and tuber crops (Oladosu et al., 2016; Viana et al., 2019). Among the several traits which are often addressed by mutation breeding are determinate versus indeterminate growth type, high yield, plant height, lodging resistance, nutritional and industrial quality attributes, days to flowering, early and uniform maturity, increased pod number, male sterility, fruit size and fruit quality, good cooking quality, seed size, head size, cob size in maize, higher seed oil or protein content, changes in flower colour and shape in ornamental crops, shattering resistance, disease and pest resistance, adaptability and tolerance to various environmental stress conditions.

13.1 Mutagens

There are two categories of mutagens, physical mutagens and chemical mutagens which are commonly employed to induce random mutations in crop plant genomes. The types of mutagen and doses applied are usually dictated by the plant species and tissues to be treated. There are biological agents which also cause random changes in gene structure and function.

13.1.1 Physical Mutagens

Physical mutagens are of two types, namely, those which give ionising radiations and those which produce non-ionising radiations. Some of the commonly used ionising radiation mutagens include X-rays, gamma rays, neutron particles, alpha particles and β-radiations. The densely ionising fast and slow neutrons and sparsely ionising X-rays and gamma rays have high penetrating ability inducing point or single gene mutations while alpha particles have less penetrating power. These radiations also create random chromosome breaks followed by random fusions resulting in chromosome aberrations and mutations affecting several genes some of which are deleterious. UV-radiation, which is the most commonly used non-ionising radiation, has very low tissue penetrating ability and is more useful for single cell protoplasts, pollen, tissue culture and cell culture. Under UV light DNA suffers damage by formation of unwanted bonds among adjacent bases. The damage caused on DNA due to UV light is often referred to as photochemical lesions. Apart from solar radiation, UV light is generated by mercury and cadmium lamps and near UV light produced by fluorescent tubes. When applied on plant materials physical mutagen radiations hit the radiation sensitive DNA within the cell nucleus causing point mutations which may be base pair substitutions, deletions, and insertions or chromosomes aberrations like inversions, translocations, duplications and deletions. The effects of radiations may even kill cells of the irradiated material depending on the mutagen and dose used. The effect of ionising radiations is enhanced by oxygen and the moisture content of seeds also influences the effects of oxygen on radiation induced mutagenesis. Chromosome translocations and inversions cause position effects due to transfer of genes from euchromatin to heterochromatin regions and vice versa.

13.1.2 Chemical Mutagens

The chemical mutagens, also referred to as genotoxic agents, are related to the DNA bases, adenine, guanine, cytosine and thymine. They include base analogues (5-bromouracil, 2-aminopurine), alkylating agents like ethyl-methane sulphonate (EMS) and ethyl ethane sulphonate (EES), acridines (proflavine, acridine orange), deaminating agents (nitrous acid) and hydroxylating agents (hydroxylamine). Some other chemical mutagens are sodium azide (SA), N-methyl-N-nitroso urea (MNU), N-ethyl-N-nitroso urea (ENU), ethylene imine (EI), hydrogen fluoride (HF), ethyl oxide (EO) and colchicine. Base analogues get incorporated into DNA during replication without affecting its replication causing specific base substitutions. For example, 5-bromouracil (5-BU) and 2-aminopurine (2-AP) are base analogues which induce transition mutations through tautomerism altering the hydrogen bonding characteristics of the bases. Base analogue 5-BU, an analogue of thymine, gets incorporated into DNA during replication, replacing thymine and causing transition as it pairs with guanine instead of adenine. Therefore, after one round of DNA replication there will be a transition change from A=T pair to G≡C pair. The other base analogue, 2-amino purine (2-AP), is an analogue for adenine and can pair with both thymine and cytosine causing a transition change from A=T pair to G≡C pair. Alkylating agents like EMS react with DNA molecule by alkylating the phosphate groups in the purines and the pyrimidines changing their pairing characteristics, resulting in mispairing of the bases. For example, EMS will act on guanine changing its pairing characteristics so that it pairs with thymine instead of cytosine. Alkylating agents and nitrous acid cause mutations in both replicating DNA and non-replicating DNA. Nitrous acid causes damage to the bases by a process called oxidative deamination and deaminates purines and pyrimidines changing their pairing properties as well. Nitrous acid acts on cytosine changing it to uracil which pairs with adenine leading to change from G≡C pair to A=T pair. It also acts on guanine changing it to xanthine which still pairs with cytosine. Nitrous acid changes adenine to hypoxanthine which pairs with cytosine leading to A=T to G≡C transition. Hydroxylamine reacts with cytosine leading to changes in its pairing properties so that it pairs with adenine instead of guanine causing G≡C to A=T transition. Acridines like acridine orange and proflavine cause frameshifts in DNA molecule by intercalating or wedging between the purines and pyrimidines, resulting in loss or addition of bases pairs in the DNA molecule, particularly during DNA replication. While most of the chemical mutagens cause point mutations colchicine is responsible for ploidy level changes. It interferes with chromosome movement and disjunction during cell division by preventing the formation of the spindle fibre leading to doubling of chromosome number.

13.1.3 Biological Mutagens

Biological mutation agents include T-DNA which is a segment of Ti-plasmid, transposons and retrotransposons which induce random mutations by insertion of DNA sequences in organism genomes resulting in insertional mutants. The T-DNA is commonly used by breeders to move target genes into plant genomes *in vitro* for the purpose of crop improvement. Excision and reinsertion of DNA segments by transposons into regions next to genes followed by DNA repair cause mutations that affect gene structure and functions as is known in *Arabidopsis*, maize, rice, wheat and barley (Viana et al., 2019; Wicker et al., 2016). Similarly, retrotransposons cause genetic changes as they insert themselves within genetic regions thereby altering the affected genes.

13.2 Plant Materials for Mutagen Treatment

In principle all plant parts can be treated with chemical and physical mutagens though certain parts should be treated with either chemical or physical mutagens only and not both. The intensity of mutations arising due to induced mutagenesis depends on the tissue treated, dosage and duration of treatment. Materials that can be treated with mutagens include whole plants, seeds, vegetative parts like tubers, stem, stolons, single cells and tissues like pollen grains, meristems, embryos, cell suspension culture, protoplast culture and other forms of tissue culture. Vegetative materials should be treated with physical mutagens because chemical mutagens do not penetrate them uniformly. The UV radiations should also be used on single celled materials like spores, pollen grains or cell tissue culture because of its low penetrating ability. Pollen grains can be irradiated with UV or gamma radiations before pollination as well. Meristems, pollen grains, protoplast and cell cultures give uniformly mutated plants which arise from single mutated cells. Whole plants can be irradiated in gamma fields or gamma rooms while seeds can be irradiated or treated with chemical mutagens as well. Seeds are the best materials for artificial mutagenesis because they don't require special conditions and culturing apart from soaking in distilled water before treatment (Oladosu et al., 2016). Seed soaking and elevated temperatures are important since they activate seed metabolism and increase sensitivity to mutagens (Oladosu et al., 2016). Seeds treated with mutagens should be planted immediately to avoid increased frequency of DNA repair

Success in mutation breeding is conditioned on obtaining materials to be treated from a carefully selected variety which is well adapted, outstanding and agronomically acceptable with high yields and resistance to most diseases, pests and abiotic stress factors but lacking in one or a few characters that require amendment through induced mutagenesis. The most desirable mutations are point mutations which have least effects on cell division and survival and are not affected by diplontic selection. Chromosome changes affect cell, organ or organism survival leading to diplontic selection.

13.3 Handling of Mutagen Treated Materials

13.3.1 Seed Propagated Crops

Approach for handling the treated seeds to recover the desired mutants depends on the breeding objectives and the target traits. In mutation breeding the objective is usually to improve an established variety with respect to one or a few attributes. Alternatively, there may be need to create variability in available breeding materials to address a number of production problems. In self-pollinated crops materials derived from seeds treated with mutagens can be handled in a similar way as segregating generations arising from hybridisation through M_1, M_2, M_3, and subsequent generations by subjecting them to the conventional breeding methods. M_0 is the material whose seed is treated with mutagens while M_1 refers to first generation of plants obtained when seed treated with mutagens are planted. When M_1 generation undergoes self-pollination M_2 seed is obtained which is planted to get M_2 generation and likewise, M_3 generation is obtained when seed harvested from M_2 is planted. Subsequent generation like M_4, M_5, M_6, etc. are similarly obtained. If irradiated pollen is used for pollination the M_1 will be self-pollinated so as to get the heterozygous mutations segregating in M_2.

The M_1 generation is usually space planted to help obtain as much M_2 seed as can be obtained. When the treated seed is planted the M_1 generation usually has several plants with abnormalities because mutagens create a lot of undesirable traits on treated materials like chlorophyll deficiency, reduced mean performances, reduced vigour and survival rate, diminutive plants and reduced height, delayed seedling emergence, whole deformed plants or plants with sectoral deformations/occurrence of chimeras, delayed flowering, failure to form flowers and flower abnormalities, sterility in normal looking flowers, increased lethality and deleteriousness, varied response to diseases, pests and abiotic stress conditions. Apart from changes on traits targeted for improvement, several morphological and physiological changes are commonly observed in the materials treated with mutagens, especially in M_1 and M_2. When seeds are treated with mutagens two types of mutations, macro-mutations and micro-mutations, arise. Macro-mutations affect Mendelian genes with major effects and include traits that are easily identifiable with distinct and discontinuous distribution in segregating populations like disease resistance, pest resistance, stature, flowering and maturity dates, shattering resistance, some rooting characteristics, plant growth type, etc. A dominant macro-mutation is noticeable in the M_1 generation and if a plant heterozygous for dominant mutation is self-fertilised some of its progenies will be homozygous dominant and homozygous recessive in addition to the heterozygous ones in M_2 generation. A recessive macro-mutant that is in heterozygous state in M_1 generation gives homozygous recessive progenies in M_2 and subsequent generations following self-pollination. In situations where there are no undesirable mutations accompanying the desired major gene mutation selection is done in M_2 and M_3. The seed of the selected mutant is multiplied and planted for yield trials and may be released as an improvement of the prevailing variety if it performs well in a number of replicated evaluations. Yield and adaptability trials can start as early as M_3 if there is sufficient seed. However, if there are other accompanying undesirable mutations, as is always the case, a few backcrosses to the mother variety with selection is necessary before yield tests are done.

Micro-mutations affect polygenically controlled quantitative traits and are detectable from M_3 and subsequent generations in terms of changes of means and variances, usually increasing genetic variances while lowering the overall mean values of metric traits. Since effects of mutations on metric traits like yield are more detectable from M_3 generation selection for such traits should not be done earlier than M_3 generation, especially for self-pollinating crops where conventional methods like pure line selection, pedigree, backcross, single seed descent and bulk breeding methods are applied. In cross-pollinating crops mutagen treated materials can be used to derive inbred lines or improve existing inbred lines through crossing and backcrossing. Products of induced mutagenesis in cross-pollinated crops can also be selected and used for improvement of existent populations through hybridisation followed by recurrent selection. In such crops dominant mutants are also recognisable in M_1 while recessive mutants can be recovered in M_2 and M_3 or even M_4 generation following enforced self-fertilisation.

Micro-mutations lead to reduced mean performances in the M_2 and M_3generations but there is increased variability among lines in subsequent generations as segregation progresses and some families with yields surpassing the parental population are obtained in later generations if well designed selection is executed as from M_3 generation. Selection among and within families in self-pollinated crops is conducted up to about M_6 at which point the best lines will have been identified (Table 13.1). By M_7 most of lines are expected to be nearly 100% homozygous at all the loci with large variations among the lines and insignificant within line variation.

13.3.2 Vegetatively Propagated Crops

Induced mutagenesis is extremely useful for creating variability in clonally propagated crops like sweet potatoes, bananas, cassava, yams, ornamental crops, sugarcane, potatoes, sisal, pineapples, etc. due to the fact that some of these crops may be difficult to cross. Mutation breeding is particularly important in the clonally propagated crops where there is no option of sexual reproduction and plants cannot produce botanical seeds at all, namely, sisal, banana and pineapples. Furthermore, for vegetatively propagated crops mutation breeding is handier and more convenient to use than conventional breeding methods because of the heterozygous and polyploid nature of their genomes which complicates selection in the sexually reproduced progenies comprising of thousands of difficult to handle segregants when selfing or hybridisation is done as is possible in crops that undergo both asexual and sexual reproduction. Some of these crops have problems of self-incompatibility and long generation time, making induced mutagenesis the most convenient and fastest way of generating selectable variability.

Table 13.1 Outline of selection for mutagen treated materials in a self-pollinated crop

Generation	Activities and observations
M_0	Seed treatment with mutagens
M_1	Mutagen treated seeds space planted immediately to give several hundreds of plants in the field. Records are taken on individual plants in regard to vigour and survival rate, abnormalities, chimeras and any observable changes. Though most observable changes are likely to be undesirable take note of any desirable major gene changes. Select a few hundreds of plants that look normal. Select plants with desirable dominant mutants for good traits
M_2	Seeds of each selected M_1 plant are planted in progeny rows. Wide spacing may be done for tillering in crops that tiller. Notes are taken for segregation in various traits and data on traits like plant height, flowering and maturity dates, tillering, branching, panicle length, determinate or indeterminate growth type, pod size, disease and pest occurrence, etc., depending on the crop and objectives. Select some plants for resistance to stress factors of concern and other favourable major gene traits. Desirable segregating mutants are noted for selection. Select vigorous normal looking plants and those that have notable desirable mutations. Selected plants are harvested individually. Compare with parental cultivar for identification of mutants
M_3	Seeds of individually selected M_2 plants are grown in rows as families in M_3. Select promising families. Early yield testing can be initiated for families if there is adequate seed. For segregating families select individual plants within families as well for planting as separate families in M_4. Promising uniform families are tested for yield in subsequent generations. Continue screening for resistance to stress factors of concern. Compare with parental cultivar for identification of mutants
M_4–M_8	Continue with between family selection so that only a few families are left by M_7. While preliminary yield trials initiated in M_3 is repeated in M_4 actual replicated yield tests is started in M_4 or initiated in M_5 using appropriate varieties as checks for comparison. Yield tests are sustained through generations so that by M_8 decision to release a promising mutant genotype as a new pure line variety is made. Useful mutants can be used in other breeding programmes through hybridisation

Both physical and chemical mutagens are useful for mutation induction in vegetatively propagated crops. Gamma rays, X-rays, neutron radiations can be applied on propagules including stem cuttings, buds, leaves, tubers, and other forms of propagules. The materials to be subjected to mutagen treatment are obtained from a clone or clones judiciously selected so that a superior cultivar with disease and pest resistance, high yielding, acceptable quality and well adapted but deficient in a particular character is subjected to induced mutagenesis for rectification.

Once the vegetative propagule has been subjected to mutagen treatment planting follows immediately and the plants obtained when such materials are planted are referred to as M_1V_1 population. Plants vegetatively derived from M_1V_1 are M_1V_2 and similarly populations grown from M_1V_2, M_1V_3 and M_1V_4 are M_1V_3, M_1V_4 and M_1V_5 respectively. When M_1V_1 is planted the activities that follow are fairly similar to clonal selection. Though very little selection is done in M_1V_1 apart from discarding poor looking and weak plants or those with albinism, mild selection is initiated in $M_1 V_2$ when the mutant characteristics can be confirmed, especially major gene mutations. In $M_1 V_2$ each of the clone plots represents a selected $M_1 V_1$ plant. Induced somatic mutants, including chimera sectors, are isolated for propagation. Superior looking vigorous clones and those with desired mutations are confirmed in $M_1 V_2$ and selected for simple distinct characters leaving out any undesirable clones or plants. Parental clone may be included for comparison to help in identification of mutants. Vegetative propagules are obtained from the selected clones and sectors to raise $M_1 V_3$ as individual clone plots for further selection and preliminary yield tests, clone uniformity and stability verification, screening for disease and pest resistance. A high selection pressure can be imposed in $M_1 V_3$ so that only superior and promising clones are used to grow $M_1 V_4$. With increased propagation materials in $M_1 V_4$ and subsequent vegetative generations, evaluations for yield and agronomic acceptability tests are done more thoroughly on the mutant clones and by $M_1 V_7$ only one or a few most promising clones should be retained as potential varieties for more elaborate replicated yield evaluations to identify one best clone. In addition to disease and pest resistance screening, tests for resistance and tolerance to other stress factors of concern can also be initiated in $M_1 V_4$.

When materials are treated with mutagens and progenies grown, it is possible to obtain complete solid homo-histont mutants which are easy to identify and select for if they have the desired mutations. The mutant may also be in form of a chimera which is a major challenge in mutation breeding because it is unstable and difficult to handle through selection. Chimera sectors which occur in progenies of both vegetative materials and seeds treated with mutagens may be periclinal involving a whole ontogenic tissue layer or mericlinal if only part of an ontogenic tissue layer is affected (Micke & Donini, 1993). Chimeras sometimes fail to express their genetic changes and may even disappear because of diplontic selection they encounter when outcompeted by surrounding unmutated tissues, making it difficult to isolate favourable mutations they may contain. Solutions to the problem of sectoral chimeras is irradiation of the gametes, zygotes before first zygotic division and adventitious bud apices in early stages giving rise to non-chimeric homo-histont mutants that can be propagated for selection and also application of methods like disbudding, grafting, using axillary buds for propagation, pruning or budding of tissues taken from irradiated explants to help isolate the mutants (Micke & Donini, 1993). Once they are isolated, chimeras are used for propagation, further testing and selection of the mutants having the target characters in subsequent vegetative generations. Problems of chimerism and diplontic selection are also resolvable by application of induced mutagenesis on shoot tips grown in vitro followed by micro-propagation, cell culture, callus culture, protoplast culture and anther culture leading to recovery of periclinal or solid mutants.

Due to high levels of heterozygousity and high ploidy levels most vegetatively propagated crops present challenges to mutation breeding because induced mutations, especially recessive alleles, may go unnoticed. If the undesirable allele, R, is dominant and the desirable allele, r, is recessive, the only quickest and easiest way of recovering the recessive mutant in a completely nulliplex mutant genotype, rrrr, in the autotetraploid is by mutagen treatment of a simplex genotype, Rrrr. In duplex,

RRrr, triplex, RRRr and tetraplex, RRRR genotypic conditions mutagenesis may affect only one copy of the dominant gene leaving the others unaffected, hence the recessive mutations remain undetected, considering the fact that in the autopolyploid crops self-pollination is difficult or impossible. An advantage of the polyploids is that they withstand deleterious macro-mutations more than the diploid.

13.4 In Vitro Techniques in Mutation Breeding

Induced mutagenesis can be used along with biotechnology methods to accelerate mutation breeding through rapid clonal propagation by exposing in vitro cultures, namely, cell culture, protoplast culture, anther culture, embryo culture, micropropagation materials, meristem culture and callus culture to UV and gamma radiations or chemical mutagens like 5-bromouracil, 2-aminopurine, ethyl-methane sulphonate, ethyl ethane sulphonate, proflavine, acridine orange, nitrous acid, sodium azide, N-methyl-N-nitroso urea, N-ethyl-N-nitroso urea, ethylene imine, hydrogen fluoride and ethyl oxide to induce random mutations for enhanced somaclonal variation followed by selection in vitro and subsequent rectification of target cultivars. Selection in vitro entails creating stress conditions and exposing in vitro cultured tissues to selective agents like pathotoxins, temperature stress, salt stress, herbicides, mineral stress so as to select cells that survive under such conditions and thereafter transferred as often as required to fresh medium for regeneration into plantlets and adult plants. In case of mutagen-treated haploids in anther culture, mutant traits can be identified in vitro by applying the selection agents and the surviving haploid cells diploidised, giving mutants in homozygous state which are regenerated into plantlets and whole plants for further propagation, selection and evaluation. Alternatively, the mutagen-treated haploids can be subjected to diploidisation then followed by selection in vitro for the desired mutants which are in turn cultured and regenerated into plantlets and adult plants. The diploid homozygotes produced from haploids have full expression of the desired recessive mutant alleles since each locus is homozygous. Mutants can be obtained in autotetraploids by mutagen treatment of male and female gametes followed by chromosome doubling and regeneration to get plantlets and adult plants with the regular tetraploid chromosome number. Micropropagation can also be employed to get plantlets in vitro for mutagen treatment and the apical buds from such treated plantlets taken into fresh medium for regeneration into new plantlets. The derived plantlets are subjected to in vitro selection by applying appropriate selective stress conditions and the surviving plantlets are transferred to fresh growth medium for regeneration into adult plants which are later multiplied and screened for the relevant characters. Acceptable mutants are multiplied to obtain sufficient planting material for further tests and selection in the field. Application of mutagens on in vitro tissue cultures involving shoot tips, micro-propagation products, single node cultures, cell culture, callus culture, protoplast culture and anther culture eliminates chimerism and diplontic selection in mutagen treated materials.

There are a number of in vitro new plant breeding tools developed as part of genome editing systems that are employed to direct required mutations at predetermined positions in the plant genome. The purpose of the new techniques for inducing mutations is to reduce the extent of bad mutations arising and improving the precision and speed of obtaining the desired genetic changes (Holme et al., 2019). Among the genome editing system techniques used to induce targeted mutations at predetermined positions are zinc finger nucleases (ZFN), transcription activator-like effector nucleases (TALENS), meganucleases (MN) and clustered regularly interspaced short palindromic repeats (CRISPR) (Holme et al., 2019; Razzaq et al., 2021). These systems employ different nucleases to introduce double strand cuts on DNA leading to stimulation of DNA repair which may be in form of homologous recombination (HR) repair system or non-homologous end-joining (NHEJ) repair system (Viana et al., 2019). During the repair process a plasmid containing a desired exogenous DNA sequence or the target PCR product can be incorporated into the plant genome at the repair site. Zinc finger nucleases cleave specific sites on DNA by binding to domains that recognise specific DNA sequences followed by HR or NHEJ repairs during which the target DNA sequence is introduced to cause the desired genetic changes at the sites (Viana et al., 2019). Also, TALENS, MN and CRISPR act similarly by using nucleases and the repair systems that can be employed to introduce the desired genetic changes into plant DNA during repair process.

Problems

13.1 How do EMS, proflavine and 5-bromouracil cause changes in the gene?

13.2 Why is autoploidy a challenge in mutation in breeding?

13.3 Why is induced mutagenesis more effective in self-pollinated crops than cross-pollinated crops?

References

Holme, I. B., Gregersen, P. L., & Brinch-Pedersen, H. (2019). Induced genetic variation in crop plants by random or targeted mutagenesis: Convergence and differences. *Frontiers in Plant Science, 10*. Article 1468. https://doi.org/10.3389/fpls.2019.01468. Creative Commons Attribution International Licence (CC BY).

Micke, A., & Donini, B. (1993). Induced mutations. In: M. D. Hayward, N. O. Bosemark, & I. Ramagosa (Eds.), *Plant breeding: Principles and prospects* (pp. 52–62). Springer. https://doi.org/10.1007/978-94-011-1524-7_5. Springer Nature No. 5742521275731.

Oladosu, Y., Rafii, M. Y., Abdullah, N., Hussin, G., Ramli, H. A., Rahim, H. A., Miah, G., & Usman, M. (2016). Principle and application of plant mutagenesis in crop improvement. A review. *Biotechnology and Biotechnological Equipment, 30*, 1–16. https://doi.org/10.1080/13102818.2015.1087333. Creative Commons Attribution International Licence (CC BY).

Razzaq, A., Kaur, P., Akhter, N., Wani, S. S., & Saleem, F. (2021). Next-generation breeding strategies for climate-ready crops. *Frontiers in Plant Science, 12*, 620420. https://doi.org/10.3389/fpls.2021.620420. Creative Commons Attribution International Licence (CC BY).

Viana, V. E., Pegoraro, C., Busanello, C., & de Oliveira, A. C. (2019). Mutagenesis in rice: The basis for breeding a new super plant. *Frontiers in Plant Science, 10*. Article 1326. https://doi.org/10.3389/fpls.2019.01326. Creative Commons Attribution International Licence (CC BY).

Wicker, T., Yu, Y., Habarer, G., Mayer, K. F. X., Marri, P. R., Rounsley, S., Cheng, M., Zuccolo, A., Panaud, O., Wing, R. A., & Rofflers, S. (2016). DNA transposon activity is associated with increased mutation rates in genes of rice and other grasses. *Nature Communications, 7*, 12790. https://doi.org/10.1038/ncommons12790. Creative Commons Attribution International Licence (CC BY).

Chapter 14
Use of Variation in Chromosome Number and Structure for Crop Improvement

14.1 Origin, Occurrence and Significance of Various Ploidy Levels

The basic chromosome number in diploid species is the basic genome for an organism and is symbolised by ×. Chromosome number for each organism represents the number of packaging units for the genome. A diploid species has $2\times = 2n$ while an autopolyploid has more than two copies of its genome present. In an autopolyploid individual the basic chromosome set (n) belongs to only one particular genome, e.g. A or B, which appears in multiples e.g., AA ($2\times = 2n$), AAA ($3\times = 3n$), $AAAA$ ($4\times = 4n$), etc., and in allopolyploids two or more genomes of different phylogenies are present in an individual, e.g., AB, AAB, ABB, $AABB$, $AAABBB$, etc. Both autopolyploids and allopolyploids exist in various crop plant species and families. The different component genomes in an allopolyploid may or may not have the same chromosome number, as in wheat there are 21 chromosome pairs with $A = B = D = 7$ chromosome pairs while in *Brassica napus* there are 19 chromosome pairs among which one genome has 10 pairs while the other genome has 9 pairs of chromosomes. Allopolyploidy arises when chromosomes of different genomes are brought together in an individual through hybridisation of different species followed by diploidisation resulting in pairs of homologous chromosomes. While a diploid may have its haploid genome represented by A, an allopolyploid may have its haploid or monoploid genome represented by AB. In the two cases the normal gamete or monoploid genome composition are A and AB while somatic genome compositions are AA and AABB respectively.

All members of a given species have the same and constant stable somatic chromosome number and the same gametic chromosome number which do not change. However, there may be variations in chromosome number among the species belonging to the same genus. Changes in the chromosome number are reflected in the phenotype in form of morphological, physiological and biochemical characters because the changes affect the genetic dosages carried by the affected chromosomes. Such

P. O. Ayiecho and J. O. Nyabundi, *Conventional and Contemporary Practices of Plant Breeding*, https://doi.org/10.1007/978-3-031-74998-8_14

phenotypic manifestations along with cytogenetic/karyological analysis make it possible to determine the effects of adding or removing individual chromosomes from the genome of an individual and alteration of whole genome by some multiple.

Changes in morphological, physiological and biochemical features arising from chromosome number changes in plants are often taken advantage of in some fruit crops, vegetable crops and ornamental plants to give more desirable products than the corresponding diploid cultivars and relatives as in cases of banana, sugarcane, sweet potato, strawberry, potato, etc. In a number of horticultural crops modified chromosome number leads to high yielding seedless cultivars. The changes in chromosome number may involve the whole genome compliment (euploidy) or only part of the genome compliment affecting only one or two or a few chromosomes (aneuploidy). Euploidy is a situation where a cell, tissue or an individual has one or more exact multiples of haploid chromosome number, and therefore refers to differences resulting from changes in whole chromosomes sets. Euploids include monoploids, diploid and polyploids. Aneuploidy is when a cell, tissue or an individual has one or more whole chromosomes of a normal chromosome set absent or represented more than once. Changes in chromosome number may involve the whole organism or only part of the organism leading to a chimera. Chimeras are also important in horticulture as regards production of ornamental crops.

14.1.1 Haploids and Monoploids

Monoploids have only one set of chromosomes or one genome, i.e., n (haploidy) and is the normal chromosome set for gametes of diploids. In many plant species the gamete is haploid while the somatic cells have diploid chromosome number. In plants the gametophyte phase (the gamete phase) with reduced chromosome number (n) due to meiosis is a brief phase in higher plants but longer in lower plants while the sporophyte phase which has the normal somatic diploid number is long and predominant in higher plants.

Though monoploids are rare in nature there are some spontaneous and experimental cases in plants which are often frail with small leaves, slow growth, low vigour and usually die early. If they survive to adulthood, they have high degree of sterility due to lack of chromosome pairing, presence of univalents and irregular segregation of chromosomes during meiosis I, leading to formation of inviable gametes. In plants haploids are known to arise spontaneously in nature or can be induced. Rare naturally occurring haploids have been noted by plant scientists in various plants including tomatoes, tobacco, potato, coconut, coffee, sugar beets, barley, pearl millet, cotton, maize, rape, wheat, etc. Rare haploids arise naturally in some plant species through semigamy, gynogenesis, androgenesis or apogamy and may be linked to polyembryony. Haploids also arise in some cases of interspecific hybridisation which result in the development of haploid zygotes that can be cultured into adult plants as is known for the cross between *Hordeum vulgare* and *H. bulbosum*. Chromosome elimination in interspecific hybrids leads to development of haploids in vivo for the cross between barley (*H. vulgare*) and bulbosum barley (*H. bulbosum*) in which the *H. bulbosum* chromosomes are eliminated from the developing embryo leading to development of haploid embryos and haploid plants when seeds from such fertilisation are planted (Kasha & Kao, 1970). Haploidy in plants is also induced through use of irradiations and alien cytoplasm introduction, anther and pollen culture. Though irradiated pollen grains are killed and become inviable they can stimulate female gamete to develop parthenogenetically leading to haploid zygotes. Anther and megaspore culture can also be used to obtain haploids. Haploids can be obtained in vitro on a large scale by use of anthers and microspores in appropriate culture medium conditions in nearly all sexually reproducing plants like rice, barley, wheat, tobacco, potato, tomato, asparagus, eggplant, pepper, maize, etc.

The haploids, also referred to as monohaploids or monoploids, derived from diploid have only one basic genome (x) with each chromosome and gene represented only once and are highly sterile with no significance in crop production. However, they can undergo spontaneous or induced chromosome doubling using colchicine or other chemical and physical agents, reinstating fertility and giving plants that are 100% homozygous at all loci. Therefore, anther culture or pollen culture followed by diploidisation of derived plantlets and regeneration into plants provides hundreds or thousands of completely homozygous plants in one or two seasons, the shortest period pure lines can be produced. Repeated self-fertilisation takes more than seven generations to obtain close to 100% but not complete homozygousity. Such 100% homozygous plants can be used as pure lines and give genetically uniform crop populations for improved selection efficiency in breeding programmes. Dihaploids, the monoploids from tetraploids, make interspecific hybridisation easier and more successful. Similarly, trihaploids which are monoploids from hexaploids, can be useful in interspecific hybridisation. Attempts have been made to use haploids by producing doubled haploids in a number of crops like maize.

14.1.2 Polyploidy

Polyploidy is the situation in which a cell, tissue or an individual has more than two sets of chromosomes (2n) or genomes in exact multiples of the chromosome set like 3n (triploid), 4n (tetraploid), 5n (pentaploid), 6n (hexaploid), 8n (octoploid), etc. Though in some plant species polyploids are not tolerated and they rarely survive, polyploidy is common in plants where several species are natural polyploids with at least one half of known plant genera containing polyploids. In several plants polyploids are desirable and compete favourably and are also of agricultural importance. Many plants of economic importance like wheat, bananas, coffee, sweet potato, tobacco, potato, strawberry, sugar beet, sugarcane, etc. are polyploids. There are two kinds of polyploids, namely, autopolyploids which have exact multiples of identical sets of chromosomes, mostly basic set of chromosomes of the same species, and allopolyploids which have chromosome sets originally from different species, hence the different chromosome sets are genetically different.

There are a number of irregularities in nature which lead to polyploidy in plants, namely sectors with somatic doubling and irregular meiosis during gametogenesis. In somatic doubling some somatic cells may undergo irregular mitosis and give rise to meristematic cells with doubled chromosome number leading to 4n cells. Affected parts of the plant may give gametes with double chromosome numbers (2n). When such rare gametes get involved in fertilisation, they give rise to polyploid ($3n$ or $4n$) zygotes which in turn develop into polyploid plants. Such plants can give polyploid plants through vegetative propagation as well. Spontaneous irregular meiosis during gametogenesis result into gametes with doubled chromosome number ($2n$) due to failure of chromosomes to segregate normally, hence both chromosome sets end up in one gametic nucleus. Asynapsis during meiosis leads to unreduced ($2n$) gametes which give autotriploids ($3n$) or autotetraploids ($4n$) when they unite with n or $2n$ gametes. Autotriploid plants also produce autotetraploids when self-fertilised. Polyploid cells that arise as calli in wounded tissues can be regenerated into polyploid plants. Polyploids also arise through interspecific hybridisation followed by doubling of chromosome number in the sterile interspecific hybrids. When chromosome number is doubled fertility may be regained. Many allopolyploid species crops arose through such mechanism. Autoplolyploidy can be induced artificially by heat treatment, decapitation, indoleacetic acid, colchicine, cyclohexane, and other chemical treatments. Heat shock causes polyploidy in plants because high temperatures interfere with meiosis. Colchicine, a highly poisonous chemical, causes doubling of chromosome number in tissues of all species, whether in somatic or gametic cells, by interfering with spindle fibre formation during cell division. Chromosomes are doubled by treating germinating seeds, seedlings or adult plants with colchicine solution. Colchicine solution or colchicine mixed with lanolin can be applied on tips of adult growing shoots and buds to achieve the such purpose. Vegetative propagules like bulbs, tubers, stem cuttings, corms and runners for clonally propagated crops can also be treated with colchicine solution by applying the solution to growing points to double the chromosome number.

Autopolyploids

Among the autopolyploidy situations which may occur naturally or be artificially induced are autotriploidy, autotetraploidy, autopentaploidy, autohexaploidy, etc. Apart from bivalent formation, there is presence of monovalents, trivalents, or tetravalents during meiosis I in autopolyploids due to some unpaired chromosomes, synapsis of three or four homologous chromosomes respectively. During meiosis I autopolyploids have multivalents and the maximum number of chromosomes coming together to form a multivalent depends on ploidy level. When more than two sets of chromosomes synapse in an autopolyploid pairing becomes irregular leading to unequal chromosome segregation and irregular disjunction in anaphase I, with consequent production of several inviable gametes and high degree of sterility. Apart from some natural autopolyploids like potatoes ($2n = 48$) whose fertility has been perfected by evolution, many autopolyploids are highly sterile because of irregular disjunction during meiosis. The daughter cells formed during meiosis have irregular chromosome number and consequently the formed gametes have varied non-haploid chromosome numbers, including aneuploid and polyploid gametes which may be inviable. This is the reason for high level of sterility of some autoploids like banana. The monoploids usually have fertility of zero or near zero while diploids are most fertile followed by autotetraploids. Autotripolids are highly sterile and give fruit crops with high quality seedless fruits like the edible bananas ($3\times = 33$) and seedless water melons. The autotetraploid bananas may produce some few seeds because autotetraploidy restores some level of fertility as compared to autotriploidy.

The comparative fertility levels of autoploids are in the following order:

$$1\times < \underbrace{2\times}_{\text{Normal fertility}} > 4\times > 6\times > 8\times > 10\times > 3\times \geq 5\times \geq 7\times, \text{etc.}$$

Most autopolyploids do not produce viable gametes and some of them have perfected vegetative reproduction through evolution such that they are strictly vegetatively propagated like bananas and some sisal species like *Agave cantala* ($3\times = 90$) and *Agave sisalana* ($5\times = 150$). Some other known important cultivated autopolyploids crops are sweet potato ($6\times = 90$), potato ($4\times = 48$) and alfalfa ($4\times = 48$).

If plants of different autoploidy levels are compared monoploids are the weakest and are of no agricultural value per se though they may help to isolate mutant genes and can be diploidised to give 100% homozygous individuals for genetic studies or breeding. Variation in chromosome number is always associated with distinct variation in the phenotype such that polyploid plants may be more deeply coloured, have larger flowers and fruits, thicker and larger underground organs (as in sweet potatoes, potatoes), larger seeds, larger cells with corresponding higher cytoplasm contents, large nuclei, larger stomatal and epidermal cells and larger pollen grains. They also have physiological features which make them possess better and wider ecological range of adaptation and tolerance making them more useful than the corresponding diploid. The doubled chromosome number in autotetraploids causes genetic duplication which is like a double genetic dose. Autopolyploids also have larger leaves, fleshy stems and roots as compared to their diploid counterparts. In some cases, tetraploids are of more commercial value than the corresponding diploid. Creation of autopolyploidy in normally diploid plants causes novel character expression in plant tissues like changed contents of alkaloids, vitamins, proteins, carbohydrates and also improves crossability of related species, especially if the species differ in chromosome number or chromosome structure. The seedless autopolyploids are good fruit crops with large fleshy fruits or ornamental crops with large beautiful seedless flowers. Sexually reproducing natural autotetraploids are cross-pollinated, highly heterozygous and suffer high levels of inbreeding depression if subjected to inbreeding and therefore, breeding of such crops must maximise and maintain heterozygosity. Most autopolyploids are vegetatively propagated.

Because of large cells and large vegetative parts many autopolyploids crops are cultivated for vegetative products instead of seed or grain as is the cases for forage crop like alfalfa, root crops like potatoes and sweet potatoes and fruits like banana and flowers in ornamental plants. Triploid watermelons are produced by crossing tetraploids as female and diploid as males. Triploid sugar beets have larger roots and higher sugar content than the diploid ones while autotetraploid cabbage and turnip have larger leaves too. Because of low or lack of fertility autopolyploids are of no importance in grain production or any form of botanical seed production but are good for forage livestock feed production due to their vegetativeness and softness, horticultural industry production of popular ornamental flowers because of extra-large flowers and large and good quality fruits. Autotetraploids are useful in gene transfer between related species by making it easier to make interspecific crosses. Through the use of dihaploids it is possible to cross autotetraploids to their diploid relatives to help transfer desired genes into them. Doubling chromosome numbers of related diploid before crossing them improves the viability of the derived interspecific hybrids.

Allopolyploids

Allopolyploids, also called alloploids, amphidiploids or amphiploids, arise through hybridisation of different species, followed by doubling of chromosome number of the progenies to reinstate fertility. They are more usable directly than autopolyploids because they have normal meiosis and high fertility. Several plants of economic importance are allopolyploids which arose naturally by bringing together more than one genome in a nucleus as a result of occasional hybridisation between closely related species or species with some degree of divergence producing interspecific F_1. Such natural interspecific F_1 hybrids are usually sterile due to irregular meiosis arising from lack of chromosome homology despite homeology and allosydentic chromosome pairing for some chromosomes of different genomes in some cases. The interspecific F_1 hybrids regain fertility when the chromosomes are doubled through occasional union of the unreduced gametes, each having complete genome of the parental species. The products of such union are allopolyploids which make new species with normal bivalent formation, disjunction and fertility because of normal pairing of chromosomes within each genome. Fertility restored in the interspecific hybrids is high if there is no homeologous pairing between chromosomes of different genomes. The usefulness of allopolyploids in plant breeding is influenced by interaction between the genomes and cytoplasmic-genome interaction. Some allopolyploids have genetic systems which hinder pairing between homeologous chromosomes of different genomes with most well-known case being the gene which prevents pairing of chromosomes of different genomes in wheat, maintaining the integrity of different genomes in the allopolyploid.

Several cultivated crop species are known allopolyploids which evolved through spontaneous hybridisation of related species followed by chromosome doubling as in wheat (2n = 42) involving *Triticum monococcum, Aegilops speltoides* and *T. tauschi* genomes (Allan, 1980) illustrated in Fig. 14.1, tobacco (2n = 48) combining genomes of *Nicotiana sylvestris* and *N. tomentosa* (Wernsman & Matzinger, 1980) shown in Fig. 14.2, strawberry (2n = 56), some Brassicas e.g., *B. carinata* (2n = 34), *B. juncea* (2n = 36) and *B. napus* (2n = 38), cotton species *Gossypium barbadense* (2n = 52) and *G. hirsutum* (2n = 52), sugarcane (2n = 80), *Coffea arabica* (segmental alloploid) (2n = 44), groundnut (segmental alloploid) (2n = 40) and other allopolyploid crops. Though North American region is the home of the progenitor species of octoploid strawberry the cultivated strawberry (*Fragaria x ananassa*), arose through hybridisation of two other octoploid species, *F. virginiana* and *F. chiloensis*, (Edger et al., 2019) as illustrated in Fig. 14.3.

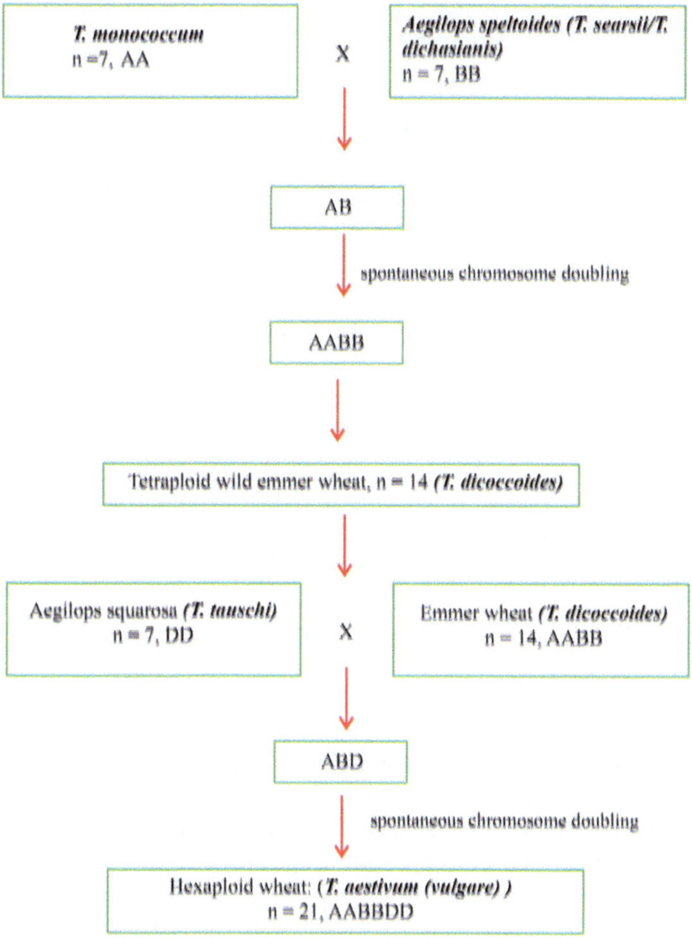

Fig. 14.1 Origin of wheat

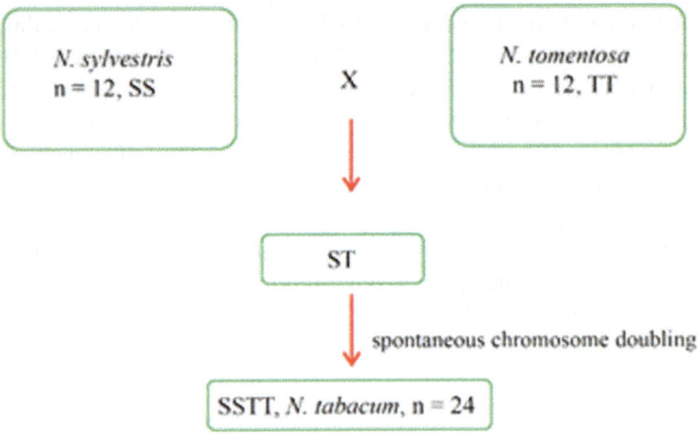

Fig. 14.2 Origin of tobacco

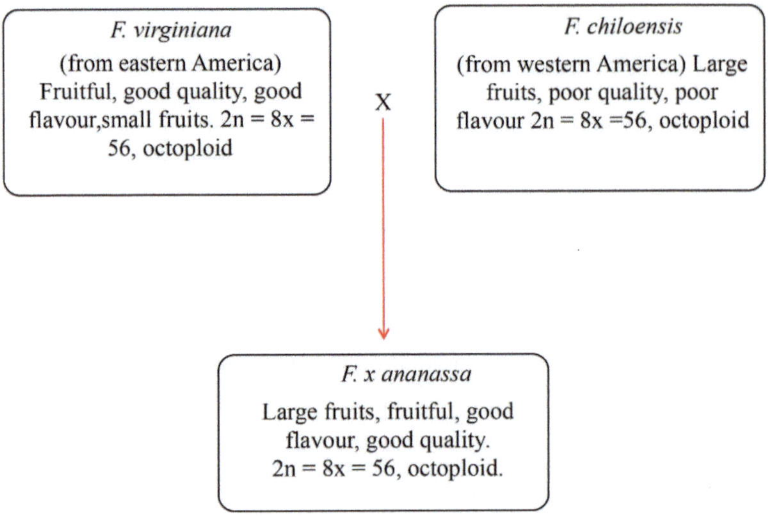

Fig. 14.3 Origin of strawberry

Apart from the allopolyploids that were obtained naturally some new species like Triticale have been synthesised by modern plant breeders as a result of artificial interspecific hybridisation of related species followed by chromosome doubling using colchicine or other chromosome duplication agents.

Table 14.1 Different aneuploid conditions

Aneuploidy	Chromosome number	Description
Nullisomy	$2n - 2$	Lack of a pair of homologous chromosomes
Monosomy	$2n - 1$	Lack of one chromosome so that one chromosome is represented only once
Double monosomy	$2n - 1 - 1$	Lack of two non-homologous chromosomes so that for each of the two affected pairs only one chromosome is present
Trisomy	$2n + 1$	Presence of one extra chromosome so that affected chromosome is represented three times instead of two times
Tetrasomy	$2n + 2$	Presence of two extra chromosomes so that affected chromosome is represented four times instead of two times
Double trisomy	$2n + 1 + 1$	Two different chromosomes represented three times each instead of two times

14.1.3 Aneupoidy

Occasional occurrence of aneuploidy is known in many plant species where different aneuploids sometimes arise that show distinct morphological features from the normal diploid plants. Aneuploids arise naturally in both diploid and polyploids due to irregular meiosis when a chromosome or a few chromosomes lag and fail to separate leading to gametes having one or a few chromosomes less or more chromosomes than normal. On fertilisation such gametes give aneuploid zygotes. Artificial or occasional natural hybridisation of autoplolyploids with the diploid parental species also leads to aneuploidy for various chromosomes. It is possible to obtain trisomics for various chromosomes in several plant species leading to distinct phenotypic expressions as known in wheat, tomatoes, tobacco, oats, chilli, rye, rice, sorghum and other plants. In addition to trisomics, other aneuploids, namely, nullisomics, monosomics, tetrasomics, double trisomics (Table 14.1) can be obtained in plant species for breeding and genetic studies purposes. In plants the extra chromosomes are often transmitted through the female gamete because the female gamete is more tolerant to chromosomal imbalance than the male gamete. A male gamete with $n + 1$ or $n - 1$ chromosomes or other aneuploidy condition is weak and often lacks viability, hence non-functional. In crop improvement programmes aneuploids obtained through interspecific hybridisation are used to transfer required genes to target cultivars in form of chromosome addition lines or they can be used to derive chromosome substitution lines and segmental substitutions. Aneuploids are also helpful in genetic analysis to locate genes on chromosomes.

14.2 Mutations in Chromosome Structure

As a natural rule, homologous chromosomes have uniform structure and order of genes. Apart from the respective allele differences at various loci, a pair of homologous chromosomes are similar in all respects and during pairing for bivalent formation their corresponding homologous segments lie side by side. However, maintenance of uniformity in the chromosome structure may be interrupted by rare structural changes known as chromosome mutations or chromosomal aberrations. Chromosome mutations involve loss or duplication of parts of a chromosome or addition of other parts to a normal chromosome or change in the order of genes. Structural changes which may be in form of deletions, duplications, inversions and translocations have distinctive configurations that are evident during meiosis I and are associated with notable phenotypic changes in the individuals which carry them and are of significance in plant breeding.

A deletion, also referred to as a deficiency, arises from the loss of a chromosome segment which may be spontaneous and rare or induced. Some agents known to cause chromosome deletions are radiations like alpha radiations, beta radiations, gamma rays, x-rays, protons and neutrons, chemicals like sodium saccharin and viruses. The agents cause breaks in chromosomes which may result in terminal or intercalary deficiencies. Terminal deletions arise where one break on the chromosome leads to a piece with a centromere and an acentric segment without a centromere which is then lost. In intercalary deletion two breaks on the chromosome are followed by loss of the section between the breaks and the union of the piece retaining the centromere and the other acentric end. Though there are cases of deletions reported in some crop plants like *Vicia faba* and maize, they are never desirable and not sustained in crop plants because they lead to reduced fertility in their heterozygous carriers and may be lethal if they are homozygous in an individual.

In a duplication a section of the chromosome is repeated in form of a tandem duplication, reverse duplication or terminal duplication. Duplications may arise from unequal crossing over or spontaneous or induced chromosome breakages caused by radiations, chemical agents and viruses followed by unequal reunion of some segments. Duplications reduce the fertility of their carriers, especially pollen viability, and have been reported in plants like maize. In plants deletions may be transmitted through the female gamete but rarely via pollen since pollen having chromosomes with deletions aborts before full development.

Inversions involve 180° reversal of the chromosome segment following two separate breaks on a chromosome with no change in the number of genes or number of chromosomes. Inversions are caused by radiations, chemicals and viruses which induce breaks in chromosomes. There are pericentric inversions in which the inverted segment contains the centromere and paracentric inversions where the inverted segment does not include the centromere. Inversions suppress crossing over and reduce fertility, especially for the male gametes. In reciprocal translocations there is exchange of parts between chromosomes that are non-homologous when breaks occur on such chromosomes. Translocations are also induced by mutagens which cause breaks on chromosomes followed by exchange of acentric fragments between the non-homologous chromosomes having such breakages. Changes in the position of genes relative to the others on the chromosome arising from inversions, duplications and translocations lead to position effects which cause alterations in the function and expression of the genes reflected in phenotype. Duplications, inversions and translocations are responsible for changes in gene position effects due to the fact that expression of a gene or a group of genes is influenced by nearby heterochromatin.

Problems

14.1 What features make polyploids have better quality horticultural crops than their corresponding diploid relatives?

14.2 How did the cultivated tobacco, wheat and strawberry arise?

References

Allan, R. E. (1980). Wheat. In W. R. Fehr & H. H. Hadley (Eds.), *Hybridization of crop plants*. America Society of Agronomy and Crop Science Society of America Publishers. Wiley, No. 5701330411481. https://doi.org/10.2135/1980.hybdizationofcrops.c15

Edger, P. P., Poorten, T. J., VanBuren, R. Hardigan, M. A., Colle, M., McKain, M. R., et al. (2019). Origin and evolution of the octoploid strawberry genome. *Nature Genetics, 51*, 541–547 (2019). https://doi.org/10.1038/s41588-019-0356-4. Creative Commons Attribution International Licence (CC BY).

Kasha, K. J., & Kao, K. N. (1970). High frequency haploid production in barley. *Hordeum vulgare* L. *Nature, 225*, 874–876. https://doi.org/10.1038/225874a0. Springer Nature, No. 5700900372196.

Wernsman, E. A., & Matzinger, D. F. (1980). Tobacco. In W. R. Fehr & H. H. Hadley (Eds.), *Hybridization of crop plants*. America Society of Agronomy and Crop Science Society of America Publishers. Wiley, No. 570133131315841. https://doi.org/10.2135/1980.hybdizationofcrops.c15

Chapter 15
Interspecific Hybridisation and Chromosome Manipulation in Interspecific Hybrids

In crop improvement programs some plant breeding challenges and production problems are resolved by mobilisation of genetic variability that exists in the species related to the crops necessitating interspecific or intergeneric crosses between the crops and their wild and weedy relatives. Hybridisations between cultivated plants and their relatives give the plant breeders opportunities to mobilise variability in alien gene pools and broaden genetic variability and diversity of cultivated plants and transfer one or a few distinct genes from wild or weedy species not available in the respective crop species. Since interspecific crosses are difficult to make and more often give no hybrids or result in sterile hybrids breeders use them only under desperate situations where there are no easier and cheaper alternatives. Some of the reasons which necessitate use of interspecific hybridisation to transfer alien genes to cultivated plants are serious pest or disease infestation or other stress conditions and climate change related problems where there are no genes to address the problems in the crop populations. Interspecific hybridisation leads to expression of new characters not found in crops and their relatives and also helps in production of alloploid species. Novel character expression not present in either of the parents arise in interspecific hybrids due to interaction of the genes of the two genomes, cytoplasm-genome interaction and possible mutations. Interspecific crosses are sometimes made to determine the phylogenetic relationship among different species. The need to use the weedy and wild relatives for crop plant improvement also depends on the need to release new varieties fast, relationship between the crop and its relative having the target gene, how desperate the situation is that necessitates the use of interspecific hybridisation, the availability of appropriate crop relatives with the desired genes, ease of using the relatives and the attitude of the breeder towards incorporation of wide crosses in crop improvement programme (Harlan, 1976). Alien genes are easy to transfer from the related species if the crop and its relative are sufficiently related as to allow the chromosomes of the two genomes to have affinity for pairing and recombination during meiosis in the interspecific hybrids.

Alien gene transfer forms pre-breeding exercise which is an important component of judicious use of weedy and wild crop relatives for crop improvement. It is an

P. O. Ayiecho and J. O. Nyabundi, *Conventional and Contemporary Practices of Plant Breeding*, https://doi.org/10.1007/978-3-031-74998-8_15

established fact that weedy and wild crop relatives and crop progenitors have genes for special attributes lacking in crop populations and that bringing alien genes into cultivated plants creates novel variation for crop plant improvement. Artificial interspecific hybridisation is a complementary strategy to conventional plant breeding for crop improvement and a number of modern crop cultivars have alien genes obtained from their wild and weedy relatives. Several characters including simply inherited traits like plant architecture traits, pest and disease resistance, abiotic stress tolerance and polygenic traits like yield, plant height, etc, have been transferred to crop plants through interspecific and intergeneric hybridisation.

15.1 Barriers to Interspecific Hybridisation

15.1.1 Phylogenetic Relationship Between Crops and Their Relatives

Through speciation various organisms have developed isolation barriers restricting natural interspecific mating and gene flow between related or unrelated species, thereby maintaining the integrity of each species. Interspecific isolation barriers limit the chances of transferring genes across species by plant breeders unless methods are devised to circumvent the barriers. However, in nature there are occasional hybridisations between species and sub-species, suggesting that introgression of genes of a particular species into populations of other related species is possible. The use of alien plant genetic variation to improve cultivated crops is determined by the phylogenetic relationship of a crop species and its relatives having the target genes, and greater the divergence between them the stronger are the interspecific isolation barriers making it difficult to transfer genes between the species. The less the divergence between any two species the easier it is to transfer the desired genes between them and the more useful are the derived hybrids. Success of obtaining useful interspecific hybrids is enhanced by absence of interspecific isolation mechanisms or presence of only weak and easy to break barriers. It is easier to transfer alien genes to polyploid species like wheat, potato and tobacco than to diploid. Based on the relationship between the crop species and their relatives and how easily the genes can be transferred, the crops and their relatives can be clustered such that those most closely related to the crop are clustered with the crop. Harlan and de Wet (1971) categorised crop species and their relatives into three gene pools as primary gene pool, secondary gene pool and tertiary gene pool. Harlan and de Wet (1971) referred to the primary gene pool as consisting of cultivated races and wild or weedy species of the same ploidy level as the crop which hybridise easily with the crop in question. The populations or species in this category are the most closely related to the crop and hybridise easily with the crop in question to give fertile F_1 hybrids that can be backcrossed to the crop making gene transfer possible. The populations in this category may be different landraces and cultivars of the crop in question and may also include species closely related to the crop with their chromosomes being homologous to those of the crop and having high affinity for pairing. For example, the cultivated *Avena sativa* and its relative *Avena sterilis;* maize *(Zea mays)* and teosinte *(Zea mays* subsp. *mexicana); Arachis hypogea* and its relative *A. monticola; Triticum turgidum* and its relative *T. dicoccoides; Hordeum vulgare* and *H. spontaneum; Sorghum bicolor* race *kafir caudatum* and its relative *S. arundinaceum* and the cultivated *Eleusine coracana* subsp. *coracana* and its wild relative *E. coracana* subsp. *africana* are close enough to give hybrids with varying degrees of fertility.

In some cases, the crop relatives may not be close enough to give fertile hybrids unless the chromosome number of the hybrid is doubled. This is the secondary gene pool category having the relatives of a given crop which cross with the crop but with varying difficulties due to barriers to interspecific hybridisation (Harlan & de Wet, 1971). The F_1 hybrids are sterile with poor chromosome pairing. Gene transfer is possible though there are problems of ploidy level differences between the crop and its relatives, genomic and cytoplasmic incompatibilities. Altering chromosome number of parental species or the hybrid, embryo rescue or other biotechnology methods may resolve some hybridisation challenges. Fertility of the hybrids is usually regained through chromosome doubling. This is the case for the relationship between cultivated tobacco and its relatives *N. glutinosa, N. debneyi and N. rustica,* wheat and its relatives *Aegilops speltoides, Ae. cylindrica* and *Ae. caudata* and *E. coracana* subsp *coracana* and its relatives E. *indica.*

There are species that are so distantly related to the crop species and are highly difficult to cross to the crop species due to strong barriers to interspecific hybridisation and any interspecific F_1 hybrids and corresponding allopolyploids are highly sterile due to genic complex effects and lack of chromosome homology between the crop and related species genomes, making it extremely difficult to mobilise genetic variation in this category for crop improvement. This is the tertiary gene pool category and the species in this category are within same genus as the crop or related genera but distantly related (Harlan & de Wet, 1971). Embryo rescue/culture, tissue culture, protoplast fusion are often employed to salvage the interspecific hybrids. Among *N. tabacum* relatives *N. repanda, N. nudicaulis* and *N. alata* belong to this category while the wild relatives of *Eleusine coracana* subsp. *coracana* that belong to this group are *E. jaegeri, and E. kigeziensis.* Chromosome translocations and genetic transformation can be employed to effect gene transfer from members of this group to the respective crop species. There is also quaternary gene pool which has no organism with any relationship to the crop in question and transfer of genes from them to the crop species is possible only through genetic transformation using indirect or direct DNA transfer methods. For example, bacterial and other organism's genes are transferable to crop plants only through transgenesis.

15.1.2 Interspecific Reproductive Isolation Mechanisms

The features preventing different species from mating to get viable progenies and achieve gene exchange are reproductive isolation barriers (reproductive isolation mechanisms) which may be pre-fertilisation/pre-zygotic isolation barriers that operate outside or within the plant tissues to prevent fertilisation or post-fertilisation/post-zygotic barriers functioning within the plant tissues preventing the survival of the hybrid (Katche, 2019; Wijerathna et al., 2022). These barrier mechanisms may operate in isolation or a combination of different mechanisms may be operational to prevent interspecific reproduction and any two species may not be isolated by one isolation mechanism only since more than one mechanism may be operational.

External barriers

Pre-fertilisation isolation mechanisms prevent the different species/populations from coming together and if they are within reach of each other, they don't mate and should they mate their gametes do no unite. The barriers that operate outside the plant tissues include the mechanisms that separate the different species in space and time, ecologically and through floral morphology and pollen dispersal (Hadley & Openshaw, 1980). Spatial isolation applies when the related species or populations are allopatric by occupying different geographical areas denying them the opportunities to meet and hybridise. Ecological isolation serves as an isolation mechanism if populations and species occupy different ecological niches within the same or different geographical areas due to strict varied ecological adaptation requirements by the populations or species. In some cases, the species isolated ecologically may have opportunities to cross-pollinate producing interspecific F_1 hybrids which fail to survive because they are not adapted to habitats for either of the parents and need special environments for their successful growth. Seasonal and temporal isolation is also a common and effective barrier which keeps plant populations/species apart even if they are within the same geographical region denying them the opportunities to hybridise because they flower at different times. Temporal isolation alone works for sympatric populations (those occupying the same geographic area) and parapatric populations (those having contact only along their borders of their ranges) if the populations strictly flower at different times or different seasons of the year. Unless it is combined with other isolation mechanisms seasonal or temporal isolation alone is easy to break. In some cases, ecological and temporal isolation mechanisms may be incomplete because of overlap of habitats in certain areas occupied by different species leading to possible contacts between them or overlap in flowering periods (Hadley & Openshaw, 1980). Isolation barriers breakdown may also happen for related populations and species not sufficiently isolated in space, especially parapatric populations and sympatric populations as is common for some crops and their weedy and wild relatives leading to possible gene flow between them. Unless other isolation mechanisms are operational, parapatric and sympatric species with some degree of sibling compatibility will form hybrid swarms or hybrid zones which is a collection of hybrids produced by local breakdown of isolation barriers between the species at points where they are in contact. Another effective external isolation barrier pertains to situations where the species differ in pollen size, style length and flower features like flower size and colour and strictly attract different insect species as pollinators. This gives no opportunities for natural cross-pollination between the different species.

Internal / post-fertilisation barriers

Prevention of fertilisation due to failure of pollen germination, pollen tube growth and penetration of the ovule and the embryo sac and all the post-fertilisation isolation mechanisms operate within plant tissues through lack of physiological/phytochemical or cytological compatibilities between the two populations or species preventing fertilisation or the interspecific hybrids from growing, developing, reproducing or leading to failure of their progenies and subsequent generations to survive and reproduce (Hadley & Openshaw, 1980). Prevention of fertilisation may also arise if pollen of one species and stigma or style of the other species interact such that pollen does not germinate or pollen tube does not grow to reach the ovule and embryo sac to effect fertilisation. During pollination the interaction of the pistil and the pollen has to provide the right environment for pollen germination and pollen tube growth. The stigmatic surface may not exude adequate water and the necessary phytochemicals for pollen germination or the style may be too long or exude chemicals that are antagonistic to the pollen tube leading to incompatibility (Kuligowska et al., 2015). The failure of pollen to germinate or pollen tube to grow down the style is attributable to female and male determinant recognition compounds in both stigma and the pollen grain and the pollen tube and stylar tissue. Pollen tube growth down the style may be hindered by deposition of callose in the pollen tube. In some cases, pollen from one species may not fertilise the other species while reciprocal fertilisation succeeds, implying unilateral incompatibility under control of incompatibility genes (Hadley & Openshaw, 1980).

In some cases, fertilisation may succeed but the F_1 hybrid or subsequent generations are inviable, weak and fail to develop properly due to genetic imbalance/unfavourable gene combinations causing physiological or developmental abnormalities. Inviability or weakness of the hybrid is manifested in form of its abortion or death at any stage, lack of vigour and failure to produce viable gametes due to sterility (Hadley & Openshaw, 1980). Some crosses give F_1 hybrids which are vigorous and survive but are completely or partially sterile due to genetic factors, hence genic sterility, or chromosomal structural differences. The hybrids may remain sterile due to gene combinations or gene complexes in the two species which fail to function harmoniously in the hybrids disrupting cell functions, cell division, gametogenesis and other cell metabolic processes. Some genes or gene combinations cause asynapsis and desynapsis interfering with distribution and frequency of chiasmata and spindle formation during meiosis I. There are cases of developmental sterility leading to abnormal development of reproductive features in F_1 and also chromosomal structural differences between the genomes of the two species causing abnormal chromosome disjunction in the interspecific F_1 hybrid resulting in segregational hybrid sterility or chromosomal sterility (Hadley & Openshaw, 1980). Structural and numerical differences between the chromosomes of the two species prevent proper pairing between chromosomes of the two genomes due to lack of homology or there may be only allosydentic pairing due to homeology. That is, chromosomes from one species may not have proper pairing partners among chromosomes of the other species, leading to breakdown of meiosis, occurrence of univalents and multivalents and consequently, random distribution of chromosomes and lagging of some chromosomes which may get lost during meiosis I, giving unbalanced inviable gametes and formation of micronuclei in the daughter cells and ultimately F_1 hybrid sterility.

Hybrid breakdown occurs in some interspecific hybridisations whereby normal and viable F_1 hybrids give inviable or sterile F_2 and subsequent generations due to genetic incompatibility between the genomes of the two parental species resulting to failure of the hybrid or even lethality in F_2 progenies (Hadley & Openshaw, 1980). This phenomenon which is genetic disharmony or genetic breakdown, affects cell division and various physiological processes, developmental and growth stages in plant tissues and organs, including development of reproductive structures and gamete production, embryo and endosperm development, meristem development, germination of seeds and growth and development of seedlings. Genetic disharmony is attributable to incompatibility arising from chromosome structural differences and unfavourable genetic interactions between the genomes of the two species which become manifested in F_2 and subsequent generations due to genetic recombination involving co-adapted and complimentary gene complexes. Such recombination interferes with stability of some complimentary genetic systems and give rise to unadapted genotypes.

Absence of harmony between the genotype of the zygote and the endosperm and the unfavourable cytoplasm-genome interaction prevent interspecific hybridisation. Incompatibility of cytoplasm of one species and the genome of the other species results in hybrid weakness and even zygote death (Bhanu et al., 2018). Differences in chromosome number in the species to be crossed is known to cause endosperm failure to provide a satisfactory environment for the developing embryo due to inappropriate balance between the endosperm and embryo genomes, leading to zygote abortion and abscission of the developing fruit (Hadley & Openshaw, 1980). Failure and degeneration of the endosperm and abortion of the zygote is related to the ploidy levels of the endosperm and the zygote since genetic dosage balance between the developing zygote, maternal tissue and the endosperm affects the development and survival of the zygote. The development of the seed is affected by the ratio of maternal and paternal genomes since under normal circumstances the endosperm is triploid and the embryo is diploid, making the ratio of endosperm genome to embryo genome ratio become 3:2. Any smaller ratio provides an unfavourable environment for embryo development. In interspecific crosses fewer disharmonies occur if the species with higher chromosome number is used as the female parent, hence, interspecific crosses are more likely to succeed if the species with the higher chromosome number is used as the maternal parent. Chromosome abnormalities like breakages and mitotic abnormalities and failure in the endosperm arising from interspecific crosses also lead to embryo-less seeds, shrivelled seeds or no seed development. In some interspecific crosses chromosomes of one species are eliminated from the developing zygotes resulting in haploid zygotes which are inviable and sterile unless chromosomes are doubled as is known for *Hordeum vulgare x H. bulbosum* cross Kasha and Kao (1970).

15.1.3 Breaking of Interspecific Isolation Mechanisms

There are a number of strategies for overcoming pre-fertilisation barriers for successful interspecific hybridisation. Once the potential populations or species to be used in the hybridisation have been identified in regard to the target traits the population or species most closely related to the crop species should be selected for making the cross. The required genes should be sought for in different populations and cultivars of the crop first. The alien gene pool and species most distantly related to crop should be used only if the required traits lack in the crop populations and species most closely related to the crop population. It is important to sample the parental populations adequately (both alien donor and recipient crop) to increase chances of success in obtaining fertile hybrids usable in the introgression process. Since incompatibility is under genetic control, and that cultivars differ with respect to this feature selection and adequate sampling of both populations or species to be crossed increases chances of capturing rare compatibility genes.

Hybridisation between related species is conditioned on successful pollination which requires that the species occupy the same place and flower at the same time. If isolation is because the species are separated by growing in different regions or different habitats or have temporal/seasonal isolation by flowering at different times/seasons attempts can be made to grow them together in appropriately controlled conditions or manipulation of planting time so that they flower at the same time to make it possible to cross them.

For sympatric and parapatric species the presence of hybrid swarms between them is evidence that the isolation mechanisms between them can be broken and the two species can be crossed to produce viable hybrids. The hybrids arising due to breakdown of isolation mechanisms lead to natural introgression by incorporation of genes of one species into the gene pool of the other one though some interspecific hybrids may be poorly adapted. If populations occupying different ecological habitats are cross-compatible giving fertile hybrids which have a restriction due to poor adaptability appropriate and special conditions can be provided for the hybrids to ensure their survival and use in the backcrossing process for introgression purposes. A number of crop plants like wheat, rice, sorghum and maize are known to hybridise naturally with their close relatives within reach of pollen dispersal forming fertile hybrids.

If mechanical barrier is the only cause for isolation interspecific hybridisation can be achieved by the breeder through artificial pollination. Incompatibility barriers preventing fertilisation are addressed by a variety of techniques, some of which are similar to the methods used to overcome self-incompatibility in crop cultivars. Correct timing of pollination overcomes the stigma pollen receptivity problem and may lead to success since the time and period when stigma is receptive varies with species. Pollination at bud stage when the stigma and style are still immature and before the female determinant incompatibility compounds appear on the stigmatic surface or in the stylar tissue or delayed pollination when such compounds are no longer active solve the problems of incompatibility. Wetting the pollen or mixing the pollen with pollen germination inducer solutions improves chances of pollination success since pollen germination is improved by humid conditions while high temperatures also favour fertilisation because temperatures of around 40 °C render incompatibility compounds inactive as known in some species like *Trifolium* (Pickersgill, 1993). Mixed pollination using mentor pollen is applied in some cases where the compatible pollen is killed using radiations, heat treatment or treatment with alcohol and then mixed with incompatible pollen and placed on stigma of intended female parent Hadley and Openshaw (1980). In the pollen mixture the killed compatible mentor pollen facilitates the germination and penetration by pollen tube of alien species due to its stimulatory effects. The killed compatible pollen induces the release of the necessary recognition compounds which suppress the incompatibility reaction and allows the alien pollen to germinate and effect fertilisation. Failure of interspecific hybridisation arising from failure of pollen germination and growth of pollen tube can also be resolved by application of appropriate growth hormones and vitamins like hexane, indole-3 acetic acid, N-m-tolphthalmic acid, 2,4-dichlorophenoxyacetic acid (2,4-D), gibberellins, auxins, 6-benzylamino purine and salicylic acid to the stigmatic surface to suppress incompatibility reaction and stimulate the alien pollen germination and growth of the fertilisation tube to effect fertilisation. In some crosses the alien pollen tube may not be adapted to crop species stylar environment and fails to grow to reach the ovule, especially in situations where the two parents have different pistil lengths. If the style is too long and the pollen tube does not reach the ovary *in vitro* pollination can be done by surgically amputating a portion of the style and applying the pollen along with sucrose solution on the remaining style stump or fertilisation is accomplished through ovule pollination by transferring pollen directly to the ovary using hypodermic syringe (Katche et al., 2019). Reciprocal crosses are also used to break interspecific barriers if interspecific hybridisation failure is caused by unilateral incompatibility and the stylar tissue incompatibility reaction is restricted to one species style only. Application of X-radiations also breaks the isolation barriers by temporarily eliminating incompatibility reaction between the species.

Alteration of chromosome numbers of the parents to be crossed improves the chances of obtaining successful interspecific crosses and compatibility is improved if the parents are mated at the same chromosome number in some cases where the related species differ in chromosome number (Hadley & Openshaw, 1980; Pickersgill, 1993) . In some cases, reciprocal crosses lead to better results if there are chromosome number differences between the species. Where species differ in chromosome numbers interspecific crosses are more successful if the species with higher chromosome number is used as the female parent. Bridging species is another method used to break interspecific isolation barriers by using three species, the crop and its two relatives. One of the crop relatives is used as a bridge between the crop species and the other relative which is the source of the gene to be transferred because the crop and the source of the target gene do not hybridise while the bridging species hybridises with both the crop and the source of the gene. The transfer is accomplished by crossing the species having the target gene to the bridging species to transfer the gene to the latter and followed by crossing the interspecific hybrid to the crop.

Where abortion or abscission is one of the ways by which the interspecific hybrid fails plant growth hormones like gibberellic acid application saves the developing interspecific zygote from aborting. Application of gibberellic acid, auxins and cytokinin to the developing interspecific cross fruit helps in retention of fruits and successful development and maturation of the hybrid embryos (Hadley & Openshaw, 1980; Pickersgill, 1993). Mixed fertilisation in the same flower helps to achieve successful development of the interspecific zygote. In this case both viable compatible pollen of the crop and the incompatible pollen of the crop relative are mixed to effect pollination and fertilisation. Due to fertilisation by both pollen types in the same ovary there will be two types of zygotes in the developing fruit: those derived from fertilisation with compatible pollen and those arising from interspecific cross. Those arising from fertilisation with compatible pollen will support those arising from interspecific cross through naturally produced growth hormones and nutrients so that the latter don't abort.

Crosses in which abortion follows interspecific fertilisation the embryos can be rescued and cultured *in vitro* on artificial medium to adult plant before it aborts or dies (Hadley & Openshaw, 1980; Pickersgill, 1993). This is a solution to endosperm failure so that the embryo can be cultured into vigorous useful plant. Embryo rescue also solves the problems of low seed set, seed dormancy and slow germination. The success of embryo culture depends on the genotype and growth conditions of the mother plant, development stage of the embryo at the time of excising, culture medium composition and culturing conditions. Since embryo should have its maximum development on the female plant before being excised for culturing growth hormones can be applied to help sustain the embryo until the rescue stage. That is, salvaging the embryo through embryo culture should be done at an appropriate stage, not too early and also not too late when the embryo is already damaged by abortion process. Embryo culture takes the form of ovary slice culture, ovule with placenta culture, single ovule culture and embryo sac culture. Interspecific compatibility and hybrid survival is improved by grafting of embryos from germinating seeds of one species into the endosperm of germinating seeds of the other. The F_1 hybrid seedling buds can be grafted to either of parent's rootstocks, especially where there is poor chlorophyll or poor root development in the hybrid (Hadley & Openshaw, 1980).

Where there is interspecific genic sterility, the barrier is broken by incorporating the right genes and inducing mutations. Interspecific hybrid sterility attributed to chromosome structural differences is resolved by doubling the chromosome number of the sterile hybrid using colchicine creating an allotetraploid thereby providing a homologue for each chromosome leading to normal chromosome pairing and disjunction during meiosis and restoring fertility of F_1 hybrid. Cytoplasmic-genic interaction is addressed by making a reciprocal cross which may give a more fertile hybrid if this interaction is the only cause for interspecific hybrid sterility. Reciprocal crosses also solve the interspecific barrier problems where maternal tissue surrounding the developing embryo fails to nourish the embryo.

Methods of biotechnology and recombinant DNA technology like protoplast fusion and genetic modification are also used to overcome incongruity by circumventing sexual barriers to interspecific hybridisation. Somatic hybridisation helps to bypass interspecific barriers and is applicable for combining both nuclear and cytoplasmic genomes of different species by bringing the diverse protoplasts together and enforcing fusion of the protoplasts using polyethylene glycol and Ca^{2+} to form somatic hybrids in the culture system. Successful regeneration and usefulness of somatic hybrids depend on phylogenetic relationships between the species involved, especially for the case of symmetrical fusion hybrids in which the contribution of both species is equal. Asymmetric protoplast fusion hybrids in which the two species have unequal genome contributions are better because they perform better in terms of rooting and regeneration, fertility and less genome conflict between the two genomes (Forsber et al., 1998). Asymmetric protoplast fusion hybrids can be obtained by fragmenting the chromosomes in the protoplasts of the donor species using UV radiations and other radiations like γ and X-ray followed by fusion with the recipient protoplasts (Cheng et al., 2004; Bruznican, 2021). The fragments get incorporated into the recipient genome and with the use of molecular markers the required integration can be selected. Use of asymmetric somatic hybrids and genome fragmentation can be used to transfer both polygenic traits and major gene traits. Genetic modification which involves engineering of identified genes from one species into another is used to transfer genes across species in one generation. Genetic transformation employing *Agrobacterium tumefaciens* or direct DNA transfer methods break interspecific hybridisation barriers, eliminate the transfer of undesirable genes and characters from the crop relatives and can be more easily employed to deliver genes between unrelated species without negatively affecting the genetic background of the recipient species.

15.2 Pre-breeding

15.2.1 Importance of Pre-breeding

Pre-breeding is useful in diversifying the genetic base of the germplasm in breeding programmes as it helps in mobilising the variability in the crop wild and weedy relatives by acting as the bridge for transfer of alien genes from such relatives to the elite breeding materials. Pre-breeding involves identification of desirable genes in non-adaptable materials, particularly wild and weedy crop relatives, for transfer to intermediate populations/lines. Pre-breeding is important in situations where wild and weedy crop relatives having the target genes are unadapted, have undesirable characteristics and are difficult to cross to the crop species. It is a necessary breeding process in which the undesirable genes are eliminated before the intermediate populations/lines developed can be utilised in breeding programmes. The process broadens the genetic base of the cultivated crops and breeding stocks in mainstream breeding programmes by providing crop materials with alien genes in more easily usable forms than if such genes are within wild/weedy relative genetic backgrounds. The process is an important way of utilising diverse germplasm resources through mobilisation of genetic variability in the wild and weedy relatives by transferring target genes from alien genomes to the materials in the mainstream breeding programme and allows for elimination of tightly linked undesired characters from the materials being developed. The resulting intermediate lines/populations having introgressed target genes from wild and weedy crop relatives are in turn used to transfer the target genes to elite breeding materials for development of superior varieties.

Pre-breeding is important where the commercial varieties have production challenges and lacking in specific attributes not resolvable by mobilising genetic variability within the crop breeding stocks alone. Catastrophic situations associated with climate change challenges necessitate diversifying the genetic base of breeding materials. Changed weather regimes have interfered with cropping patterns and more frequent severe droughts and increased rainfall storms leading to floods and waterlogged soils have become serious production challenges which must be addressed through broadening genetic base of crop plants. Emerging pathogen races and insect biotypes that were of no significance previously like the recent incidences of maize lethal necrosis disease (MLND) caused by maize chlorotic mottle virus (MCMoV) in conjunction with other viruses (2011) and fall army warm (2017) in Kenya and known established production problems not yet adequately addressed, e.g., *Striga* menace on cereals within the tropics and subtropics, necessitate introduction of greater diversity in stocks at the breeders' disposal.

15.2.2 Pre-breeding Strategies

An important component of pre-breeding is collaboration between breeders and gene bank personnel in multiplication, evaluation and access to any other useful information about the relevant collections in stock for identification of the germplasm resources with the required genes. During multiplication and screening of the existing germplasm resources wild and weedy relatives having the target traits like disease and pest resistance, *Striga* resistance, resistance and tolerance to heat stress, salinity, drought stress, mineral stress and other abiotic stress factors are identified for use in hybridisation with the crop. Parents are chosen from populations most closely related to the crop by giving preference to populations in the primary gene pool, followed by secondary and tertiary gene pools. Germplasm in the secondary gene pool and tertiary gene pool categories are resorted to in situations where no populations or species having the required genes are found in the primary gene pool. Crop parent should be of be desirable agronomic characters except for lack of the target trait. The selected parents are used in ♀ crop x ♂ crop relative interspecific hybridisations unless there is known unfavourable cytoplasm-genome interaction or unilateral incompatibility making it necessary to use the crop relative as the female parent in the initial cross and the crop as the female parent in the final backcross. The hybrid progenies which are backcrossed to the crop parent are used as the female parents while the recurrent parent is used as the pollen parent. While making the hybridisation the presence of any possible hybridisation barrier as dictated by diversity of relationship is taken into account and resolved. Many interspecific F_1 hybrids are sterile and may have to be subjected to chromosome doubling for improved fertility if the wild or weedy parent is from the secondary gene pool category. The hybrid with doubled chromosome number and some amount of fertility is backcrossed to the crop parent giving progenies with a wide range of chromosome composition. Selection is made among the segregating progenies in every generation of backcross for the target trait and high intensity of expression for the recurrent crop parent characters. Through generations of backcrossing substitution lines, addition lines and segmental substitution lines involving the chromosomes bearing the targeted gene may arise. Incorporation of the target gene from the alien genome into the crop chromosomes may be achieved through homoeologous chromosome recombination or segment translocation, sometimes enhanced by application of radiations. Molecular mapping and sequencing are used in identifying molecular markers associated with the target genes/QTLs in crop wild and weedy relatives to be transferred so as to fast-track and improve the efficiency of the process and eliminate linkage drag and unwanted alien genome in the selected progenies. Biotechnology methods like embryo rescue, somatic hybridisation and genetic transformation methods are employed in difficult hybridisations involving populations and species in secondary and tertiary gene pools.The products of pre-breeding are intermediate populations or lines for easy and fast use in mainstream breeding programmes to develop superior crop varieties combining the alien target genes and the good characters of the crop.

15.2.3 Challenges for Pre-breeding Efforts

Wild, weedy and various forms of crops relatives which are often used as target gene donors in interspecific crosses usually lack important agronomic, yield, quality attributes and have too many undesirable genes which must be eliminated through several generations of backcrossing. The relatives have undesirable genes that are too tightly linked to the target genes and there are also barriers to hybridisation between them and crops to be improved. Some crop relatives are unadapted to certain conditions and will need special conditions to grow and flower.

Backcross progenies originating from interspecific crosses have wide ranging chromosome numbers and variable fertility, with a large fraction being highly sterile due to irregular chromosome segregation and have to be subjected to chromosome manipulation process to obtain usable materials in form of segmental substitution/translocation lines, chromosome addition and substitution lines. Segmental substitutions and translocations are sometimes difficult to obtain. Where interspecific hybridisation barriers are controlled by gene complexes that are not easy to break induction of mutations, incorporation of appropriate genes and genetic transformation are used as alternatives to resolve such challenges.

15.3 Chromosome Manipulation in Interspecific Hybrids

Interspecific crosses are expensive to make and are resorted to in desperate situations when some factors pose major threats to crop plants necessitating the transfer of specific genes to address the production challenges which may be in form of abiotic or biotic stresses. The target genes for transfer may be for characters like disease and pest resistance, tolerance or resistance to drought, cold conditions and high temperatures, mineral and salt stress conditions. Since wild and weedy crops relatives used as target gene donors in interspecific crosses lack important agronomic, yield and quality attributes and have too many undesirable genes, alien genome transfer to address the production problems at hand should move only one or a few necessary genes from the genome of the crop relative and ensure that the genome of the recipient crop species remains stable, with no undesired effects on crop yield, quality attributes and other and agronomic parameters.

The usefulness of interspecific hybrids depends on the extent of their fertility which is also determined by closeness between the crop and its relative. If the interspecific hybrid has normal fertility, it is simply backcrossed to the crop parent for the number of generations necessary to reinstate the traits of the crop parent. In the initial cross the crop is used as the female parent to retain the crop cytoplasm unless the genome-cytoplasm interaction dictates that the crop relative be used as the female parent. During backcrossing process, the recurrent crop parent is used as the male parent to avoid any pollen sterility problems due to chromosome imbalance in the hybrid and backcross progenies. However, in the final backcross generation when nearly all the alien genome has been eliminated except for the gene(s) under transfer, the crop parent is used as the female parent. While backcrossing is done selection is conducted for the trait under transfer to ensure that it is not lost while at the same time the undesired alien traits are eliminated by intense selection against them. Use of flanking molecular markers or markers located within the target genes/QTLs and molecular markers for the recurrent crop parent traits improve the efficiency and speed of recovery of the traits of the recurrent parent and elimination of the alien genome while ensuring retention of the target traits. If the chromosomes of the two species have affinities due to homology or homeology allowing genetic exchange during backcrossing a segment of the alien chromosome bearing the target gene may be transferred to the crop chromosome through crossing over, giving a segmental substitution line.

More often the interspecific hybrids have limited fertility or not fertile at all due to lack of pairing between the chromosomes of the two genomes or sterility imposed by unfavourable gene combinations, especially if the crop relative belongs to secondary or tertiary gene pools. The interspecific F_1 hybrid may be sterile due to formation of univalents that are randomly distributed resulting in micronuclei formation in the daughter cells or some chromosomes may lag and get lost during meiosis I. In some interspecific crosses the chromosomes of the two species have homeology leading to allosydentic pairing thereby forming heteromorphic bivalents instead of homomorphic bivalents. Fertility of such interspecific F_1 hybrids can be regained by doubling the chromosome number using colchicine or other chromosome doubling agents to create an amphidiploid which is a synthetic allotetraploid. Backcrossing of the amphidiploid to the crop parent and subsequent backcrosses involving the crop parent and backcross progenies may result into segmental substitution/translocation lines or chromosome substitution or addition lines (Fig. 15.1).

15.3.1 Segment Substitutions/Translocations

If hybrid sterility is due to chromosome structural differences the diploidised interspecific F_1 have normal fertility with normal bivalent formation and normal chromosome assortment in meiosis I and can be backcrossed to the cultivated parent to produce a sesquidiploid. When sesquidiploids are backcrossed to the crop species the progenies obtained have varying degrees of fertility attributable to variable chromosome numbers ranging from diploid to sesquidiploids with various aneuploids of chromosome additions and even substitutions. In the process of backcrossing selection is done for the traits of cultivated parent and the alien gene(s) being transferred. There may be rare occurrence of simple genetic crossing over due to homeologous pairing or translocations between the crop and alien chromosomes. Radiations like X-rays can be used to facilitate such translocation and segment exchange events but this has the problem of creating non-homeologous exchanges, deficiencies and duplications (Prieto, 2020). Substitution cases have been reported in cereals like wheat, oats and rye. Segmental substitutions/translocations can be induced in monosomic substitution lines or addition lines followed by selection (Fig. 15.2). Selection is made for the progenies having the least of the alien traits and alien chromosomes with the help of molecular markers. Progenies having the target alien traits but no alien chromosome and least of alien traits have the relevant segment recombination (segment substitution) or translocation and are of the desired genotype. The success of translocations is determined by recipient ability to tolerate genetic deletions and duplications in their genomes, a situation tolerated better by polyploids than diploid.

15.3.2 Addition Lines

Chromosomes in interspecific hybrids can be handled through addition or substitution lines with the aim of obtaining translocation and segmental recombination genotypes carrying the target genes. Monosomic (single chromosome) or disomic (one homologous pair) addition lines of the relevant alien chromosome often arise among the backcross generations, especially in crosses of closely related species with

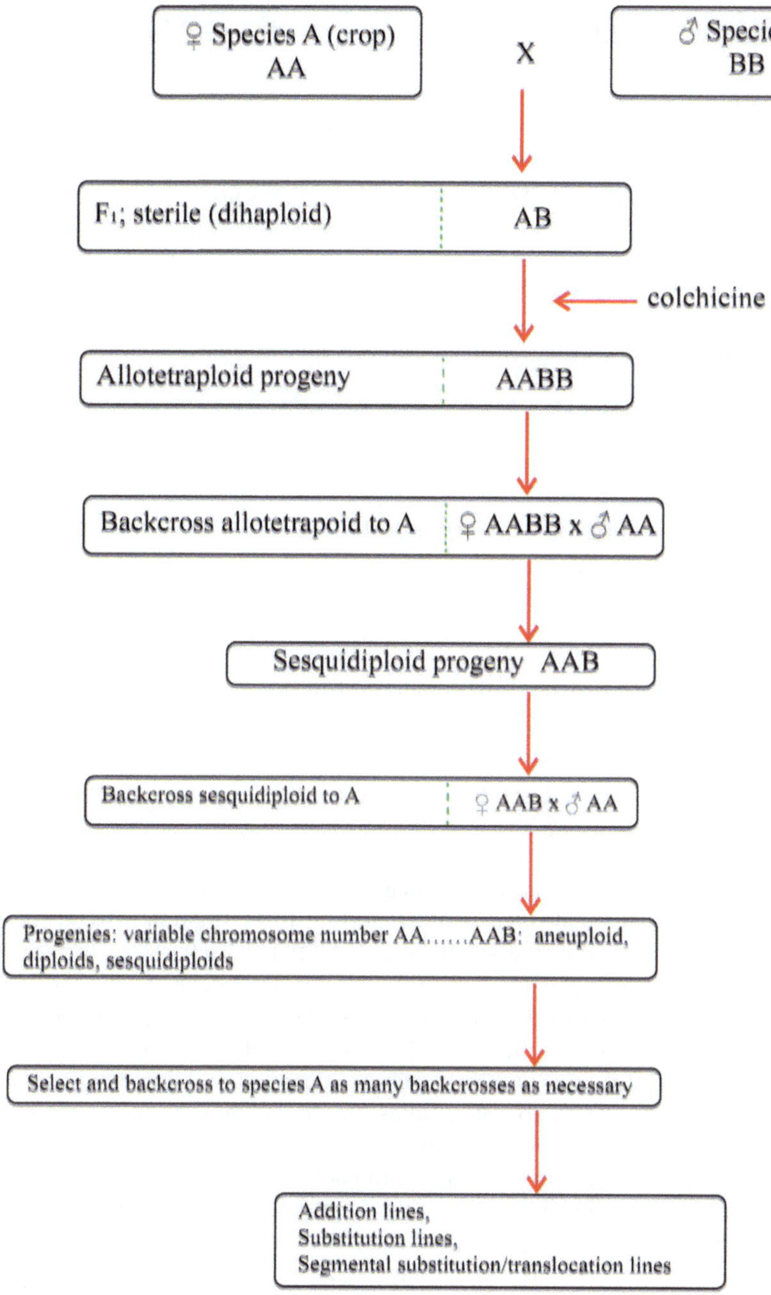

Fig. 15.1 Outline for chromosome manipulation in interspecific hybrids

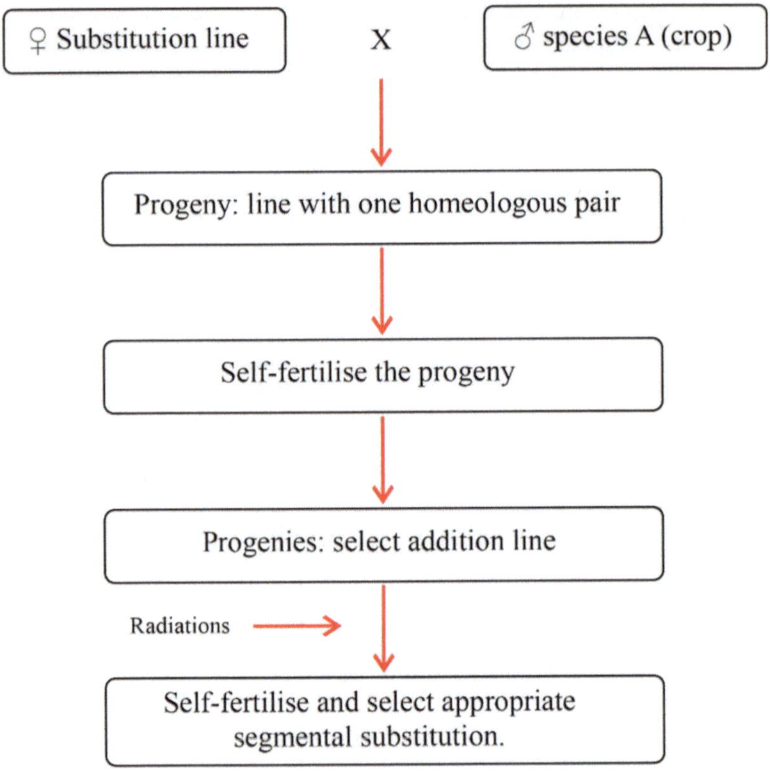

Fig. 15.2 Outline of getting segment substitution/translocation lines from substitution lines

homeologous pairing. To obtain an addition line the hybrid is crossed to the cultivated parent which is used as the male recurrent parent (Fig. 15.1). If a trait is to be transferred from species B to A, then the amphidiploid, AABB, is backcrossed to species A to obtain a sesquidiploid progeny with genome AAB. Despite the formation of A bivalents, the presence of B univalents during meiosis I makes the sesquidiploid have reduced fertility as compared to the AA, BB and the allotetraploid AABB parents. However, some fertile progenies with varied chromosome constitution ranging from AA to AAB with various addition aneuploids are also obtained. The aneuploids are mainly for the chromosomes of B genome which are additions to the normal diploid A genome constitution in the products of backcrossing. There are also a few cases of B chromosomes substituting for some A chromosomes, especially if there is high level of homeology. If it is possible to carry out karyotypic selection based on cytological meiotic configurations and chromosome banding along with phenotypic selection and molecular markers progenies having least number of B chromosomes but containing the trait to be transferred from B can be identified for selection.

Selection for the target characters in the absence of chromosome parts exchange between the chromosomes of the two species leads to isolation of monosomic or disomic addition lines in the backcross progenies having added B chromosomes to the A genome. If the cross between crop species A and species B gives the relevant B chromosome substitution line such a line is crossed as a female parent to a normal diploid A. The hybrid from the cross may be subjected to self-fertilisation. Among the progenies derived from self-pollination it is possible to obtain an addition line for the B chromosome with the target trait. Biotechnology *in vitro* methods are also applicable whereby microprotoplasts are used to transfer one or a few chromosomes from donor species to the crop species to create chromosome addition lines without involvement of sexual reproduction route.

Extra chromosomes in addition lines, especially in monosomic additions, have the disadvantage of forming micronuclei during meiosis I, chromosome instability and reduced fertility. The additional alien chromosomes are usually not stable in the cytoplasm and genome background of the crop parent and often get lost. Self-pollination of addition lines often leads to several euploid progenies arising because of the alien chromosome loss during gamete formation and after a few more generations the addition line totally disappears as the progenies become euploid. The addition lines may be more chromosomally stable if they are disomic addition lines but this causes too much genetic dosage of unwanted genes. Because the additional alien chromosome brings several genes responsible for undesirable phenotypic changes, addition lines are agronomically inferior to regular crop cultivars and are unacceptable as commercial varieties.

Since a single or a pair of B chromosomes will have too many genes from species B the addition lines of the relevant chromosome can be subjected to X-rays or other radiations for possible translocation of the alien segment carrying the target gene to one of the A chromosomes followed by selfing and selection. The progenies with desired segmental substitution or translocation having the target character, nearly all A traits and least B traits are selected for further breeding purposes to develop a desired variety.

Two types of translocations are obtainable from addition lines, intercalary and reciprocal translocation. The most desired translocation is an intercalary small segment containing the desired gene which is broken from the alien chromosome and inserted in the crop chromosome. Reciprocal translocations involve exchange of homeologous segments. Because the reciprocal translocation involves exchange of chromosome segments it amounts to the substitution of a foreign segment for crop chromosome segment and loss of some crop genes. Loss of crop traits through segmental translocation may not be desirable as homeologous segments may not necessarily replace one another sufficiently, hence intercalary translocation of a small segment of the alien chromosome bearing the target gene is the best alternative if it is obtainable. A more precise approach is genetic approach to promote crossing over through chiasmata formation between homeologous chromosome segments as known in wheat where absence of the Ph gene on 5B chromosome which suppresses homeologous pairing allows allosydentic pairing leading to precise crossing over involving homeologous segments (Thomas, 1993).

15.3.3 Substitution Lines

To produce a substitution line, the relevant alien (B) chromosome disomic addition line is crossed to the crop (A) genotype that is monosomic for the chromosome to be replaced. Among the progenies there will be the substitutions in which one A chromosome is substituted for by the B chromosome. If such monosomic substitutions are self-pollinated some disomic substitution lines in which a pair of A chromosomes is substituted for by a pair of B chromosomes may be obtained. Though the disomic substitution lines are agronomically inferior to the addition lines they have better fertility and chromosome stability than the addition lines. However, they have much lower fertility as compared to the parental cultivated crop. Like addition lines, the substitution lines are rarely commercialised because of loss of the several crop species genes and presence of too many undesirable genes from the crop relative which do not adequately compensate for the lost crop genes. Substitution lines are more tolerated in polyploids like wheat than in diploid and that the chromosomes can appropriately substitute for those to which they have homeologous relationships (Thomas, 1993).

15.3.4 Bridging Species

Chromosome manipulation using bridging species involves three different species, the crop (species A) and its two relatives, species B and C with genomes AA, BB and CC respectively. One of the crop relative species, B, is used as a bridge between the crop species, A, and the other relative species, C, which is the source of the desired gene (Fig. 15.3). The crop and the source of the desired gene are not cross-compatible while the bridging species hybridises with both the crop and the source of the gene. To transfer the gene a cross is made between the species C having the target gene and the bridging species B so as to transfer the gene to the latter and followed by backcrossing to parent B. Backcrossing is done for a few generations to reinstate the characters of B while ensuring that trait under transfer is not lost. In the cross, species B is used as the female parent. Once a satisfactory line of B having the trait under transfer has been obtained through backcrossing the derived line is crossed to the crop species A. The resulting interspecific hybrid is backcrossed to the crop parent with selection for traits of the crop and the trait being transferred. Addition or possible translocation or segmental substitution lines having the target gene is selected for further breeding. It is important that the genotype selected as having the required characteristics of the cultivar being improved and the target gene has the cytoplasm of the crop cultivar. In the backcrosses appropriate molecular markers are employed to fast-track the transfer of the target gene, eliminate linkage drag and recovery of the genome of the recurrent parents.

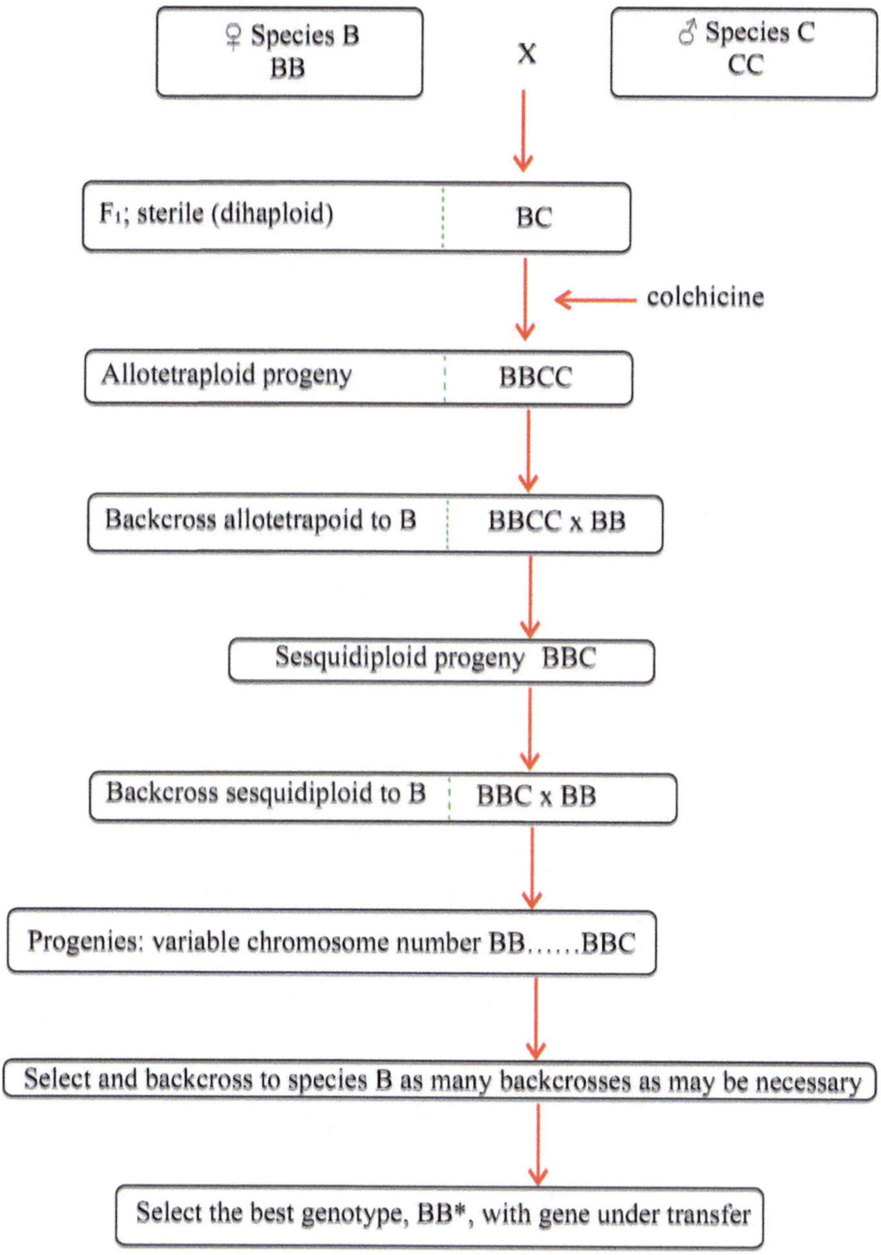

Fig. 15.3 Outline for use of bridging species in interspecific hybrids (from C to A via B)

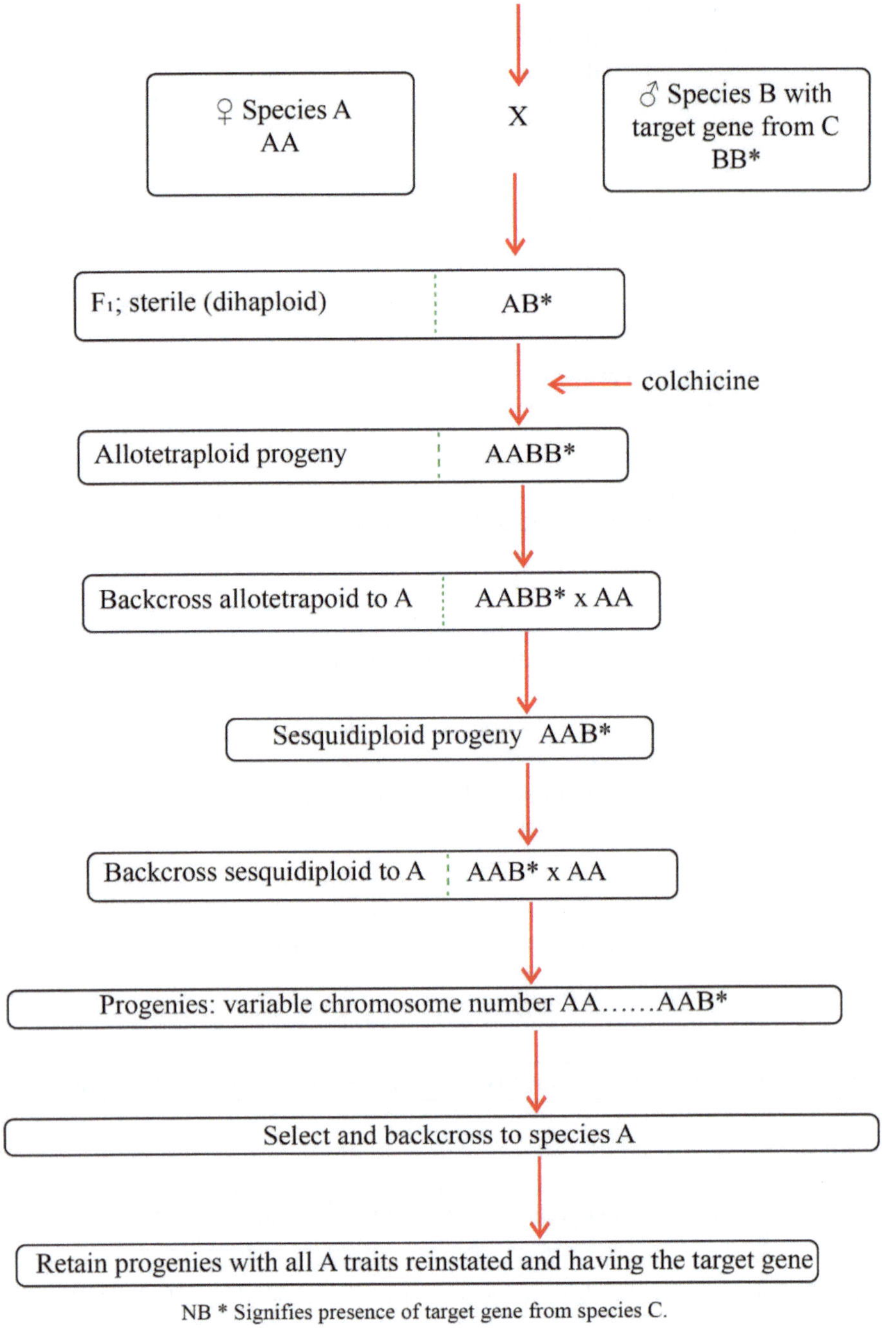

Fig. 15.3 (continued)

15.4 Cytoplasmic Conversion

Interspecific hybridisation can be used to transfer the alien cytoplasm to the crop species where the desired characters are controlled by cytoplasmic factors or if there is favourable interaction between the crop genome and the alien cytoplasm. Allospamic conversion or cytoplasmic conversion helps to broaden the cytoplasmic diversity of crop plants without having adverse effects on agronomic and quality traits. Cytoplasmic conversion is also done within species where the cytoplasm of one crop cultivar is used to replace the cytoplasm of the other one.

In addition to influence on pest or disease resistance and susceptibility as known in susceptibility to southern corn leaf blight for maize genotypes having T-cytoplasm, genes within the cytoplasm control other traits, namely, male sterility, growth and agronomic traits in several crops like groundnuts, maize, sorghum, wheat and oats. To conduct allospamic conversion a cross is made using the source of the desired cytoplasm as the female parent and in backcrosses that follow to recover the traits of the crop cultivar the recurrent crop parent is used as the pollen source. Backcrossing is done for as many generations as necessary to reinstate the traits of the recurrent parent apart from the cytoplasm which is from the donor parent. The fastest and easiest method of cytoplasmic replacement uses biotechnology where two cells from different species or cultivars are fused with one cell having its nuclear genome eliminated while the other one has its cytoplasm incapacitated, hence the new cell will have nuclear genome of species A or cultivar A and the cytoplasm of species B or cultivar B.

Problems

15.1 Which are the different gene pools for crops and their relatives?

15.2 How do various external interspecific hybridisation barriers operate?

15.3 Give any four ways of overcoming

(a) Pre-fertilisation barriers?
(b) Post-fertilisation barriers?

15.4 Why are substitution and addition lines not used as commercial varieties in crops?

15.5 Under what conditions do breeders find it necessary to undertake allospamic conversion?

References

Bhanu, A. N., Singh, M. N., & Srivastava, K. (2018). Crossability studies of interspecific hybridization among *Vigna* species. *Biomedical Journal of Scientific & Technical Research, 2*(5). https://doi.org/10.26717/BJSTR.2018.02.000818. Creative Commons Attribution International License (CC BY).

Bruznican, S., Eeckhaut, T., van Huylenbroeck, J., De Keyser, E., De Clerq, H., & Geelen, D. (2021). An asymmetric protoplast fusion and screening method for generating celeriac cybrids. *Science Report, 11*, 4553. https://doi.org/10.1038/s41598-021-83970.y. Creative Commons Attribution International License (CC BY)

Cheng, A., Xia, G., Zhi, D., & Chen, H. (2004). Intermediate fertile Triticum aestivum (+) Agropyron elongatum somatic hybrids are generated by low doses of UV irradiation. *Cell Research, 14*, 86–91. https://doi.org/10.1038/sj.cr.7290207. Springer Nature No. 5745350087525

Forsberg, J. D., Dixelius, C., Lagercantz, U., & Glimelius, K. (1998). UV dose-dependent DNA elimination in asymmetric somatic hybrids between *Brassica napus* and *Arabidopsis thaliana*. *Plant Science, 131*, 65–76. Elsevier, No. 5802440550240.

Hadley, H. H., & Openshaw, S. J. (1980). Interspecific and intergeneric hybridisation. In *Hybridization in Crop Plants*. (Fehr, W. W. R., & Hadley, H. H. (eds.). American Society of Agronomy Inc. and Crop Science Society of America, pp. 133–159. https://doi.org/10.2135/1980hybridizationofcrops.c7. John Wiley and Sons, Nos. 5701350712785, 5701360433179, 5744921244497 and 5744930151795.

Harlan, J. R. (1976). Genetic resources in wild relatives of crops. *Crop Science, 16*, 329–333. https://doi.org/10.2135/cropsci1976.0011X001600030004x. John Wiley and Sons No. 5744750383132.

Harlan, J. R., & de Wet, J. M. (1971). Towards a rational classification of cultivated plants. *Taxon, 20*, 509–517. https://doi.org/10.2307/1218252. John Wiley and Sons No.5744800480911.

Kasha, K. J., & Kao, K. N. (1970). High frequency haploid production in barley, Hordeum vulgare L. *Nature, 225*, 874–876. https://doi.org/10.1038/225874a0. Springer Nature No. 5700900372196

Katche, E., Quezada-Martinez, Katche, E. I., Vasquez-Teuber, P., & Mason, A. S. (2019). Interspecific hybridization for Brassica crop improvement. *Crop Breed, Genet. Genome*, e190007. https://doi.org/10.20900/cbgg20190007. Creative Commons Attribution International License (CC BY).

Kuligowska, K., Lutken, H., Christensen, B., Skovgaard, Ib., Linde, M., Winkelmann, T., & Muller, R. (2015). Evaluation of reproductive barriers contributes to the development of novel interspecific hybrids in the Kalanhoe genus. *BMC Plant Biology, 15*, 15. https://doi.org/10.1186/s12870-014-0394-0. Creative Commons Attribution International License (CC BY)

Pickersgill, B. (1993). Interspecific hybridisation by sexual means. In: M. D. Hayward, N. O. Bosemark, & I. Ramagosa (Eds.), *Plant Breeding: Principles and Prospects*. Springer Dordrecht, pp. 63–78. https://doi.org/10.1007/978-011-1524-7_6. Springer Nature No. 5745270696889.

Prieto, P. (2020). Chromosome manipulation for plant breeding purposes. *Agronomy, 10*, 1695. http://creativecommons.org/licences/by/4.0/. https://doi.org/10.3390/agronomy10111695. Creative Commons Attribution International License (CC BY).

Thomas (1993). Chromosome manipulation and polyploidy. In: M. D. Hayward, N. O. Bosemark, & I. Ramagosa (Eds.), *Plant Breeding: Principles and Prospects*. Springer Dordrecht, pp. 63–78. https://doi.org/10.1007/978-94-001-1524-7_7. Springer Nature No. 5746360205585.

Wijerathna-Yapa, A., Ramtekey, V., Ranawaka, B., & Basnet, B. R. (2022). Application of in vitro tissue culture technologies in breeding and genetic improvement of wheat. *Plants, 11*(17), 2273. https://doi.org/10.3390/plants11172273. Creative Commons Attribution International License (CC BY)

Chapter 16
Biotechnology for Crop Improvement

16.1 Tissue Culture

In addition to conventional breeding, interspecific hybridisation and mutation breeding methods there are strategies for creating genetic changes in plants without pollination and fertilisation but involve genetic manipulation at the cellular and molecular levels. Instead of dealing with the whole plant the breeder uses in vitro cultures for tissues like callus, meristems, anthers, embryo, cells or protoplasts to create genetic variation and conduct selection at such levels (Fig. 16.1). The tissue cultures are also used for propagation, application of artificial mutagenesis, recombinant DNA technology and genetic modification.

The idea of plant tissue culture was first presented in 1902 by German botanist Gottlied Haberlandt who cultured leaf palisade cells in knob's salt solution with sucrose. Plant tissue culture research was thence conducted in the early decades of twentieth century and by 1950s it was realised that plantlets and whole plants could be regenerated from tissues or callus cultures by altering concentrations of growth hormones like auxins and cytokinin in the culture media. In 1960s plant biotechnology had a major step forward when single cell totipotency, i.e., ability of a single cell to develop into a whole plant, was demonstrated in tobacco and carrot. Plant cells can give rise to any undifferentiated cell type of the mature plant and differentiated cells can also change to embryonic state and re-differentiate to new cell types. The discovery of enzymes that digest polysaccharides like cellulose, making protoplast culture possible, and the demonstration that embryos could be formed directly from anthers were also major developments in plant biotechnology in 1960s. In 1970s it was further discovered that plant cell cultures could be subjected to microbial genetical manipulation and that protoplasts from two different sources could be fused to form somatic hybrids.

Developments in plant tissue culture and use of modern biotechnology methods have helped overcome some limitations of conventional plant breeding methods which include the use of whole plants as selection units that makes it difficult to

P. O. Ayiecho and J. O. Nyabundi, *Conventional and Contemporary Practices of Plant Breeding*, https://doi.org/10.1007/978-3-031-74998-8_16

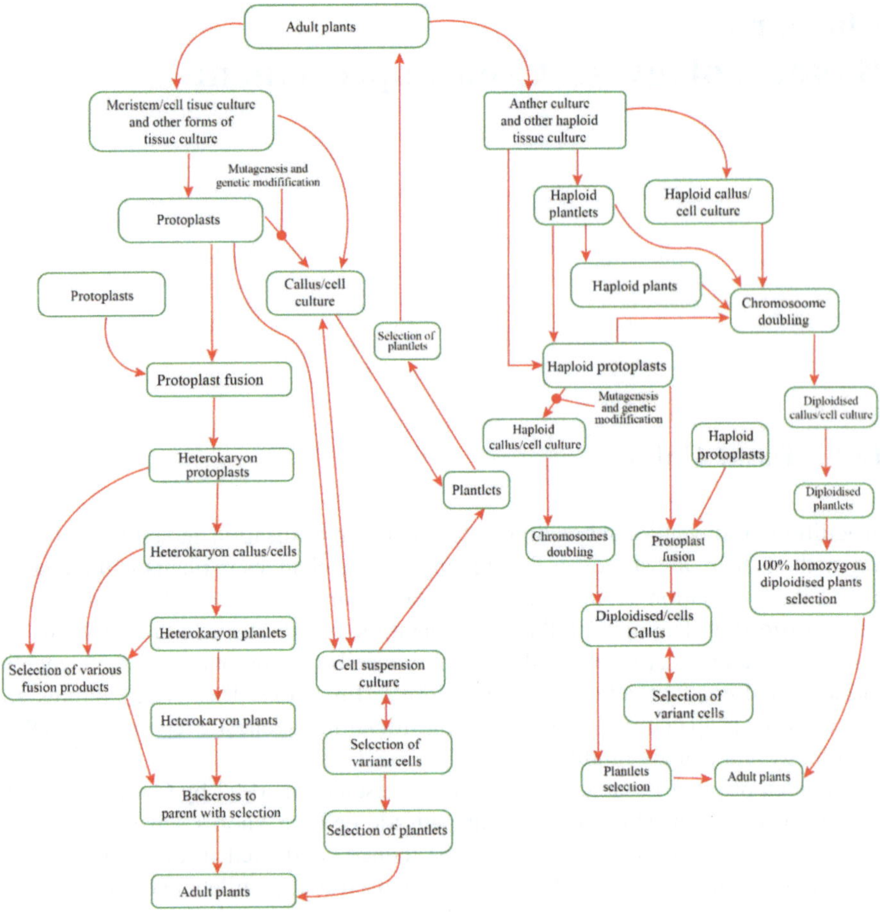

Fig. 16.1 Some in vitro methods for crop improvement

visualise genetic variability, low seed production in some plant species, difficulty or failure of some plant species to flower and long generation periods for some plant species hindering breeding through sexual approach. There are challenges of long periods needed to develop improved crop varieties by conventional methods, inefficiency of selection in diploid or polyploids due to masking of recessive gene expression in heterozygotes, inbreeding depression, costs and periods involved in inbred line production and maintenance and barriers to interspecific hybridisation. Use of conventional methods to transfer a gene or a few genes from a donor cultivar to an established variety of an annual crop takes at least five years since the process entails making a cross followed by several generations of backcrossing and selection to reinstate the characters of the recurrent parent and eliminate undesirable traits of the donor cultivar. Application of plant tissue culture aids in fast production of mature plants, mass propagation of desired cultivars, regeneration of genetically modified plants, production of disease-free propagules, elimination of plant viral infection and conservation of germplasm in vegetatively propagated plants and micropropagation of superior quality propagules (Bednarek & Orlowska, 2020).

16.1.1 Regeneration in Tissue Culture

Explants can be obtained from plant organs like cotyledons, leaves, petioles, hypocotyls, stems, roots, anthers, embryos, meristems, etc., for culturing in artificial media and the various tissues that are often cultured include embryo culture, meristem culture, callus culture, cell suspension culture, anther culture, protoplast and culture bud culture. Most plant tissue culture media components are vitamins (e.g., thiamine, pyridoxin, nicotinic acid, etc.), amino acids like arginine, asparagine etc., mineral micronutrients (manganese, iron, molybdenum, zinc, copper and boron), mineral macronutrients (nitrogen, sulphur, phosphorous, potassium, magnesium and calcium), carbon source (lactose, maltose, galactose, raffinose and cellobiose), growth hormones (auxins, cytokinin, abscisic acid, gibberellins, naphthalene acetic acid, indoleacetic acid and zeatin) and antibiotics (streptomycin, kanamycin, etc.) whose proportions are altered depending on species, the tissue being cultured and purpose. The culture media may be liquid or solid with pH adjustment usually between 5.4 and 5.8. Commonly used commercial culture media formulations are Murashige and Skoog (MS) medium, Nitsch medium, Gamborg medium, White's, Linsmaier and Skoog (LS) medium, Woody Plant (WPM) medium, DRC medium and Schenk Hilderbrandt (SH) medium.

A number of factors which influence the performance and successful regeneration of cultured tissues into plants include genotype of plant used as source for culture material, growth conditions and history for plant used as tissue source, type and part of organ from which the explant is obtained. Though plant tissues and cells can easily be induced to regenerate by wounding, the tissues from different genotypes and species differ in their response to wounding (Nic-Can et al., 2015). Other factors influencing tissue culture performance are developmental stage, health and age of the tissue, age of the culture, cytological and biochemical changes, composition and physical conditions of culture medium, phytohormones present, sub-culturing manipulation and culture environment in terms of pH, gaseous environment, osmotic pressure, temperature and lighting conditions (Bidabadi & Jain, 2020). Immature embryos, mature seeds, anthers, young leaves and roots usually perform well in regeneration.

When tissues are excised from plants, sterilised using bactericides and fungicides to remove contaminants and placed in appropriate culture medium the cells in the tissues undergo cell division, differentiate and grow to form a callus which form plantlets when transferred to a fresh medium having appropriate growth hormones like kinetin. Transfer to fresh culture, adjustment of nutrients and growth hormones is done as required for development of shoots and other organs. Plant regeneration from tissue culture can take place through somatic embryogenesis or shoot-root morphogenesis referred to as organogenesis (Bidabadi & Jain, 2020). Regeneration through somatic embryogenesis occurs when a cell or a group of cells from somatic tissue form embryos which develop into plantlets and then into whole plants as determined by origin of explant, culture medium, and in vitro culture conditions (Nic-Can et al., 2015). Somatic embryogenesis can take place directly from cultured tissue without callus formation or it occurs indirectly through callus proliferation that develops into embryoids or pro-embryos forming globular and heart-like structures, regenerating into embryo, then into plantlets and finally into whole plants (Nic-Can et al., 2015; Bidabadi & Jain, 2020). Somatic embryos can be used for germplasm conservation using cryopreservation technique and artificial seed production for commercial production and genetic transformation. In organogenesis cells form an aggregation into a callus from which organ primordia arise followed by root and shoot development and ultimately giving a plant. Organogenesis takes place by development of roots, shoots and leaves directly on explant tissue or on callus and may be induced by manipulation of plant hormone content in the culture medium. Manipulation of auxin and cytokinin levels in the medium determines the formation and differentiation of roots, shoots or embryos from explants. Low auxin to cytokinin ratio leads to shoot formation while high auxin to cytokinin ratio promotes root formation. Balance between auxins and cytokinin leads to callus growth with the morphology being influenced by plant species and medium composition. Gibberellins also help to enhance cell growth and elongation.

16.1.2 Meristem Culture

In meristem culture meristem tissue is excised from the plant, sterilised and placed in artificial culture medium for proliferation. Meristems may be derived from shoot tips, axillary and adventitious buds and even from tips of primary and lateral roots. Proliferating meristems are transferred to fresh medium and regenerated into plantlets, and the plantlets are subjected to other rounds of culture media as may be required or transferred directly onto soil for regeneration into whole plants. Meristem culture helps in multiplication of single plants into large populations fast and the method can be used to propagate species which have very low seed yields and those having long and infinite juvenile periods without flowering or species which produce no seed as in triploid bananas, pineapple and some aneuploids. Inbred lines can also be maintained using meristem culture without sexual inbreeding to avoid excessive inbreeding depression and in some species this method can be used to multiply superior hybrids and other types of superior varieties indefinitely. Meristem culture is used to obtain disease free plants, especially in case of virus diseases.

16.1.3 Anther Culture

Guha and Maheshwari (1966), while studying meiosis process in *Datura innoxia,* discovered that cultured anthers formed embryo-like structures which arose individually from pollen grains. It was later realised that such structures could be diploidised and cultured into plantlets for regeneration to give to give 100% homozygous whole plants. Since Guha and Maheshwari discovered that gametocytes could be used to generate sporophytes in vitro through anther culture the method has been used to generate plants in several plant species including tobacco, tomato, egg plant, wheat, barley, rice, maize, triticale, Brassicas, rubber, tobacco, cotton, soybean, pepper, sorghum, sugarcane, sugar beet, soybean and many others.

Anther culture which is applied to generate sporophytes in vitro from gametocytes involves culturing of pollen grains or anther containing microspores obtained from closed flower buds using appropriate culture media. In this culture method anthers are excised from closed flowers, sterilised and placed on the medium for proliferation of the microspores into haploid plantlets. Regeneration of the microspores into plantlets can take direct androgenesis route or indirect androgenesis path. In direct androgenesis the plantlets are regenerated from plated anthers directly as the anthers undergo embryogenesis by developing into pro-embryos that regenerate into plantlets and then into whole plants while indirect androgenesis takes the route of callus production which develops into plantlets and ultimately into whole plants (Papenfuss & Carman, 1987). In either case the plantlets are transferred to fresh medium as required and plantlets are ultimately transferred from the culture medium to a soil growth medium for development into whole plants. Haploids can also be obtained through in vitro culture of unpollinated ovaries and ovules.

If the haploid plantlets are treated with colchicine fertile doubled haploid plants which are 100% homozygous are obtained, thereby giving stable diploid 100% pure line plants in a single generation and reducing number of generations required to obtain homozygous lines through conventional breeding. With conventional breeding it takes up to 6 generations to achieve about 98.44% homozygosity through self-pollination. Homozygous lines derived from different cultivars through anther culture can be crossed to give superior hybrid cultivars combining different traits. In the diploidised plants there is no heterozygosity and therefore no genetic masking of recessive alleles through dominance, making it easy to identify dominant and recessive genes controlling various traits and also makes phenotypic selection for both qualitative and quantitative becomes more effective in such plants. Anther culture can be used in mutation breeding, genetic modification and selection in vitro by subjecting microspores or the haploid plantlets to induced mutagenesis by use of mutagens like UV-radiations and thereafter the plantlets are diploidised. Selection for the desired genetic changes is easily done on the mutant plantlets in vitro or on the regenerated plants since recessive traits will be clearly expressed in doubled haploids. Haploids derived from anther culture can be used to obtain monosomics, nullisomics and also used in protoplast fusion to generate diploid, triploids and other forms of polyploids. Anther culture can also be used to mobilise alien genetic variation into cultivated crops.

16.1.4 Callus Culture

Callus culture refers to any tissue of a mass of unorganised dedifferentiated cells exhibiting unrestrained growth in vitro. Cell dedifferentiation involves reversing of some changes that occur during development and specialisation of cells and tissues with respect to morphology and metabolism. Callus growth is induced from explants of meristems, somatic tissues, microspores, macrospore mother cells, embryos, single plant cells, cell suspension and protoplast cultures when such tissues are sterilised and placed on an appropriate agar solidified or liquid medium with required levels of relevant nutrients and phytohormones. Callus cultures are maintained by periodical transfer to fresh media as required. Callus culture can be made of densely aggregated cells or friable with loosely associated cells. Callus tissues exhibit morphological and physiological variability and also differ with respect to ability to withstand various stress factors, growth rate and regeneration capacity.

Plantlet formation and plant regeneration from various forms of in vitro tissue cultures often pass through callus formation. Regeneration is induced by manipulation of auxins and cytokinin. Callus culture can be used for cloning genotypes in short periods, molecular and cell-based selection for resistance to various stress factors like disease, herbicides, temperature, salt and mineral resistance by introducing selection agents into the culture medium

16.1.5 Embryo Culture

Embryo culture is applied mainly in situations where the embryo is rescued and cultured to save it from abortion and is particularly useful where there is incompatibility between the embryo and the maternal tissue interrupting embryo development prematurely within the developing seed. The developing immature embryo is excised from the developing seed before it aborts and transferred to an appropriate culture medium. The embryo is transferred to fresh culture medium periodically and as frequently as required with the necessary adjustment in the medium composition depending on the age and developmental stage of the embryo until it develops into a viable seedling. The seedling may develop directly from the embryo or through formation of callus followed by root and shoot development.

Embryo culture is employed to break post-zygotic hybridisation barriers between distantly related species so as to allow successful interspecific hybridisation for gene transfer. In vitro culture of excised embryos makes it possible to salvage and grow interspecific hybrids which would otherwise abort in embryo stage if not rescued. Embryo culture makes it possible to undertake conservation and propagation of forest species and rare and endangered species. Embryo culture is applied in speed breeding and other rapid generation advancement methods. Embryo culture is also useful in situations of seed dormancy.

16.1.6 Cell Suspension and Cell Cultures

When mechanical agitation is applied to friable callus culture in liquid medium the loosely associated cells are dispersed giving cell suspension cultures which are usually comprised of friable cultures with basic growth units of up to 50 cells. The cultures are transferred to fresh liquid medium as frequently as required for efficient growth and eliminate build-up of toxic by-products or exhaustion of nutrients in the medium. Since suspension culture can have a large population of cells of up to 10^8 cells per 100 ml of liquid medium it helps to increase the number of selection units that can be handled and makes it possible to apply the tools of molecular genetics to higher plants as selection can be done on traits expressed by individual cells. Suspension cultures are cultured into calli and then regenerated into plantlets and plants.

Cell suspension culture and cell culture make it possible to treat higher plants like microorganisms since selection is conducted at the cell level for traits like resistance to biotic and abiotic stress conditions and improved nutritional quality. Mutations can also be induced in the cells by treating cell suspension and cell cultures with mutagens like gamma and UV radiations, nitrous acid and EMS. Induction of mutations is followed by selection in which the cell cultures are subjected to selective stress environments in the medium where the cells would not survive unless they had the relevant mutational changes. The cells are cultured in selective media having selective agents like pathotoxins (fungal toxins), specific herbicides like atrazine, pollutants, undesired specific mineral concentrations and saline conditions, undesirable temperature levels, etc. This helps to eliminate the cells lacking resistance genes while the variant cells having resistance genes will survive. Cells may also be selected for amino acid production, particularly the essential amino acids like methionine, lysine, valine, tryptophan, etc., by introducing into the culture the respective amino acid analogues that would stop the synthesis of particular amino acids in the cell culture. This helps to select the cells that continue to produce the respective amino acid despite the negative feedback mechanism for improved protein quality. The presence of the amino acid analogue starves most of the cells to death due to lack of the respective amino acid while any surviving cells are mutants which do not respond to the negative feedback mechanism attributable to the analogue and overproduce the target amino acid.The variant cells that survive such selection processes are presumed to have acquired genetic changes that make them resist or tolerate the respective stress conditions and are selected for regeneration into plantlets and then into whole plants. Cells in culture are also useful in transgenesis since target alien DNA can be transferred into the genomes of the cells using various DNA transfer methods.

16.1.7 Protoplast Culture and Protoplast Fusion

The application of protoplasts in plant manipulation came from discoveries of the use of cellulase to obtain protoplasts from plant cells (Cocking, 1960), regeneration of plants from protoplasts (Takebe et al., 1971) and use of polyethylene glycol (PEG) for fusing protoplasts (Kao & Michayluk, 1974). Protoplasts are obtained from various plant tissues and organs like leaves, roots, tubers, flower parts, fruits, root nodules, microspores, tumours, callus culture, cell suspension and cell cultures. The protoplasm with the plasmalemma is isolated from the cell wall by subjecting the cells to plasmolysis and appropriate enzyme mixtures having cellulase, hemicellulase and pectinase to degrade the cell wall by digesting cellulose, hemicellulose and pectin leaving the protoplast intact. The isolated protoplasts are purified, washed to remove the enzymes and other chemicals used in the isolation and transferred to appropriate culture medium that may take the form of liquid agar, droplet, and bead cultures which have to be changed periodically as required. Culture media used for protoplast culture and other tissue cultures like cell culture, anther culture, etc. are usually similar though the one for protoplast culture may have mannitol or sorbitol added so as to provide the needed osmotic potential. Once the cell wall has formed the protoplasts in culture are regenerated into plantlets and plants either through organogenesis or embryogenesis depending on the species and culture conditions. Somaclonal variation or variation obtained through induced mutagenesis in protoplast culture is exploited to identify variant cells with desired traits for selection at the cellular level.

Though the idea of protoplast fusion was proposed as early as 1909 by Kuster it was not until 1960s and 1970s that it was done practically for animals and plants respectively. The cultured protoplasm can be fused easily with another protoplasm, leading to somatic cell hybridisation without the necessity of going through the sexual cycle. It is possible to do protoplast fusion between protoplasts from different sources, even from different species followed by culturing of the derived hybrids in appropriate medium for formation of the cell walls, mitotic cell division, regeneration into callus, plantlets and plants.

Since protoplasts do not fuse automatically because they are negatively charged and repel each other they are induced to fuse by neutralising the plasmalemma negative charge using Ca^{2+} in a solution of high pH of about 9.0–10.5 containing polyethylene glycol or electrofusion which brings protoplasts together in an electric field of brief high voltage pulse (Kao & Michayluk, 1974; Melchers & Labib, 1974; Bates & Hasenkampf, 1985). Polyethylene glycol facilitates the fusion of protoplasts by agglutinating and fusing the protoplasts, altering the cell membrane properties under high pH of 9.0–10.5 and high Ca^{2+} concentration in buffer solution.

Somatic cell protoplast fusion makes it possible to mix two genomes and cytoplasm from different parents into one heterokaryon cell creating a somatic hybrid. The heterokaryons can be isolated from the protoplast mixture, which also contains homokaryons (unfused single protoplasts) and multiple cell fusion products, by help of markers in either of the parents where one marker is in one parent and another marker is in the other parent, e.g., genes for antibiotic resistance, amino acid analogues and herbicide resistance. If parents have different genes for some of these traits the fusion products should have all the genes from both parents. Such recombinants are selected by adding the selective agents such as antibiotics, amino acid analogues, herbicides, etc. to the culture medium. Fusion products having all the relevant genes from the two protoplasts survive in the selective medium due to genetic complementation while the other cells which do not have the needed combination of genes are killed. While isozyme and DNA markers can also be used to identify the hybrids, chromosome number determination helps to identify heterokaryons as well since such a hybrid should have chromosomes from the two parental protoplasts. In regenerated plants the hybrids will exhibit some morphological characters from either of the parents or have intermediate features.

Somatic hybridisation is useful for circumventing interspecific sexual incompatibilities in difficult to cross species. Some interspecific somatic hybrids are difficult to handle unless they have some level of fertility since the breeder still has to manipulate the variability generated by fusion using conventional breeding methods. Heterokaryons produced by protoplast fusion may behave as normal somatic tissues if the two nuclei fuse during mitosis resulting in some chromosome stability within the hybrid cell with unreduced chromosome compliment from either of the genomes (Power et al., 1976). Mixing of cytoplasm from two sources creates a diverse cytoplasmic base and opportunities for extra-chromosome manipulation so that favourable cytoplasmic hybrids are taken advantage of in breeding. Somaclonal variability arising from fused protoplast cultures also give opportunities for selection of desirable variants. While it is possible to use somatic hybrids as bridges for distantly related species, interspecific gene transfer using heterokaryons can be done by radiating one species' protoplast and then fusing the radiated protoplast with the recipient species protoplast (Pelletier, 1993) followed by in vitro selection for the hybrid cells having appropriate chromosome fragment translocation using selective media and culturing them into plantlets and plants. Through asymmetric chromosome elimination where more chromosomes of one parent are lost than the other parent chromosomes creating aneuploids, it is possible to have cells that have only the required chromosome added which can be regenerated into plants for backcrossing to the parent whose genome is retained to help transfer genes across species. Apart from being used in bringing diverse genomes together through protoplast fusion protoplast culture is extremely useful for genetic transformation since protoplasts take up DNA making it easy to insert the target genes directly into the recipient crop genome.

While mitosis can take place in somatic hybrids between distantly related species regeneration may be difficult due to problems associated with chromosome structural and numerical changes which arise during culturing of the interspecific protoplast hybrid cells. In some diversely related cases the nuclei may fail to fuse and integrate due to genetic incompatibility. Though protoplast fusion breaks the hybridisation barriers between different species in cases of extreme divergence the heterokaryon cell may not regenerate into callus and plantlets or the chromosomes of one or both parents may be lost. There is great variability in somatic hybrids due to genetic instability, mitotic recombination, cytoplasmic segregation, chromosome elimination resulting in sterile aneuploid somatic hybrids and endo-mitosis, leading to the fusion products having ploidy levels higher than tetraploidy (Power et al., 1976). Most somatic hybrids have problems of sterility making it difficult to cross them to either of the parents. Chromosome substitution or additions lines obtained through asymmetric chromosome changes in the interspecific heterokaryons are unstable, have problems of loss of a lot of genes and too many alien genes added. If such lines are regenerated into plants, they have problems of reduced yield, lack of acceptability, poor chromosome stability and low fertility. It may be necessary to induce mitotic recombination or use radiations on protoplasts to create chromosome breakages that lead to intergenomic chromosome segment transfer.

It is possible to get cybrids which are cells with one nuclear genome and a mixture of cytoplasm from two sources by incapacitating the nucleus of one cell and enforcing the fusion of the two cells, that is, fusing one cytoplasm with protoplasm from another source. Fusion of cytoplasm from different sources, known as allospamic hybridisation or cybridisation, is used to fuse cytoplasm and protoplasm of related or unrelated species. This is achieved by using X-rays to inactivate nuclear genome of one parent before protoplast fusion with another parent (Evans & Flick, 1983). In some cases, the somatic hybrids undergo mitosis with segregation and total loss of chromosomes of one genome, leading to cybrids. Allospamic hybrids are useful where certain cytoplasmic genes like male sterility, disease and pest resistance factors desired for a commercial species are found in another species which does not cross easily with the commercial species. Allospamic hybrids can be regenerated into callus, plantlets and whole plants.

The process of protoplast fusion can be modified to achieve allospamic conversion or cytoplasmic conversion where the cytoplasm from one parent and the genome from another parent are brought together in one cell. This is done by inactivating the cytoplasm in the protoplasts of one parent and at the same time inactivating the nucleus of protoplasts of the other parent and followed by fusing the two products. Chemicals like iodoacetate or iodoacetamide and irradiation can be used to eliminate cytoplasmic organelles of one parent while the nucleus of the other parent is also subjected to irradiation followed by fusion of the two so that one parent provides active cytoplasm while the other parent provides active nucleus (Pelletier, 1993).

16.2 Somaclonal Variation

16.2.1 Occurrence of Somaclonal Variation

When using callus culture, cell culture or other forms of in vitro tissue culture for cloning or micropropagation it is expected that the regenerated plants are genetically and phenotypically identical to the plants from which the tissues are derived. However, cells in culture, plantlets and plants regenerated from callus, cell suspension, anther, protoplast cultures and other forms of in vitro cultures have a wide range of spontaneous genetic, cytoplasmic organelle and epigenetic changes which arise due to culturing and regeneration in abnormal and stressful environment attributable to in vitro culture conditions, medium composition, sub-culturing, level of growth hormones, origin of explant and genotype of explant, state of growth medium and duration of culturing leading to variation referred to as somaclonal variation (Larkin & Scowcroft, 1981). Variation among clones derived from the cultured tissues is attributable to point genetic changes, aberration in chromosome number and structure leading to varied polyploidy levels, aneuploidy, chromosome deletions, translocations and duplications in the cultured tissue cells and passed on to the regenerated plantlets and plants (Krishna et al., 2016). Somaclonal variation may be protoclonal variation (among regenerants or cells from protoplast cultures), calliclonal variation (among regenerants from callus cultures) and gametoclonal variation (regenerants from gametic cultures) which may be androclonal variation (among regenerants of anther and pollen culture) or gynoclonal variation (among regenerants of ovary or ovule culture). Apart from the spontaneous changes attributable to the abnormal and stressful culture conditions, pre-existing variability in the parental explant source also contributes towards somaclonal variation (Bednarek & Orlowska, 2020).

Under tissue culture conditions chromosomes undergo endo-reduplication during interphase of the cell cycle leading to changes in ploidy levels attributable mainly to failure of spindle formation or other forms of spindle abnormalities during mitosis associated with endopolyploidy and even polyteny in some cases (Leva et al., 2012). The occurrence of mixaploidy (presence of all ploidy levels like haploids, diploid, triploids, aneuploids, etc. together in a tissue or an individual), aneusomaty (presence of euploid and various aneuploid cells in an individual) along with chromosome fragments, deletions, translocations, inversions, acentric, telocentric, dicentric and multicentric chromosome pieces, ring chromosomes and micronuclei during mitosis are common features of cultured tissues and regenerated plantlets and plants (Larkin & Scowcroft, 1983; Krishna et al., 2016). Aberration in chromosome number and structure are responsible for chromosome loss and abnormalities in cell division and presence of multivalents during meiosis I in regenerated plants. Point genetic changes arising in tissue culture are often associated with gene amplification and de-amplification, DNA methylation, DNA base sequence changes, mitotic recombination, activation of transposable elements, silencing of some genes, reactivation of silent genes in multiple families, histone modification due to epigenetic changes, and cytoplasmic alterations (Sato et al., 2011). DNA sequences may undergo amplification or de-amplification while organelle mutations like mitochondrial DNA (mtDNA) and chloroplast DNA rearrangements are also detectable in regenerants and cells in culture in form of phenotypic variations. Mitotic recombination (somatic crossing over) arises due to rare synapsis between chromosomes in somatic tissue cells. Some changes taking place in tissue culture are not genetic or chromosomal but are due to physiological response of the cells to the culture environment and epigenetic events reflecting abnormal and altered levels of gene expression. Epigenetic changes are expressed phenotypically in the cultured tissues, regenerated plantlets and plants and even in their progenies. They are non-heritable, reversible and are reflections of altered gene expression like inactivation of some genes and physiological responses.

Point genetic mutations may comprise of dominant, recessive or codominant Mendelian genes with recessive genes remaining unexpressed and unnoticed where there is heterozygosity but they show up in homozygous state following segregation if C_0 plants are self-pollinated to produce C_1 progenies and C_1 plants are in turn self-pollinated to produce C_2. Plants regenerated from cell culture irrespective of tissue are referred to here as C_0 (primary somaclones) and progeny arising from selfing of these and subsequent generations are respectively referred to as C_1, C_2, C_3, C_4, C_5, etc. Polygenic traits are also affected by changes responsible for somaclonal variation.

Genetic variability among somatic hybrids is common and is comparable to variability among plants regenerated from other tissue cultures (Larkin & Scowcroft, 1981; Evans & Flick, 1983). They display increased phenotypic variability and several useful traits can be obtained from such hybrids as in sexual hybrids. Causes of variation among the somatic hybrids are nuclear incompatibility and instability, mitotic recombination, cytoplasmic and organelle segregation and elimination, genetic point changes, chromosome number changes, chromosome rearrangements and chromosome segment exchanges (Evans & Flick, 1983). Nuclear incompatibility often arises if distantly related species protoplasts are fused leading to chromosome elimination and aneuploidy in somatic hybrids and cybrids. Mitotic recombination or somatic crossing over arises due to rare synapsis between chromosomes in somatic tissue cells, resulting in reciprocal recombinant daughter cells with translocations and mixing of genetic information, hence intergenomic mitotic recombination. In the somatic hybrids such exchange may take place between chromosomes of the different genomes. Cytoplasmic segregation in somatic hybrids takes place such that genetic information in one of the cytoplasm may be lost leading to organelle segregation. Organelle segregation accompanied by somatic recombination causes a wide range of nuclear-cytoplasmic genetic combinations and variation among somatic hybrids not achievable with sexual hybrids. Organelle segregation happens for chloroplasts, mitochondria and other cytoplasmic controlled traits like male sterility.

16.2.2 Factors Influencing Somaclonal Variation

The extent of somaclonal variation is determined by a number of factors, including type of culture medium, culture conditions, medium composition, timing and frequency of sub-culturing, level of growth hormones in culture medium, age of the culture, duration of culturing and pre-existing variation and explant origin, i.e., plant genotype from which the tissue is derived and the kind of tissue (Leva et al., 2012). It is possible to get higher somaclonal variation from certain particular tissues but less from other tissues of the same individual plant. Explants from differentiated tissues like stems, leaves, roots, give more variation than explants from meristems like auxiliary buds and shoot tips. Increased number of sub-culturing and prolonged culturing increase the extent of variation because long culturing duration makes the tissues not to be cytokinin requiring, a phenomenon referred to as cytokinin habituation (Meins, 1982). The extent of occurrence of aneuploidy increases with the increasing age of culture and also depends on the organ from which the explant is obtained. Genotypic differences regarding the extent of somaclonal variation are explainable by the fact that different genotypes vary in their responses to stressful culture medium conditions and that genetic stability is linked to genetic make-up of the parent genotype (Leva et al., 2012). The influence of the type of culture medium on somaclonal variation is related to sugar and hormonal contents of the medium since auxins in cultures of unorganised cells lead to increased genetic variation due to DNA methylation and unbalanced concentrations of auxins and cytokinin is known to induce ploidy level changes and DNA methylation (Israel et al., 1995; Leva et al., 2012). Point genetic changes arising in tissue culture are often associated with gene amplification and de-amplification, DNA methylation, DNA sequence changes, mitotic recombination, activation of transposable elements, reactivation of silent genes in multiple families and cytoplasmic alterations (Bednarek & Orlowska, 2020). Copies of particular DNA sequences may vary due to amplification or de-amplification. Organelle mutations like mitochondrial DNA (mtDNA) and chloroplast DNA rearrangements are also detectable in regenerants and cells in culture some of which are responsible for albino variants. Reduced levels of growth regulators lower the extent of variation. The number of subculturing and their durations also elevate the extent of somaclonal variation in cell suspension and callus cultures (Bairu et al., 2006). Tissues under culture suffer from oxidative stress which raises the levels of reactive oxygen species (superoxide, hydrogen peroxide, hydroxyl radical) inflicting damage to the cells in culture through interference with DNA integrity.

16.2.3 Exploitation of Somaclonal Variations

The broad spectrum of somaclonal variation affecting quantitative and qualitative characters controlled by nuclear genes and traits influenced by cytoplasmic factors is often stable and exploitable by selection for crop improvement with respect to resistance to biotic and abiotic stress factors, agronomic and quality traits. Variation derived through tissue culture can be detected and confirmed by morphological and physiological traits, resistance to various stress factors, cytogenetic/karyological analysis, biochemical analysis and use of molecular markers like RAPDs, RFLPs, AFLPs, simple sequence repeats (SSRs), ISSRs, methylation sensitive amplified polymorphism (MSAP), etc. Somaclonal and gametoclonal variation relate to a variety of changes of agronomic importance in crop plants and may include pigmentation, yield, vigour and size of plant, leaf and flower morphology, essential oils, fruit size, fruit solids, fatty acids, amino acids, proteins, disease and pest resistance, tolerance to abiotic stress factors changes, etc. Such variation has been observed in most crops, e.g., wheat, tomato, rice, maize, banana, sugarcane, oats, etc. Induced mutagenesis can be combined with somaclonal variation to increase the range of selectable variation. In gametoclonal variation both recessive and dominant mutants are expressed directly in haploids and doubled haploids. The doubled haploids regenerated from anther and pollen cultures are used as 100% homozygous lines for development of new pure line varieties following selection. Somaclone selection in vitro using selective media on tissue cultures and regenerated plantlets or selection of regenerated plants has led to development of lines that are resistant to herbicides, pathotoxins, diseases and pests and abiotic stresses in various crops like potato, maize, banana, rapeseed, tomato, tobacco, sugarcane and other crops. Protein amino acid composition can also be selected for by use of amino acid analogues which limit production of certain amino acids for improved efficiency of amino acid production leading to recovery of cell lines efficient in production of the desired amino acids.

The best strategy to exploit somaclonal variation to improve the agronomic performance of a particular variety is to select the variety with favourable qualities and but lacking in one or a few attributes and use its tissues in culture to generate somaclonal variation and focus on the improvement of the traits it is lacking. Selected genetically stable regenerants can be backcrossed to the parental cultivar where necessary or can be evaluated as potential varieties. Somaclonal variation also has an advantage in that unlike interspecific hybridisation and interspecific protoplast fusion it doesn't bring in alien genomes because the variation, whether undesirable, neutral or desirable, arises from the genome of the cultivar targeted for improvement. As compared to induced mutagenesis, somaclonal mutations are mostly solid with very low amounts of chimerism, especially in vegetatively propagated crops. Somaclonal variation can also be applied to handle interspecific hybrids and protoplast fusion involving diverse populations or different species. Subjecting tissues from products of interspecific hybridisations and interspecific protoplast fusion to in vitro culture can help in the recovery of useful somaclonal variants with various chromosomal (whole or segmental) substitutions or additions.

16.2.4 Challenges Associated with Somaclonal Variations

Variations arising in tissue culture are a handicap since they are random and undesired where clonal uniformity is required like in vitro micropropagation of crop cultivars, inbred lines, germplasm conservation and cases of genetic engineering. The breeder has no control and cannot direct the genetic changes which are also too many to be handled conveniently. In these variations some favourable point changes may have their expression altered by chromosome aberrations, unfavourable DNA amplification and de-amplification and the movements of transposable elements may also lead to further changes. While there are certain genetic changes that are expressed in individual cells and not in multicellular plant tissues or organs or whole plants there are some changes which are epigenetic and not inherited or may be attributable to cytoplasmic factor changes.

16.3 Micropropagation for Mass Multiplication and Artificial Seed

Use of biotechnology helps in rapid mass propagation of plantlets to be grown out as propagation materials by farmers for commercial production. Tissue culture is applicable in micropropagation for rapid cost-effective mass multiplication of superior genotypes of vegetatively propagated crops, where highly uniform disease-free planting material is required, for crops in which fecundity is low and in woody plants that take too long to mature or are difficult to multiply through conventional sexual methods.

Artificial seed or synthetic seed is encapsulated tissue which may include somatic embryo, shoot tip/bud, axillary bud or any other meristematic tissue that imitates natural seed and is used for sowing to obtain plants (Rihan et al., 2017). Other tissues which can be encapsulated to give artificial seeds are in vitro-derived propagules like cell aggregates, nodal segments, micro-shoots, embryogenic calli, protocorm-like bodies and other micro-propagules. Success of micropropagation depends on genotype, age, origin, health and physiological condition of the explants, media composition and growth hormones like indole-3-butyric acid (IBA), α-naphthalene acetic acid (NAA), indole-3-acetic acid (IAA) and 2,4-dichlorophenoxyacetic acid (2,4-D). Explants from shoot tips and meristem tips, hypocotyls, embryos, cell suspension or nodal axillary bud, floral meristems, leaf discs, callus culture, etc., are suitable for this purpose. To produce somatic embryos the somatic cell has to be induced into embryogenesis through cell division initiation and establishment of polarity to create apical and basal meristems which give the shoot and the root respectively (Meucci et al., 2024). Embryogenesis induction is achieved by use of growth regulators, especially auxins (e.g., 2,4-D, NAA) and cytokinin. Following induction a bipolar somatic embryo is formed, starting with globular heart-shaped structure then torpedo-shaped embryo which is allowed to undergo maturation then subjected to encapsulation along with artificial endosperm comprising of nutrients, growth regulators, fungicides, bacteriocides, pesticides, bio-controllers and bio-fertilizers (Abbas et al., 2022). The endosperm is improvised by adding these components to the encapsulation matrix for improved germination capacity and viability of the seed. In the encapsulation process the somatic embryos are placed in sodium alginate solution and the droplets of the mixture allowed to drop into calcium chloride solution to form round firm beads (Meucci et al., 2024). Artificial seed resembles the natural seed by virtue of consisting of the somatic explant material representing the zygotic embryo and the capsule comprising of artificial coat and artificial endosperm. Apart from desiccated synthetic seeds which may be encapsulated (coated in polyoxyethylene glycol) or naked (uncoated), there are also hydrated encapsulated (coated in hydrogel) synthetic seeds (Rihan et al., 2017; Abbas et al., 2022). Desiccated synthetic seeds are produced for desiccation tolerant somatic embryos only while hydrated synthetic seeds are for somatic embryos which do not withstand desiccation, i.e., recalcitrant.

Artificial seeds are important for large scale mass propagation of promising and established cultivars and are particularly useful for vegetatively propagated species which do not produce seeds (like banana, pineapple, seedless watermelon), species with long juvenile periods (fruit trees and forest species) and polyploid species difficult to propagate sexually due to undesirable segregation. These seeds are also important where there are problems of sterility like in products of interspecific hybridisation, somatic hybridisation and transgenesis. Furthermore, aseptically produced artificial seeds are disease free and prevent spread of diseases, are cheaper to produce, more reliable and durable as regards storage and easier to handle during transportation and planting (Rihan, 2017). Since artificial seeds have no problems of heterogeneity and segregation, they are useful in preservation of genetic uniformity. They can be subjected to storage for 6 months at 4°C and long-term cryopreservation storage for germplasm conservation, especially for recalcitrant seed producing species. They also help in maintenance of parental lines in hybrid crop production and preservation and multiplication of elite genotypes. Regeneration is achieved through direct organogenesis or somatic embryogenesis using 2, 4-D, NAA or other hormones for growth induction and differentiation of tissues and providing opportunities for mass production of propagules. One major challenge for commercial production of propagules using tissue culture is lack of genome stability due to somaclonal variation. Uniformity in micro-propagated plants can be ascertained by use of DNA molecular markers. Micropropagation has the challenge of labour for frequent transfer of materials from one vessel to another, acclimatisation and possible contamination in the laboratory. Production of synthetic seeds has been attempted in several crops like tomato, banana, pineapple, potato, sweet potato, cassava, orange, lemon, cauliflower, broccoli, raspberry, celery, grape, garlic, capsicum, carrot, breadfruit, asparagus, cucumber, eggplant and other crops.

Problems

16.1 Distinguish between somatic embryogenesis and organogenesis in regard to tissue culture.

16.2 Which factors influence regeneration of cultured tissues?

16.3 How do cytokinin and auxin levels in the culture medium affect cultured tissues?

16.4 What role does polyethylene glycol play in cultured tissues?

16.5 What is responsible for somaclonal variation in cultured tissues?

16.6 What are the advantages of synthetic seeds?

References

Abbas, M. K., Mahood, H. E., & Alhasan, A. S. (2022). Production of synthetic seeds in vegetable crops: A review. In *IOP Conference Series: Earth and Environmental Science* (vol. 1060, p. 012099). https://doi.org/10.1088/1755-1315/1060/1/012099. Creative Commons Attribution International License (CC BY).

Bairu, M. W., Fennel, C. W., & Van Staden, J. (2006). The effect of plant growth regulators on somaclonal variation in Cavendish banana (Musa AAA cv Zelig). *Scientia Horticulturae, 108*, 347–351. https://doi.org/10.1016/j.science.2006.01.039. Elsevier, No. 5750701219807.

Bates, G. W., & Hasenkampf, C. A. (1985). Culture of plant somatic hybrids following electrical fusion. *Theoretical and Applied Genetics, 70*, 227–233. https://doi.org/10.1007/BF00304903. Springer Nature, No. 5626360434826.

Bednarek, P. T., & Orlowska, R. (2020). Plant tissue culture environment as a switch-key of (epi)genetic changes. *Plant Cell, Tissue and Organ Culture, 140*, 245–257. https://doi.org/10.1007/s11140-019-01724-1. Creative Commons Attribution International License (CC BY).

Bidabadi, S. S., & Jain, S. M. (2020). Cellular, molecular and physiological aspects of in vitro plant regeneration. *Plants, 9*, 702. https://doi.org/10.3390/plants9060702. Creative Commons Attribution International License (CC BY).

Cocking, E. C. (1960). A method of isolation of plant protoplasts and vacuoles. *Nature, 187*, 962–963. https://doi.org/10.1038/187962a0. Springer Nature, No. 5750281131739.

Evans, D. A., & Flick, C. E. (1983). Protoplast fusion: Agricultural applications of somatic hybrid plants. In Kosuge, T., Meredith, C. P., & Hollaender, A. (Eds.), *Genetic engineering of plants: An agricultural perspective* (vol. 26, pp. 271–288). University of California, Davis, USA, Boston, MA: Springer. https://doi.org/10.1007/978-1-4684-4544-2_19. Springer Nature, No. 5750361283437.

Guha, S. R., & Maheswari, S. D. (1966). Cell division and differentiation of embryos in the pollen grains of Datura in vitro. *Nature, 221*, 97–98. Springer Nature, No. 5747730405044.

Israel, Y., Lahav, E., & Reuveni, O. (1995). In vitro culture of bananas. In Gowen, S. (Ed.), *Bananas and plantains world crop science series* (pp. 147–178). Dordrecht: Springer. https://doi.org/10.1007/978-94-0011-0737-2_6. Springer Nature, No. 5750711190407.

Kao, K. N., & Michayluk, M. R. (1974). A method for high frequency intergeneric fusion of protoplasts. *Planta, 115*, 355–367. https://doi.org/10.1007/BF00388618. Springer Nature, No. 5699230520444.

Krishna, H., Alizadesh, M., Singh, D., Singh, U., Chauhan, N., Eftekhari, M., & Sadh, R. K. (2016). Somaclonal variations and their applications in horticultural crop improvement. *3 Biotech* (6), 54. https://doi.org/10.1007/s13205-016-0389-7. Creative Commons Attribution International License (CC BY).

Larkin, P. J., & Scowcroft, W. R. (1981). Somaclonal variation—A novel source of variability from cell cultures for plant improvement. *Theoretical and Applied Genetics, 60*, 197–214. https://doi.org/10.1007/BF02342540. Springer Nature, No. 562670004990.

Larkin, P. J., & Scowcroft, W. R. (1983). Somaclonal variation and crop improvement. In Kosuge, T., Meredith, C. P., & Hollaender A. (Eds.), *Genetic Engineering of Plants: An agricultural perspective* (vol. 26, pp. 289–314). University of California, Davis, USA, Boston, MA: Springer.https://doi.org/10.1007/978-1-4684-4544-2_20. Springer Nature, No. 5750390139427.

Leva, A. R., Petruccelli, P., & Rinaldi, L. M. R. (2012). Somaclonal variation in tissue culture: A case study with olive. *Recent Advances in Plant in Vitro Culture*. https://doi.org/10.5772/50367. Creative Commons Attribution International License (CC BY).

Meins, Jr., F. R. (1982). Habituation of cultured plant cells. In Schell, J., & Kahl, G. (Eds.), *Molecular biology of plant tumours* (pp. 3–31). https://doi.org/10.1016/B978-0-12-394380-4.50007-4. Academic Press Inc., No. 5803090831909.

Melchers, G., & Labib, G. (1974). Somatic hybridisation of plants by fusion of protoplasts. Selection of light resistant *Thybrid* of haploid light sensitive varieties of tobacco. *Molecular Genetics and Genomics, 135*, 271–294. https://doi.org/10.1007/BF00271144. Springer Nature, No. 5626361424846.

Meucci, A., Ghelardi, C., Chietera, G., & Mensuali, A. (2024). Synthetic seed production and slow growth storage in in vitro cultured plants of Iris pallida Lam. *Horticulturae, 10*(3), 272. https://doi.org/10.3390/horticulturae10030272. Creative Commons Attribution International License (CC BY).

Nic-Can, G. I., Galaz-Avalos, R. M., De-La-Pena, C., Alcaza-Magana, A., Wrobel, K., & Loyola-Vargas, V. M. (2015). Somatic embryogenesis: Identified factors that lead to embryogenic repression. A case of species of the same genus. *PLoS ONE, 10*, e0126414. https://doi.org/10.1371/journal.pone.0126414. Creative Commons Attribution International License (CC BY).

Papenfuss, J. M., & Carman, J. G. (1987). Enhanced regeneration from wheat cultures using dicamba and kinetin. *Crop Science, 27*, 588–593. https://doi.org/10.2135/cropsci1987.0011183X002700030o35x. John Wiley and Sons, No. 5747751340109.

Pelletier, G. (1993). Somatic hybridisation. In Hayward, M. D., Bosemark, N. O., & Ramagosa, I. (Eds.), *Plant breeding principles and prospects* (pp. 93–106). https://doi.org/10.1007/978-94-011-1524-7_8. Springer Nature, Nos. 5750310996089 & 5802981408968.

Power, J. B., Freason, E. M., Hayward, C., George, D., Evans, P. K., Berry, S. F., & Cocking, E. C. (1976). Somatic hybridisation of *Petunia hybrida* and *P. parodii. Nature, UK 263*, 500–502. https://doi.org/10.1038/263500a0. Springer Nature, No. 5802650663130.

Rihan, H. Z., Kareem, F., El-Mahrouk, M. E., & Fuller, M. P. (2017). Artificial seeds (principles, aspects and applications). *Agronomy, 7*, 71. https://doi.org/10.3390/agronomy7040071. Creative Commons Attribution International License (CC BY).

Sato, M., Hosokawa, M., & Doi, M. (2011). Somaclonal variation is induced *de novo* via the tissue culture process: *Sainpaulia. PLoS ONE, 6*(8), e23541. https://doi.org/10.1371/journal.pone.0023541. Creative Commons Attribution International License (CC BY).

Takebe, I., Labib, G., & Melchers, G. (1971). Regeneration of whole plants from isolated mesophyll protoplasts of tobacco. *Naturwissenschaften, 58*, 318–320. https://doi.org/10.1007/BF00624737. Springer Nature, No.5750280446157.

Chapter 17
Recombinant DNA Technology and Genetic Modification

Recombinant DNA (rDNA) technology refers to the technology of gene splicing *or* joining pieces of DNA from different sources into one DNA molecule, a process which involves the application of endonucleases and ligases, isolation, joining and cloning of specific genes or sets of genes. Recombinant DNA techniques make it possible to identify and isolate genes that encode various enzymes and structural proteins and splice two or more heterologous DNA fragments from different organisms to form recombinant DNA in vitro creating new DNA molecules that do not exist under natural conditions. This is facilitated by availability of methods that help break DNA molecules into fragments using endonucleases which recognise specific target nucleotide sequences in the DNA molecule where they create cuts on both strands. Recombinant DNA production process requires purified DNA from cells, endonucleases used for breaking the DNA into fragments and the ligases for joining the fragments to other DNA molecules which serve as vectors or carrier molecules (Khan et al., 2016). DNA fragments can be joined into DNA pieces smaller than DNA molecules in the chromosomes for easy genetic analysis and gene transfer. If two DNA fragments which are products of the same restriction endonuclease like EcoR1 are mixed together under appropriate conditions, the DNA fragments from the two sources form recombinant molecules through hydrogen bonding of their sticky ends catalysed by ligase. The recombinant DNA formed is usually amplified or subjected to cloning by producing several copies through normal DNA replication employing DNA replication enzymes. Gene cloning entails in vitro incorporation of the gene (target DNA segment) or recombinant DNA piece of interest into a small self-replicating vector chromosome belonging to a phage like lambda or a plasmid or the recombinant DNA may be incorporated into a host cell where it will replicate into many identical copies. The vector and the inserted DNA fragment is then moved to a host cell in which the recombinant DNA molecule replicates using host cell DNA polymerase producing several identical copies or clones. The cloned recombinant DNA is isolated from host cells, purified for genetic analysis or used in genetic modification or it can be transcribed, its RNA translated and the gene product isolated.

P. O. Ayiecho and J. O. Nyabundi, *Conventional and Contemporary Practices of Plant Breeding*, https://doi.org/10.1007/978-3-031-74998-8_17

17.1 Vectors—Carriers of DNA Fragments

Fragments of DNA produced by restriction enzymes need vectors or vehicles as carriers to enter bacterial cells or other cells for cloning. A vector or transfer vehicle for DNA fragment transfer into a recipient host cell is a small DNA molecule capable of autonomous replication along with the foreign DNA fragment joined to it and should also allow easy recovery of the hybrid DNA from the host cell. The appropriate vector DNA has the respective restriction enzyme cleavage sites that are cleaved by the same restriction endonuclease producing the DNA fragments making insertion of the foreign DNA fragments possible. The vector also has a distinct selectable marker not found in the host cell helping distinguish host cells having the recombinant DNA from those lacking it. Some commonly known vectors include various types of naturally occurring plasmid DNA plasmids like SC101, Ti plasmid, bacteriophages like lambda phage (λ-phage) and vectors based on plant viruses which are usually used to introduce target foreign DNA into the bacteria, plant and animal cells (Agrawal & Rami, 2022).

17.2 Recombinant DNA Production

Cloning experiments use site specific restriction endonucleases which cleave DNA molecules at specific palindromic nucleotide sequence restriction sites. Palindromes are nucleotide-pair sequences that read the same forward or backward from a point in the double stranded DNA. A palindromic recognition sequence for a restriction endonuclease may read the same in the 5' to 3' direction on one strand and as it reads in the 5' to 3' direction on the complimentary strand from a point of symmetry. A restriction nuclease cleaves a palindromic sequence of the DNA two strands at different points producing DNA segments with complimentary single stranded ends, called sticky ends. It is possible to get complimentary sticky ends for DNA from different sources so long as the palindromic sequences are similar and restriction products are of the same restriction enzyme activity. It also feasible to join DNA fragments from different sources like viruses, plasmids, bacteria and eukaryotes when the complimentary sticky ends anneal through hydrogen bond formation between the pairing bases in presence of DNA ligase. In some cases, DNA digestion by certain restriction endonucleases leads to DNA with blunt ends if the two DNA strands are cut at the same point. Under such situations the enzyme terminal transferase helps to synthesise complimentary single stranded tails on 3'ends of DNA fragments to act as sticky ends for joining DNA molecules and fragments from different sources.

17.3 Genetic Transformation Methods

The technology of adding specific exogenous DNA or sets of genes to the genome of a target organism, often referred to as genetic modification, genetic transformation, genetic engineering, or transgenesis, uses recombinant DNA methods to manipulate the genetic constitution or genotype of an organism making it possible to mobilise diverse gene pools, especially the alien genes into crop plant species. Genetic transformation goes back to late 1920's when the process was discovered by Fredrick Griffith in *Diplococcus pneumoniae* (*Streptococcus pneumoniae*) and in 1940's when Oswald Avery, C. M. McLeod and M. McCarty identified the transformation principle as DNA.

Genetic transformation technology makes it possible to transfer alien DNA segments or genes into genomes of otherwise superior varieties to improve them with respect to specific traits in which they are deficient in a single step or within the shortest time possible without introducing other undesirable genes. Transformation saves on time, costs, labour, space and resources since the operation is done in vitro at the cellular level within a small laboratory space and eliminates the time-consuming conventional hybridisation and backcrossing approach. Biotechnology and genetic engineering make it possible to transform plants genetically to obtain transgenic crops by substituting or adding whole genes into plant cells for the desired traits and then evaluate the millions of cells, each of which is a potential plant. Genetic transformation overcomes interspecific hybridisation barriers by allowing for transfer of genes across different organisms without the need for sexual reproduction and effectively widens the gene pool for germplasm resources. Genetic transformation is used to transfer genes easily across species in one generation without need for crossing, thereby breaking interspecific hybridisation barriers. Among the several traits that are transferred to plants through transgenesis are resistance to diseases, pests, nematodes and herbicides, male sterility, fruit ripening and quality parameters. The transgenic plants are also developed with improved yield and tolerance to other abiotic stresses like drought, heat, cold, salinity, mineral stress and other agronomic characters.

Use of recombinant DNA technology for transferring genes directly or indirectly from one organism into the DNA of another one was first done in 1973 and it was not until 1982 when products of genetic engineering were first commercialised in form of medicinal drugs followed by first commercialisation of genetically modified crops 1996 in USA. Other countries which followed immediately to commercialise genetically modified crop varieties include Japan, Brazil, Argentina, Paraguay, China, Australia, Canada, South Africa and some European Union countries.

There are two types of gene transfer systems for purpose of transformation, namely, vector dependent (indirect) biological gene transfer using plasmids and direct gene transfer methods which are based on physical or chemical penetration of the cell wall and the plasma membrane. Genetic transformation is conditioned on successful transfer and stable integration of target alien DNA segment into the genome of the host plant, replication of alien DNA segment along with host DNA and expression of the transferred gene in the host genetic background and tissue/organ. Apart from being integrated into the nuclear genome of the recipient plant, the transferred gene must also undergo regular Mendelian inheritance and must not have undesirable effects on growth and development, yield and product quality of the host plant.

17.3.1 Indirect Transfer: Agrobacterium-Mediated Transfer

The components of indirect DNA transfer include the DNA of interest, the transfer vehicle and the host. The DNA of interest is the isolated cloned recombinant DNA. The target foreign DNA is inserted into the vehicle DNA and the two pieces of joined DNA are moved into the host cell. The delivered foreign DNA should express itself and change the phenotype of the host cell without the influence of vehicle DNA. When a target foreign DNA is spliced onto a bacterial plasmid the resulting recombinant plasmid is amplified in the bacterial cell under controlled conditions as the bacterium multiplies leading to millions of copies of the foreign gene before being introduced into the target plant protoplasts or cells in culture for assimilation of the foreign DNA into cell or protoplast chromosome and subsequent regeneration into plantlets and whole plants.

The most commonly used indirect DNA transfer for crop plants employs *Agrobacterium tumefaciens.* The *A. tumefaciens*-mediated transformation helps in incorporation of any gene of interest into the genomes of crop plants e.g., genes for disease, pest and herbicide resistance, quality aspect genes, genes for tolerance to abiotic stress factors like drought, cold, heat, mineral toxicity, alkalinity and mineral deficiency. The success of the technology is dependent on cultivar genotype, *Agrobacterium* strain, regeneration system and appropriate wounding response behaviour by the host plant because *A. tumefaciens* infection requires wounding which stimulates the plant cells to switch from resting stage to dividing stage (Potrykus, 1993). This transformation approach can be applied for any dicotyledon plant species, e.g., celery, *Arabidopsis,* sugar beet, oil seed rape, *B. oleracea*, musk melon, cucumber, strawberry, soybean, cotton, sunflower, walnut, lettuce, flax, tomato, apple, alfalfa, tobacco, *Petunia*, garden pea, eggplant, potato and many others. The technology has been extended to monocotyledons like maize, wheat, rice and others as well.

Agrobacterium tumefaciens is a gram-negative soil-borne bacterium which infects plants to cause crown gall. Though there are other *Agrobacterium* strains like *Agrobacterium rhizogenes*, the most commonly used strain in plant genetic modification is *A. tumefaciens.* This bacterium has Ti plasmid which is responsible for crown gall formation in plants and is a well-known and frequently used vector plasmid with natural genetic transformation system that is taken advantage of for genetic modification of crop plants.

When *A. tumefaciens* infects a plant cell a piece of its tumour inducing (Ti) plasmid DNA, called T.DNA, is transferred into infected plant cell genome at the wound point followed by crown gall tumour formation as T.DNA genes which encode enzymes responsible for synthesis and overproduction of auxins (indoleacetic acid) and cytokinin leading to cell proliferation and tumorous gall formation, are expressed in the plant tissue (Agrawal & Rami, 2022; Narusaka et al., 2012). The other T-DNA genes include those which help in the synthesis of opines, octopine and nopaline utilised by the bacterium as carbon source and virulence (vir) genes that enhance the transfer of T-DNA into the plant genome (Agrawal & Rami, 2022; Ameen Rashid & Dhahir Lateef, 2016). The vir genes are not transferred with the T-DNA into the plant genome. When the bacterium attaches to the wound site, mediated by the vir genes, a trans-membrane forms between it and the host cell to facilitate the transport of T-DNA/protein complex into the host cell, a process that is comparable to bacterial conjugation mechanism. This biological phenomenon is used by breeders to transfer target DNA into plant protoplasts or cells in vitro or other tissues in culture like callus, leaf discs or root sections by engineering the target DNA segment onto T-DNA and the arising recombinant DNA is put back into *A. tumefaciens* to be cultured together with plant cells in culture medium (Agrawal & Rami, 2022; Khan, 2009). If the sterilised explant tissues are co-cultured with *A. tumefaciens* carrying the plasmid with the target foreign gene the bacteria bind to plant cells and the process of gene transfer to accomplish transformation is initiated. The cells that become genetically modified by acquiring the target foreign genes in the culture are isolated by use of appropriate selective media to kill the bacteria and the non-transformed cells and thereafter cultured through organogenesis or embryogenesis into plantlets and consequently into plants. Apart from appropriate antibiotics to eliminate the bacteria the selective medium has other selective agents to select the transformed cells, depending on the target gene incorporated or the marker gene being used, e.g., herbicide resistance, pathotoxin resistance, resistance to kanamycin, etc. Before it is transferred T.DNA is modified such that it does not lead to the gall disease but only transfers the target gene into the plant cells. T.DNA modification is achieved by eliminating its tumour formation inducing genes and replacement of unnecessary sequences by the DNA of interest without interfering with genes responsible for its transfer and integration into the plant genome (Agrawal & Rami, 2022). Disarming the Ti plasmid by removing the genes responsible for tumour formation makes the non-oncogenic or disarmed plasmid a useful tool in genetic engineering because it can be used to move target genes to plant cells without causing the tumour disease. Apart from plasmids, plant viruses like cauliflower mosaic virus (CaMV) which has double stranded DNA, Gemini virus with single stranded DNA, wheat dwarf virus which infects a number of cereal crops, brome mosaic virus (BMV) which infects monocotyledons and tobacco mosaic virus (TMV) are also used as vector vehicles for the recombinant DNA transfer. Viruses that cause plant diseases are genetically modified so that they don't cause diseases on infection for transfer of target DNA.

17.3.2 Direct Transfer Methods

A number of direct DNA transfer methods which are based on physical or chemical agents to enhance the entry of DNA into the cells are in use for monocotyledonous plants like maize, wheat, rice, sorghum, barley, rye, oats and dicotyledonous plants as well. Direct transfer methods have the advantage of not requiring vectors for DNA delivery and transformation and at the same time solve the limitations associated with *Agrobacterium*-mediated transformation such as limited host range, tedious processes involving modification of Ti plasmid and viruses used as vectors, incorporation of target genes into the plasmid genomes, reintroduction of plasmid into the bacterium and co-culturing the bacterium with the plant tissues to accomplish the transfer. Furthermore, direct physical DNA delivery is faster, more efficient, simpler, easier to conduct and allows for simultaneous delivery of more than one gene into a host genome. Some of the direct DNA transfer methods include protoplast DNA uptake, polyethylene glycol (PEG) and liposome-mediated transfer, electroporation, biolistic transfer, microinjection, macro-injection, micro-targeting, silicon carbide-mediated transfer, laser microbeam-mediated transfer, ultrasound-mediated transfer, sonication-assisted *Agrobacterium*-mediated transformation (SAAT), pollen and embryo sac transformation and use of florets or flowers.

Protoplasts DNA uptake and use of liposomes and PEG

Protoplasts are ideal units for transgenesis because they absorb DNA directly through the plasmalemma and are easily transformed by adding the target foreign DNA to the protoplast culture medium for purpose of incorporating the DNA into the nuclear genome. The foreign DNA added to the protoplasts in culture will be taken up by protoplasts directly and incorporated into the nuclear genome without the support of any vector and such transformed protoplasts are transferred to appropriate selective medium where the transformed cells are selected using markers and regenerated into plantlets and then into plants (Potrykus, 1993). Naked DNA uptake by protoplasts is induced by adding polyethylene glycol (PEG) and calcium (Ca) cations which destabilise the plasmalemma and modify pore size on the plasmalemma thereby allowing DNA molecules to enter the protoplasts more easily for incorporation into the plant protoplast genome (Khan, 2009; Potrykus, 1993). Once this is accomplished PEG is removed and the protoplasts are cultured into cells and for in vitro selection of transformed cells and plantlets. DNA can also be encapsulated in liposomes to facilitate its entry into the protoplast. Encapsulation is achieved by mixing the DNA solution and liposomes in an organic solvent. The liposomes carrying the DNA are then incubated with protoplasts in a medium having PEG and Ca cations to induce and facilitate the fusion of liposomes and protoplasts so that the foreign DNA is incorporated into protoplast nuclear genome (Khan, 2009; Shruti et al., 2024).

Electroporation

In electroporation the cells or protoplasts in culture to which solution of the target foreign DNA has been added are subjected to a short duration of high voltage electric pulses of 200–600 V/cm in an electroporator making the plasmalemma temporarily permeable to DNA (Khan, 2009). Electric pulses produce charge anomalies in the plasmalemma resulting in pore formation allowing penetration of the target DNA into the cytoplasm and the nucleus and subsequent incorporation onto the chromosomes (Ameen Rashid & Dhahir Lateef, 2016). The protoplasts having alien DNA is isolated by culturing in selective media and then regenerated into plantlets and whole plants. Electroporation method is fast, efficient, simple and causes less damage to the protoplasts as compared to use of PEG and has been applied to develop transgenic plants in several crops like rice, wheat, maize, oilseed rape and several other crops.

Biolistic transformation

Biolistic (biological ballistics) transformation, also known as particle bombardment or particle gun, is used for delivering target DNA into cells of intact organs, callus culture, meristem culture, protoplast culture, anther or pollen culture, zygotic embryos, cell suspension or other cultured tissue by micro-projectile bombardment of heavy metal (tungsten or gold) micro-particles using a particle gun for incorporation of the alien DNA into the cell genome (Potrykus, 1993). The target DNA is coated on tungsten particles or gold particles which are loaded onto the particle gun and then directed into the target tissue at high velocity by the particle gun (Ameen Rashid & Dhahir Lateef, 2016; Khan, 2009). This makes the particles coated with target DNA get into the cell nuclei and incorporate the foreign DNA into the plant. The transformed tissue is subjected to selective media and surviving plant material which has taken up the target foreign DNA is cultured for regeneration into plantlets and whole plants. Biolistic transformation is a highly versatile and simple method and has been applied to develop transgenic cultivars in both dicot and monocot crops like rice, soybean, maize, wheat, barley, sugarcane and groundnuts. This method has the advantage of being simple, efficient, is applicable on cells from different tissues and can be used to transform several cells in one shot (Khan, 2009).

Micro-injection

Micro-injection is a direct DNA transfer method which employs microscopic facilities and fine glass micro-capillary injection pipettes with orifices and micromanipulator to deliver the target DNA precisely into protoplasts, microspore-derived embryos or somatic cells (Neuhaus & Spangenberg, 1990). When introducing exogenous DNA by micro-injection protoplasts are immobilised and a micro-pipette is used to inject the target DNA solution through the plasmalemma onto the chromosomes within the nucleus (Khan, 2009). Though it has a limitation of being labour intensive and that only one cell can be handled per injection and needs more skill with precise instrumentation, micro-injection is superior to other methods because only small amounts of the target DNA is delivered avoiding wastage and it has high precision and predictability (Potrykus, 1993). This method has been used to produce transgenic plants in several crops like to sunflower, maize, wheat, rice, barley, cotton.

Pollen and embryo sac transformation through DNA uptake

Gene transfer can be accomplished by using pollen pathway whereby the styles are cut after pollination and target DNA solution applied at the tip of the style stump. The DNA follows the path of the pollen tube to reach the ovule to accomplish transformation (Khan, 2009; Luo & Wu, 1988). Alternatively, gene transfer is achieved by treating pollen of the plant to be transformed with target exogenous DNA and the treated pollen is used for regular pollination by placing the pollen on the stigma (or silk in case of maize) for fertilisation. As fertilisation takes place the target DNA will be incorporated into the genome of the endosperm and embryo and the effect of such transformation will be observed after seed from such fertilisations are harvested, grown and derived plants screened for the gene under transfer. The target DNA solution may also be applied directly into the inflorescence allowing the DNA to get incorporated into the gametes or zygote as pollination and fertilisation take place. The DNA solution is applied to the stigma and immediately followed by pollination. When planted the seed derived from such fertilisation gives some plants that reflect the expected phenotype attributed to the transferred exogenous DNA. De-La-Pena et al. (1987) developed macro-injection transformation method which they applied for delivering exogeneous DNA to developing inflorescence of rye plants using standard hypodermic needle and syringe. After pollination and fertilisation, the developed seeds are harvested and some plants obtained after planting the seed may have the phenotype attributed to the exogenous DNA that was injected into the floral organ, suggesting effective transformation. In macro-injection the applied alien DNA follows the pollen tube pathway during fertilisation into the ovule and is taken up directly leading to transformation (Khan, 2009). The seed having the alien gene following such fertilisation matures with others on the mother plant. Macro-injection approach is also employed to introduce target foreign DNA into embryo sac, meristems and germinating pollen through injection for in vivo transformation (Shruti et al., 2024). Macro-injection can be used to transfer genes easily across species in one generation without need for crossing, thereby breaking interspecific hybridisation barriers. Plants raised from seeds arising from such genetic modification exhibit phenotypic changes due to the DNA transferred to the ovary or style. Advantages of pollen and embryo sac macro-injection transformation methods are that there is no somaclonal variation problem and there is also no need for tissue culturing since transformed plants having the target gene are obtained through regular sexual reproduction and seeds. These methods also solve the biotechnology procedure problems associated with *Agrobacterium*-mediated transformation.

Silicon carbide-mediated transfer and Sonication transformation method

Use of silicon carbide fibres for plant transformation, often referred to as silicon carbide-mediated transformation (SCMT), was proposed in 1990 by Kaeppler et al. (1990). The procedure employs small needle-like silicon carbide fibres with high tensile strength to puncture the cells without killing them but allow for DNA transfer into the cells in tissue culture like cell, immature embryos, callus cultures, etc. This is achieved by placing silicon carbide fibres in a solution containing target DNA and the tissue culture followed by shaking or application of vortex making the cells collide with the fibres creating small pores on the cells and the DNA coated fibres transfer DNA into the cells as they puncture the cells (Kaeppler et al., 1990; Shruti et al., 2024). The transformed cells are selected and regenerated into plantlets and plants. This method is cheap and simple but has the disadvantage of damaging the cells.

Sonication transformation method uses ultrasonic waves for gene transfer which cause sonoporation by rupturing cellular membranes making it possible to introduce exogenous DNA into the protoplasts and other in vitro tissues for incorporation into the nuclear genome (Khan, 2009; Shruti et al., 2024). Sonication is also used along with *Agrobacterium* so that in sonication-assisted *Agrobacterium*-mediated transformation (SAAT) the plant tissues, which may be leaf tissues, somatic and zygotic embryos, shoot tips, meristematic tissues, whole seedlings and other tissues, are subjected to a brief period of ultrasound while placed in a suspension of *Agrobacterium* having recombinant plasmid with the target DNA (Shruti et al., 2024; Trick & Finer, 1997). The treatment produces openings on the tissues through which the bacterium gains access into the cell leading to transformation. Though this method has been used to transform a number of dicotyledonous and monocotyledonous crops it is not much used because of damage it causes to the cell membrane.

Laser microbeam and other methods

Laser microbeam-mediated transformation creates small self-healing pores in the cell wall and plasmalemma (Shruti et al., 2024). Target DNA solution present in the culture medium gains entry into the punctured cells and gets incorporated into the nuclear genome. The pores in cell wall close immediately and get healed. Dry seeds or germinating seeds and seedlings can also be treated with target DNA for the purpose of transformation.

17.3.3 Genome Editing and Gene Silencing

Genome editing

Genome editing is a molecular genetic technology for improving plants by creating genetic changes in the genome through insertions, deletions and substitutions of nucleotide base pairs. Genome editing is used for cheap and fast development of new varieties by making precise specific nucleotide changes and also diversify alleles, even in multiple gene traits. Through genome editing changes are created that rectify the cultivars with respect to the characters in which they are deficient. Target qualitative and quantitative trait nucleotides are subjected to genome editing and alleles with undesirable effects subjected to correctional process in as fast as one generation. A number of in vitro plant breeding tools developed as part of genome editing systems that are employed to direct required mutations at predetermined positions in the plant genome, reduce the extent of bad mutations arising and improve the precision and speed of obtaining the desired genetic changes include Zinc Finger Nucleases (ZFNs), Transcription-like Effector Nucleases (TALENS), Meganucleases (MN) and Clustered Regular Interspaced Short Palindromic Repeats (CRISPR-Cas9) (Holme et al., 2019; Razzaq et al., 2021).

These gene editing techniques are applied to help in rapid molecular breeding for crop improvement by altering a few nucleotides in the target gene, knock out genes, modify expression of genes by inserting transgenes at certain locations through homologous recombination or disrupting gene promoter sequences and creating new alleles. The gene editing systems employ site directed nucleases that cleave specific sites on DNA by binding to domains which recognise specific DNA sequences followed by single or double cuts on DNA, methylation, deamination and acetylation (Viana et al., 2019). This also causes base changes that can be employed for gene editing, gene silencing and alteration of gene expression. In the cellular DNA repair process of the double breaks that follows there is homologous repair (HR) and non-homologous end joining repair (NHEJ) systems (Razzaq et al., 2021; Viana et al., 2019). During the repair process a plasmid containing a desired exogenous DNA sequence or the target PCR product can be incorporated into the plant genome at the repair site. The DNA repair allows small base insertions or deletions (indels) changes in the DNA molecule. There may also be large deletions and translocations. Transcriptional activator-like effector nucleases (TALENS) and Zinc Finger Nucleases (ZFNs) have been applied but they have problems of poor specificity, need for cloning and vector construction and have less effective delivery systems as compared to Clustered Regularly Interspaced Short Palindromic Repeats (CRISPR associated system) (Razzaq et al., 2021). Though CRISPR has at least 30 alternatives, the most extensively employed genome editing method is CRISPR-Cas9 (Clustered Regularly Interspaced Short Palindromic Repeats-Associated Protein 9) which is being applied in crop improvement efforts and has been used to create genetic changes in a number of crop plants like tomato. CRISPR-Cas9 is a system with higher target specificity, precise nuclease activity, efficient delivery and a broad range of targets and has been used to create drought tolerance, salt tolerance, osmotic stress tolerance, disease resistance and virus disease resistance in a number of crops (Razzaq et al., 2021). Genome editing is applicable in domestication of genes of wild relatives by editing identified genes to create specific genetic changes in crop wild relatives to improve the alien alleles and even convert unadapted materials to superior varieties. Genome editing can be used in conjunction with speed breeding to develop crop varieties that have the required agronomic traits and cope better with biotic and abiotic vagaries associated with climate change.

Gene silencing

In some cases of virus pathogen attack on the plant host, the plant resists virus pathogen attack through silencing the virus genome by using its RNA interference

(RNAi) system for self-defence. When the virus infects the plant, it uses the plant system to replicate itself through formation of double stranded RNA (dsRNA) which is recognised by the host plant and degraded by the host Dicer enzyme into short virus-derived small interfering RNAs (vsiRNAs or siRNA) or microRNA (miRNAs) (Hussain, 2015; Zamore, 2004). These are in turn used by the effector RNA-induced silencing complex (RISC) to activate RNase for disruption of the viral genome and protein synthesis; hence the virus genome is silenced and the host shows resistance (Hussain, 2015; Zamore, 2004). On the other hand, the virus pathogen produces RNAi suppressor or may even use the host RNAi to silence the host resistance genes for successful pathogenicity. The RNAi gene silencing system is known to work against other pathogens like bacteria, fungi and nematodes that attack plants. The host RNAi gene silencing system can be taken advantage of to develop disease resistant cultivars through genetic modification where transgenic plants are armed with siRNA or miRNA to silence the pathogen genome.

17.4 Confirmation of Gene Transfer

Successful transfer of the target genes is confirmed by use of selective agents for markers like antibiotic resistance (e.g., resistance to kanamycin, hygromycin, neomycin), herbicide resistance, quality aspects like amino acid analogues and other appropriate markers. When the selective agents are introduced into the culture only the transformed cells or tissues will survive and grow despite the presence of the substance. Some marker genes are referred to as reporter genes and involve testing of the products of transformation by assessing gene expression without necessarily involving plant tissue destruction.

Apart from selection using selective media, gene transfer can be confirmed by use of molecular markers like RAPD, RFLP, AFLP, ISSR and others which help to ascertain the presence of specific sequences in the plantlet and plant samples to confirm the presence of transferred genes and the number of copies of the transgene. This is done for different generations, e.g., the primary transformed material (plants regenerated from the transformed tissue) and progenies derived by selfing of these in subsequent generations. Transformation is also confirmed by expression of the transferred traits in the regenerated plants.

17.5 Bt Gene and Trypsin Inhibitor Genes

One of the major developments of genetic transformation is the transfer of Bt gene obtained from the gram-positive soil borne bacterium, *Bacillus thuringiensis* into a number of crops like cotton, tobacco, maize and tomato for resistance to Lepidopteran insects. The transfer is achieved through *Agrobacterium*-mediated transformation or any of the direct transformation methods. In *B. thuringiensis* Bt gene is responsible for synthesis of insecticidal crystalline proteins (ICP), also referred to as Bt crystals, that have toxic effects on Lepidopteran and Dipteran insects and not to other organisms. If this gene is engineered into plants it leads to production of the same crystalline proteins and when the larvae of the Lepidopteran insects ingest the proteins from plant tissues, the proteins dissolve in their midgut and are converted into toxic polypeptides that bind to the midgut epithelial cells (Convents et al., 1990; Hussain, 2015). This disrupts the gut cell membranes, making the epithelial cells lyse resulting in the death of the larvae. The crops into which Bt gene has been engineered are resistant to damage by caterpillars of a number of Lepidopteran pests like pink bollworm, cotton bollworm, tobacco budworm, and fall army worm. There are also ICP's from *B. thuringiensis* which can be used to target Dipteran and Coleoptera insects. Tobacco, maize, cotton, tomato varieties having Bt gene for resistance against such pests have been commercialised with millions of hectares across the globe. Some Bt crop varieties have both Bt gene and herbicide tolerance gene stacked together as Bt/HT varieties with consequent reduced costs, high yields and high benefits to the farmers. Cowpea trypsin inhibitor gene is another gene which has been transferred to genomes of crops like tobacco for resistance against a number of Coleoptera and Lepidoptera insects. This gene normally protects certain cowpea varieties from the storage pest *Callosobruchus maculatus* through synthesis of high levels of trypsin inhibitor protein.

17.6 Genetically Modified Crop Cultivation

The first genetically modified crop was developed in 1983 and the first commercial production of genetically modified crops was done in 1996. The global hectarage was 1.7 million ha in 1996 followed by a fast rise so that by 2001 there were 52. 8 million hectares globally which climbed to 160 million hectares and 179.5 million hectares in 2011 and 2015 respectively (Clive, 2015). In 2018 the area under genetically modified crops was 191.7 million hectares with USA having the highest area followed by Brazil and Argentina (ISAAA, 2018). The most dominant crops were soybean, maize, cotton, alfalfa, canola (oilseed rape) and sugar beets.

Genetically modified cultivars have been developed in many crops like soybean, maize, cotton, papaya, canola, tomato, cassava, sugarcane, oranges, sunflower, potato, rice, sweet pepper, squash, sweet potato and other crops (Table 17.1). The target traits for which genetically modified crops are developed are insect pest resistance, nematode resistance, herbicide tolerance, drought tolerance, male sterility, fertility restoration, disease resistance, improved quality like elevated oil content in crops like canola, delayed fruit ripening and other traits (Clive, 2015). The challenges to the development and adoption of transgenic crops include availability of genes to be transferred, the efficiency of gene transfer methods, regulatory and legal requirements related to genetically modified organisms (GMO's) in many countries and slow acceptance rate by many countries, especially the developing countries which have greatest need to address food insecurity.

Table 17.1 Some crops in which genetically modified cultivars have been commercialised (Source of information—Clive, 2015)

	Crop	Traits incorporated by transformation
1	Maize	Insect resistance (Bt), virus resistance, herbicide tolerance (HT), drought tolerance, male sterility
2	Canola (*B. napus*)	HT, fertility restoration, modified oil content
3	Cotton	Insect resistance (Bt), HT
4	Alfalfa	HT
5	Sugar beet	HT
6	Papaya	Virus resistance
7	Poplar	Insect resistance (Bt)
8	Carnation	HT, modified flower colour
9	Potato	Insect resistance, virus resistance, quality
10	Soybean	HT, modified oil content, insect resistance (Bt)
11	Squash	Virus resistance
12	Tomato	Delayed ripening, insect resistance
13	Petunia	Virus resistance
14	Sweet pepper	Virus resistance
15	Rose	Modified flower colour + flavonoid biosynthetic pathway
16	Wheat	HT
17	Tobacco	HT, nicotine reduction
18	Rice	HT, insect resistance
19	Melon	Delayed ripening
20	Plum	Virus resistance

Problems

17.1 How does T.DNA help in genetic transfer?

17.2 What is the role PEG and Ca ions in genetic transformation?

17.3 How do electroporation and biolistic processes operate to effect genetic modification?

17.4 What are advantages of direct DNA transfer over *Agrobacterium*-mediated DNA transfer?

17.5 Indicate any three genome editing systems.

References

Agrawal, S., & Rami, E. (2022). A review: Agrobacterium-mediated gene transformation to increase plant productivity. *The Journal of Phytopharmacology, 11*, 111–117. https://doi.org/10.31254/phyto.2022.11211. Creative Commons Attribution International License (CC BY).

Ameen Rashi, A., & Dhahir Lateef, D. (2016). Novel techniques for gene delivery into plants and its applications for disease resistance in crops. *American Journal of Plant Sciences, 7*, 181–193. https://doi.org/10.4236/ajps.2016.71019. Creative Commons Attribution International License (CC BY)

Clive, J. (2015). Global status of commercialised Biotech/GM crops: 2015. In *ISAAA Brief No. 51*. ISAAA. ISAAA Director Permission.

Convents, D., Houssier, C., Lasters, I., & Lauwereys, M. (1990). The *Bacillus thuringiensis* delta-endotoxin. Evidence of a two-domain structure of the minimal toxic fragment. *Journal of Biological Chemistry, 265*(3), 1369–1375. https://doi.org/10.1016/S0021-9258(19)40023-9. Creative Commons Attribution International License (CC BY).

De La Pena, A., Lorz, H., & Schell, J. (1987). Transgenic rye plant obtained by injecting DNA into young floral tillers. *Nature, 325*, 274–276. https://doi.org/10.1038/325274a0. Springer Nature, No. 5753731267365.

Holme, I. B., Gregersen, P. L., & Brinch-Pedersen, H. (2019). Induced genetic variation in crop plants by random or targeted mutagenesis: Convergence and differences. *Frontiers in Plant Science, 10*. Article1468. https://doi.org/10.3389/fpls.2019.01468. Creative Commons Attribution International Licence (CC BY).

Hussain, B. (2015). Modernization in plant breeding approaches for improving biotic stress resistance in crop plants. *Turkish Journal of Agriculture and Forestry, 39*, 515–530. https://doi.org/10.3906/tar-1406-176. Creative Commons Attribution International License (CC BY)

ISAAA. (2018). Global status of commercialised Biotech/GM crops in 2018. Biotech crops continue to help meet the challenges of increased population and climate change. In *ISAAA Briefs No. 54*. ISAAA. ISAAA Director Permission.

Kaeppler, H. E., Gu, W., Somers, D. A., Rines, H. W., & Cockburn, A. E. (1990). Silicon carbide fiber-mediated DNA delivery into plant cells. *Plant Cell Reports, 9*, 415–418. https://doi.org/10.1007/BF00232262. Springer Nature, No. 5753791388926

Khan, K. H. (2009). Gene transfer technologies in plants: Roles in improving crops. *Recent Research in Science and Technology, 1*, 116–123. ISSN 2076-5061. Creative Commons Attribution International License (CC BY).

Khan, S., Ullah, M. W., Siddique, R., Nabi, G., Manan, S., Yousaf, M., & Hou, H. (2016). Role of recombinant DNA technology to improving life. *International Journal of Genomics, 2016*. Article ID 2405954. https://doi.org/10.1155/2016/2405954. *Creative Commons Attribution International License (CC BY).*

Luo, Z. X., & Wu, R. (1988). A simple method for the transformation of rice via the pollen tube pathway. *Plant Molecular Biology Reporter, 6*, 165–174. https://doi.org/10.1007/BF02669590. Springer Nature, No. 5754220633537

Narusaka, I., Narusaka, M., Yamasaki, S., & Iwabuchi, M. (2012). *Methods to transfer foreign genes to plants*. Transgenic Plants-Advances and Limitations: InTech. https://doi.org/10.5772/32773. Creative Commons Attribution International License (CC BY)

Neuhaus, G., & Spangenberg, G. (1990). Plant transformation by microinjection techniques. *Physiologia Plantarum, 79*, 213–217. https://doi.org/10.1111/j.1399.3054.1990.tb05890.x. Wiley, No. 5754250548062

Potrykus, I. (1993). Gene transfer to plants: Approaches and available techniques. In M. D. Hayward, N. O. Bosemark, & Ramagosa, I., et al. (Eds.), *Plant breeding principles and prospects*. Wiley. https://doi.org/10.1002/9780470514009. Nos. 5752520711397, 5752661124106, 5752230389193.

Razzaq, A., Kaur, P., Akhter, N., Wani, S. S., & Salem, F. (2021). Next-generation breeding strategies for climate-ready crops. *Frontiers in Plant Science, 12*, 620420. https://doi.org/10.3389/fpls.2021.620420. Creative Commons Attribution International License (CC BY)

Shruti, Tripathi, S., & Shukla, S. (2024). Methods of genetic transformation: major emphasis to crop plants. *Journal of Microbiology, Biotechnology and Food Sciences, 13*(4), e10276. https://doi.org/10.55251/jmbfs.10276. Creative Commons Attribution International License (CC BY).

Trick, H. N., & Finer, J. J. (1997). SAAT: Sonication assisted Agrobacterium-mediated transformation. *Transgenic Research, 6*, 329–336. https://doi.org/10.1023/A:1018470930944. Springer Nature, No. 5754240624491

Viana, V. E., Pegoraro, C., Busanello, C., & de Oliveira, A. C. (2019). Mutagenesis in rice: the basis for breeding a new super plant. *Frontiers in Plant Science, 10*. Article 1326. https://doi.org/10.3389/fpls.2019.01326. Creative Commons Attribution International Licence (CC BY).

Zamore, P. D. (2004). Plant RNAi; How a viral silencing suppressor inactivates siRNA. *Current Biology, 14*, R198–R200. https://doi.org/10.1016/j.cub.2004.02.021. Elsevier, No. 5804700933211

Chapter 18
Breeding for Resistance to Biotic Stress Factors

The biotic factors reducing yields for which resistance or tolerance is developed are parasitic weeds, diseases and pests that attack crops in the field and those that destroy produce after harvest. In the daily agricultural activities, the farmer has to contend with one or more of stress factors which limit crop performance causing serious economic losses to the farmer. The stress conditions are always expensive or sometimes very difficult to alter for profitable farming undertaking since efforts to control or alter the conditions often lead to serious reduction in marginal returns to the farmer.

To address the production problems associated with the stress factors for improved crop yield and product quality the environment has to be manipulated to suit the crops or the crop genotype has to be altered so that the crop can cope with the prevailing environment better. To reduce crop losses due to stress it is cheaper and more beneficial to the farmer if crops that withstand or resist the stresses are developed and availed to the farmer for cultivation. Through genetic manipulation of crop plants, the genotypes resistant or tolerant to the biotic and abiotic stresses and better adapted to the prevailing environmental conditions are developed for maximisation and stabilisation of yield and improved quality of the major crops the world depends on.

Cultivation of stress resistant or tolerant genotypes is the best approach for reducing the cost of production for the farmers since cultivation of high yielding resistant or tolerant varieties eliminates or substantially lowers the need to purchase and use pesticides, resulting in increased returns to the farmers. Resistant varieties cultivation is environment friendly with no adverse effects on the environment while application of pesticides to protect susceptible varieties pollutes the environment and is not safe to the farmer, farm workers, the general public and consumers because of produce contamination and it generally precludes the use of predator insects in biological control of insect pests. Use of resistant varieties is compatible with biological control and integrated control systems for sustenance of high production. The resistant varieties must combine the resistance genes with genes for high yield, adaptability,

P. O. Ayiecho and J. O. Nyabundi, *Conventional and Contemporary Practices of Plant Breeding*, https://doi.org/10.1007/978-3-031-74998-8_18

quality and other desirable traits for them to be acceptable. Despite their desirability, the resistant varieties take long to develop using conventional breeding methods and that resistance genes tend to be pest biotype or disease race specific, especially for major gene resistance. Molecular marker breeding, genomic selection, genome editing, transgenesis and speed breeding help to fast-track the development of stress resistant and tolerant varieties.

18.1 Breeding for Disease Resistance

Breeding for disease resistance is one of the most important objectives of plant breeding because of the devastating impact diseases have on crop yields and produce quality and that application of plant resistance is the cheapest and most effective approach for crop plant disease control. Crop plant diseases are never seen before they attack the host plants except for some insect-spread pathogens where the presence of the respective insect pest is a sign that the disease spread by that particular insect is likely to occur. Development of strategies to achieve the objective of disease resistance in crops necessitates proper understanding of the dynamics and genetics of host-pathogen interaction. Relationship between the pathogen and the plant host is such that the pathogen strives to cause disease by gaining access into the host, avoid, suppress or overcome the host's resistance system and take advantage of the host for its nutrition while the host responds by recognising the presence of the pathogen and restricts the pathogen using its resistance system to suppress the growth and the establishment of the pathogen in its tissues (Leach et al., 2014). The dynamics of host-pathogen interaction is easily explained using gene-for-gene model which helps to understand the genetics of different types of resistance and genetics of host-pathogen relationship.

18.1.1 Classification of Resistance

Plant disease resistance can be classified into three categories, namely, specific or vertical resistance, general or broad (field or horizontal) resistance and tolerance. Genetic classification also identifies three categories for resistance as oligogenic/monogenic resistance, polygenic resistance and cytoplasmic resistance (Parlevliet & Zadoks, 1977). Vertical resistance is equated to oligogenic (major gene) resistance while horizontal resistance is similar to polygenic (minor gene).

Vertical resistance

Vertical resistance, also known as oligogenic resistance, single gene or monogenic resistance is attributable to a specific major gene or a few major genes with discontinuous variation and is the most common and mostly used resistance. The genes for resistance in the host and those for virulence in the parasite have large effects leading to discontinuous distribution and distinct categories of resistant and susceptible genotypes and where a cross is made between a cultivar having one or two or three or four genes for vertical resistance and a susceptible cultivar there will be Mendelian monofactorial, bifactorial, trifactorial or tetrafactorial segregation ratios respectively in F_2. However, there is no generalisation regarding such relationship since there are cases in which resistance genes at different loci have epistatic and complimentary gene action giving modified segregation ratios such as 9:7, 15:1, 13:3, etc. instead of 9:3:3:1 for two loci. There are several cases of multiple allele resistance genes and situations of linkage involving multiple genes that confer resistance to a particular pathogen. Some resistance genes are also linked in blocks. Where there is clear dominance-recessive relationship host resistance tends to be dominant over susceptibility while the pathogen virulence is usually recessive to avirulence. However, there some deviations from this generalisation.

Vertical resistance is race specific since the responsible gene or genes condition resistance to a specific race or races of a particular pathogen and corresponds to a specific virulence and avirulence locus in the pathogen. A cultivar having one or a few resistance genes will be resistant to one or a few races of the pathogen only but remains susceptible to other races. Vertical resistance is unstable and is only effective before the new races to which it does not confer resistance arise. If a new disease race comes up such resistance fails and another gene for resistance against the new pathogen race has to be identified for incorporation into the variety by crossing and back crossing. If the pathogen does not have high mutation rates for pathogenicity genes or if new diseased materials are not introduced vertical resistance can keep the crop cultivar protected for a long time. Wheat resistance to leaf rust (*Puccinia recondita* f. sp. *tritici*) and yellow rust (*P. striiformis*) are examples of major gene resistance with variable pathogen races making varieties with monogenic resistance highly vulnerable because new mutations in the pathogens' pathogenicity genes can cause serious disease outbreaks if conditions are favourable. Others also include resistance of barley to yellow mosaic virus and resistance of soybean to soy bean mosaic virus. Vertical resistance is also associated with hypersensitivity against specific pathogen races.

Horizontal resistance

In horizontal resistance, also referred to as polygenic resistance, partial, or non-specific resistance, individual polygenes controlling host-pathogen interaction have small quantitative effects, both from the host and the pathogen, resulting in quantitative or continuous variation in expression of both resistance and pathogenicity. The resistance is pathogen species specific but is not specific to any particular race of the pathogen. Apart from gene effects like dominance and epistatic relationships, this type of resistance is also influenced by the environment though the impact of environment is not as dramatic as for the more complex traits like crop yield and yield related parameters. This resistance is highly stable, slows down disease spread and epidemics and is durable due to genetic homeostasis in parasite populations because any genetic changes in the pathogen races will have only small effects which are not dramatically felt (Parlevliet & Zadoks, 1977). Since polygenic resistance is attributable to several genes the pathogen will need several independent virulence minor gene mutations to cause any significant change in the resistance of the host with polygenic resistance. Some examples of known horizontal resistance in crop plants are resistance of maize to grey leaf spot, maize resistance to anthracnose stock rot, resistance of barley to leaf rust, resistance of rice to bacterial blight and resistance of maize to leaf rust.

Mixed major gene and polygene resistance

Vertical resistance and horizontal resistance are complimentary and some genes that act vertically to certain pathogen races can act horizontally to some other races. There are cases where genes known to be responsible for major gene resistance also exhibit horizontal resistance. The expression of resistance genes may vary with genetic background such that a gene will have major effects in a particular genetic background but behave as a minor gene in a different genetic background and there are cases where major genes for resistance modify or mask the effects of minor resistance genes while minor genes may also mask the expression of certain major genes in regard to disease expression (Nelson, 1984). In some cases, both major genes and minor genes influence resistance and pathogenicity for a particular disease as known in maize resistance to *Puccinia sorghi*. In maize resistance to *Puccinia sorghi* is controlled by both major and minor genes since some maize genotypes are known to have at least two major resistance genes with large effects and other genotypes are known to have one major gene with large effects and some other genes with minor effects (Kim & Brewbaker, 1977).

Cytoplasmic resistance

Cytoplasmic genes are also known to determine the inheritance of host-pathogen interaction for certain disease expression. Oats stem rust (*Puccinia graminis* f. sp. *avenae*) virulence in some oat varieties is known to be maternally inherited and is believed to be controlled by a single plasma-gene. In maize the inheritance of resistance to southern corn leaf blight, *Bipolaris maydis* (*Cochliobolus heterostrophus* in teleomorph state), is also known to be under the control of cytoplasmic genes.

18.1.2 Gene-for-Gene and Effector-Receptor Models for Pathogen-Host Relationships

For diallelic resistance locus in the host with dominant R and recessive r being resistance and susceptibility alleles respectively, and in the pathogen the corresponding locus has dominant V and recessive v as avirulence and virulence alleles respectively,

Table 18.1 Host reactions to different pathogen genotypes

Pathogen (alleles V and v)	Host (alleles R and r)	
	rr	RR and Rr
VV and Vv	Susceptible	Resistant
vv	Susceptible	Susceptible

the expected disease reaction between the host and the pathogen genotypes is as in Table 18.1.

This interaction between the pathogen and its plant host is explained by gene-for-gene model proposed by Flor (1956). In major gene resistance there is an allele for resistance in the host and a corresponding avirulence allele in the pathogen such that for each gene that conditions resistance in the host there is a corresponding virulence gene in the pathogen that breaks the resistance of the host. Resistance occurs when the host has resistance allele and the pathogen has avirulence alleles at the corresponding locus while disease expression occurs when the host does not have resistance allele or the parasite has virulence allele in homozygous condition. Resistance allele in the host protects the plant only if there is avirulence allele at the corresponding locus in the pathogen. However, if the pathogen has virulence alleles at the corresponding locus the resistance allele in the host does not express itself as it is overcome by the corresponding gene for virulence in the pathogen. There is a direct and specific relationship such that a gene for race-specific resistance can only be overcome by a corresponding gene for virulence in the pathogen. In cases of dominant resistance alleles and recessive virulence alleles there is no disease occurrence when both genes in the host and the pathogen are homozygous dominant or heterozygous but if in either the host or the pathogen or both the alleles are homozygous recessive susceptibility occurs and there is disease expression in host plant. The individual polygenes for resistance in the host also have corresponding virulence polygenes in the pathogen with which they interact on a gene-for-gene basis with each polygene contributing minor effects marred by the environmental effects.

Molecular interpretation of resistance and pathogenicity

Both the pathogen and the host produce proteins that help in causing pathogenicity and defence respectively. Molecular interpretation of gene-for-gene concept requires the presence of avirulence (Avr) genes, also referred to as effector genes, which encode effector proteins (virulence factors) in the pathogen that suppress the plant proteins to cause disease. In the host resistance is under the influence of race-specific R (resistance) genes which encode surface immune receptors (e.g., receptor-like kinase, i.e., RLK) or intra-cellular immune receptors (e.g., nucleotide-binding site leucine-rich repeat proteins, i.e., NLRs) that recognise the effectors/avirulence (Avr) proteins from the pathogen (Piquerez et al., 2014). The R-mediated resistance is mostly qualitative race-specific gene-for-gene resistance which breaks down easily when new races arise through mutation of the Avr genes.

The pathogen releases its toxins and effector proteins into the host cells to manipulate and defeat the plant host immunity system while complex proteins encoded by the host's R genes help to identify pathogen effectors (Jones & Dangl, 2006). During host attack by the pathogen the host R genes are induced to have plant immune receptors synthesised while pathogen-associated molecular patterns (PAMPS) and pathogen effectors are produced by the pathogen when it gains entry into the host. When the pathogen gains entry into the host's tissues it produces PAMPS and the host in turn responds to defend itself through a PAMP-triggered immunity (PTI) which helps it to identify foreign bodies in its tissues (Andersen et al., 2018; Deng et al., 2020). The host identifies the foreign bodies by employing surface receptor protein molecules called pattern recognition receptors (PRR) which recognise and bind PAMPS so that the plant defence response is induced (Piquerez et al., 2014). The receptor protein is activated when its conformation changes on recognition of the pathogen's effectors or effector-modified plant protein and interaction with other proteins while effectors have the role of overcoming host PTIs and avoiding recognition so that there is successful growth of the pathogen in the host with the pathogen taking advantage of the host for its nutrition (Bonardi et al., 2012; Piquerez et al., 2014). Hence, the pathogen injects effectors into the host to suppress PTI leading to effector triggered susceptibility (ETS) if host immunity fails. Since the pathogen effectors may cause evasion or suppression of the host PTI the host produces another set of receptors called effector triggered immunity (ETI) proteins which are encoded by R genes to activate the defence response (Andersen et al., 2018). The interactions lead to prevention of further infection through hypersensitive response resulting in localised cell death and several other reactions by the plant. Hypersensitive reaction is due to specific interaction arising between products encoded by the resistance gene and avirulence gene such that if either one of the products is absent there is no compatibility and disease expression occurs. In crops R genes are mostly responsible for this system of immunity to various pathogens. The ETIs lead to activation of systemic acquired resistance. The PAMP-triggered immunity and ETI are R-mediated resistance responses which are conditioned on activation of resistance proteins in the presence of the respective Avr proteins and pathogen detection by the host defence system that depends on recognition of molecules unique to the respective pathogen. Some R genes also encode PRRs [e.g., RLKs, receptor-like proteins (RLP) and cell wall associated protein kinases (WAKs)] leading to race-specific resistance too (Andersen et al., 2018; Deng et al., 2020). In some cases, RLKs and WAKs are also encoded by quantitative trait loci. The R encoded proteins also play the role of guarding the effector target proteins referred to as guardee and that any modification of the target proteins by the pathogen proteins results in activation of the R proteins and resistance reaction in the host (Andersen et al., 2018). The receptor-effector model for R genes-Avr genes relationship suggests that R proteins and Avr proteins have direct corresponding relationship since R genes have direct recognition of pathogen avirulence genes to cause resistance. This specific R-Avr relationship corresponds to the gene-for-gene resistance. The ETI is involved in major gene resistance while

PTI-triggered resistance may be attributable to PTI, with some overlap between the two mechanisms. The PTI and ETI are interrelated and share signalling pathways and ETI depends on PTI to cause resistance to the pathogen.

Gene silencing

When a virus particle infects the plant through a wound or deposited into the plant tissue by a vector it is recognised by the plant defence system and the receptors encoded by the dominant R genes, are synthesised to counter the attack. The virus avirulence system also produces effectors. This interaction localises the virus at the point of infection because of programmed cell death, creating a necrotic lesion as evidence of hypersensitive reaction. In some cases of virus infection, the plant uses its RNA interference (RNAi) system to defend itself by cutting the RNA of the virus pathogen into small pieces thereby silencing the virus pathogenicity gene (Hussain, 2015; Zamore, 2004). The virus protein synthesis is disrupted and the plant resistance is expressed.

18.1.3 Pathogen Race Variability and Host Differentials

As a general rule most pathogens are highly variable for pathogenicity, especially the pathogenic fungi cases for which there is major gene host-pathogen relationship. Hosts are equally variable for resistance and susceptibility to the pathogens. Both hosts and pathogens have inherent variation which is reflected in their interaction resulting in disease occurrence or absence of disease expression in the host. Variability of the pathogen which is a product of mutations, sexual and mitotic recombination, especially in fungi and bacteria, is responsible for occurrence of races in pathogen populations. The more variable the pathogen is or more frequently mutable its genes at the virulence/avirulence loci are, the more often it will generate new races with different virulence alleles that are likely to break the defence system of the resistant host varieties. Some pathogens, namely, *Puccinia graminis, P. recondita*, powdery mildews and downy mildews are known to be highly variable having different pathogenic races due to mutation and recombination forces.

Due to gene-for-gene race-specific resistance relationship different races of a particular pathogen are usually distinguished according to their virulence capabilities using a set of genotypes of the host crop. Each host genotype has a single and unique major gene for resistance against a specific pathogen race which also has the corresponding avirulence major gene. There is a direct and specific relationship such that a gene for race-specific resistance can only be overcome by a corresponding gene for virulence in the pathogen. Since each host genotype has a gene that confers resistance against a particular race the genotypes are used to identify resistance genes, differentiate the various races of the pathogen, study the extent of variation in a given pathogen, detect occurrence of new mutant races and monitor the race trends for disease pathogens.

18.1.4 Breeding Methods for Disease Resistance

Success in resistance breeding depends on adequate knowledge of reproductive system of the crop species, technology and resources at the disposal of the breeder, the inheritance pattern and type of resistance and the ease with which it can be transferred from the donor to the recipient cultivar. Resistance should be stably expressed under varied conditions and not tightly linked to undesirable characters in the donor parent. In major gene resistance there should be distinct simple Mendelian inheritance with full expression of resistance and susceptibility in the relevant crop genotypes and adequate protection of the plant having the resistance genes. With clear cut major gene inheritance, it is easy to develop resistant cultivars using conventional methods of crossing and backcrossing while it is difficult to develop polygenic resistance using such strategies unless the methods used for quantitative traits are applied. The requirements for resistance breeding are pathogen inoculum, resistant germplasm resources for screening and resistance evaluation. Screening for resistance is done using appropriate germplasm, parasite inoculum, field nurseries, greenhouse, growth chamber and laboratory facilities to discriminate between the resistant and susceptible genotypes at the right stage of the host life cycle. Artificial infestation of genotypes for screening under controlled conditions is usually better than selection under natural infestation in uncontrolled field conditions. Sometimes a known susceptible spreader genotype which helps in building up the pathogen inoculum in the field is used under field screening conditions where it is planted side by sidc with the genotypes under resistance screening. The pathogen is expected to spread from the spreader crop to the cultivars being screened. Screening is also done by subjecting the plants or their tissues in vitro at the right stage to the appropriate pathotoxin extracted from the pathogen where the toxin is associated with disease expression following infection as is the case for toxin producing pathogens.

Sources of resistance

Regular germplasm resources from which the resistance genes are obtained include established varieties, introduced cultivars, obsolete varieties and landraces, wild and weedy species related to the crop in question. Availability of diverse germplasm is not only important for identification and selection for resistance genes but is also a safeguard against pathogen races which may arise and break the resistance of established resistant varieties. Different resistance gene donor parents in which resistance is well expressed should be assembled and it is important that the breeder works in collaboration with the gene banks to identify good resistance source germplasm since most germplasm materials are held in national and international gene banks and breeding institutions. The breeder identifies a source of clear resistance from among the established varieties first followed by obsolete cultivars, landraces and other populations, all within the primary gene pool. The search is extended to the secondary gene pool having the weedy and wild relatives for the crop if the desired genes are not obtained in the primary gene pool. Crop relatives in the tertiary gene pool should the last resort because of difficulties in transferring genes from them to the crops. A number of disease resistance genes have been obtained from wild, weedy relatives and landraces and are being used in crops like soybean, wheat, rice tobacco, potato and other crops against various diseases.

Advances in genomics and phenomics helps in allele mining using landraces, wild and weedy relatives and identification of novel alleles or different alleles of various candidate genes with the aim of identifying superior genes in the germplasm. In addition to allelic variation detection, allele mining helps to identify and analyse haplotypes concerned with resistance, identifying the relationship between the germplasm accessions and identification of molecular markers for the resistance alleles. Successful allele mining is conditioned on genetic information from the materials at hand, the diversity of materials used in the screening, genome and gene sequence data and bioinformatics tools.

Disease resistance breeding requires that the available diverse germplasm be subjected to screening using efficient methods and appropriate high-throughput inoculation procedure and identification of R genes using genome sequencing and mapping techniques. Field nursery has the advantage of multiple infection which, along with artificial inoculation, helps to identify resistance genes responsible for broad spectrum resistance. Apart from the use of the conventional breeding procedure, the identified R genes are cloned to be used for transfer of resistance to the target cultivars through molecular marker assisted breeding methods, genomic selection and transformation. De novo domestication is also used to broaden resistance while genome editing can be employed to create mutations in susceptibility genes for strengthening resistance in the cultivars under cultivation and elite breeding materials.

Other sources of germplasm that are important in breeding for resistance are products of mutation breeding, biotechnology and transgenesis. Induced mutagenesis can be done in vivo and in vitro for target cultivars or elite breeding materials followed by selection for resistance to the disease in question and crossing and backcrossing to the established cultivar where necessary. In vitro selection is most suitable to selection for resistance to necrotroph pathogens where the pathotoxins are used as selection agents. Somaclonal variation created in in vitro cultures is a rich source of variation for possible resistance genes. If various tissue cultures are subjected to in vitro selection some variants that survive are likely to be mutants having genetic changes conferring resistance and are then regenerated into whole plants. In tissue culture toxins of fungi like *Fusarium, Phytophthora, Alternaria, Helminthosporium*, etc. and bacterial pathogens like *Pseudomonas, Xanthomonas, Erwina* are used for screening in vitro to develop disease resistant genotypes. In vitro selection may not be applicable for resistance to biotrophic pathogens which do not produce toxins but resistance is based on features of intact and differentiated tissues.

Resistance gene deployment

Single gene vertical resistance breeding

Because major gene monogenic or oligogenic resistance has simple Mendelian inheritance it is easily transferred and manipulated through crossing and backcrossing, especially in the presence of molecular markers. Major gene resistance is deployed through use of a gene for a period and then replaced when it is no longer effective. In self-pollinated crops once resistance gene has been identified it is introduced into the recipient desirable established variety by crossing the variety to the donor cultivar followed by backcrossing to the variety as the recurrent parent for a number of generations. In every generation selection is made for resistance trait and the characteristics of the recurrent parent. Use of DNA molecular markers improves the efficiency and fast-tracks the gene transfer process and recovery of the genome of the recurrent parent through marker-assisted backcrossing. A common strategy is to develop and release superior resistant varieties with major resistance genes as new pathogen races arise and when their resistance breaks down newly identified resistance genes are again incorporated by crossing and backcrossing. This strategy of using one resistance gene at a time leads to development of several varieties at great costs, many of which remain abandoned if the pathogen is variable with new races which keep coming up frequently. A major gene protects the plant against the pathogen until a new virulent race to which it does not confer resistance arises to break the resistance and become established. An R gene can also be used for some time and then replaced with another R gene when it breaks down and brought back later into use when the race which made it breakdown is no longer prevalent. In cross-pollinated crops inbred lines for hybrid development are improved by crossing to resistance gene sources followed by backcrossing to the inbred line.

Where a cultivar with good agronomic traits and fairly good quality attributes is the donor for resistance to another well-established cultivar only a few generations of backcrossing with selection for the traits of the recurrent parent will be needed. Introduction of resistance from agronomically unfavourable cultivars with low yields, poor quality, poor adaptability and even related species, necessitates several generations of backcrossing to eliminate undesirable genes from the donor of resistance genes.

Development of varieties with multigenic resistance for major genes and durability of resistance

Resistance stability is reflected in its durability and how long it remains unbroken by the pathogen and how widespread the resistant variety is cultivated in a region or across regions. The durability is determined by the number of resistance genes in a cultivar, pathogen host range, variability of the pathogen population which is in turn determined by the mutability of its pathogenicity genes and how fast the pathogen races spread in a region. Where the pathogen has a wide host range resistance lasts longer than if the pathogen has a restricted host range (Ribeiro do Vale et al., 2001). A new pathogen race will cause disease on a previously resistant variety only if it is virulent and overcomes all of that variety's resistance genes. Resistance breakdown applies more to single major gene resistance and resistance of hypersensitive nature than to the more stable multigene or polygenic resistance because the pathogen may need only one or a few gene mutations to unlock the resistance. Inherent variability in the pathogen and frequent mutation of the pathogen genes leading to new races is a major threat to the single major gene resistance and jeopardises production of cultivars having single major gene resistance.

Varieties with two or more resistance genes against a particular disease have lower probability of becoming susceptible to new races than the varieties possessing single resistance genes. Multigenic major gene resistance needs several independent pathogenicity gene mutations so that the pathogen can break resistance in the host as compared to the monogenic vertical resistance where the pathogen needs only one such mutation to break the host resistance. Since probability that several mutations will occur in the same pathogen at the same time is generally low the likelihood that a race will become homozygous for two or more new virulence genes is much less than it will do so for one gene. If a variety has three resistance genes for a given disease at three different loci and the pathogen also has avirulence genes at the corresponding pathogenicity loci the variety has a very low probability of succumbing to the disease unless the pathogen has races with virulence alleles at the three loci. If the pathogen genes at the three loci have only forward mutation to recessive virulence alleles at rates u_1, u_2 and u_3 respectively, the probability that at each locus the pathogen will be homozygous for the new mutant virulence allele is u_1^2, u_2^2 and u_3^2 respectively. The probability that the pathogen will be homozygous for virulence at the three loci at the same time as a result of mutation is $(u_1^2)(u_2^2)(u_3^2)$. Therefore, taking into account that natural mutation rates are always low and are usually less than 1×10^{-5} a variety with multigenic resistance will maintain its resistance for longer periods than varieties with single resistance genes. Varieties with multigenic resistance can be developed by pyramiding of genes, regional gene deployment and multiline varieties. Multiline cultivars (see Sect. 11.4) have been used to reduce disease level, increase the diversity, stability and durability of major gene resistance in self-pollinated crops like wheat, barley, oats, rice, etc.

(i) *Gene pyramiding*
Gene pyramiding involves stacking of several major genes at different loci or several $QTLs$ in a cultivar to confer durable resistance against a particular pathogen. Since different R genes encode equally different nucleotide-binding leucine-rich repeat proteins pyramiding of several different *R* genes against a particular pathogen in one genetic background gives a more stable and longer lasting resistance than a single R gene. Combination of R genes and $QTLs$ responsible for different resistance components against a particular pathogen provides the plant with a much more stable and durable resistance. This is achieved through identification of different sources of resistance genes, multiple parent hybridisation, backcrossing and selection and use of more than one pathogen race in screening. Bringing genes together through pyramiding is done efficiently and faster employing molecular marker assisted backcrossing with the help of flanking molecular markers tightly linked to the target genes or QTLs or molecular markers located within the genes or QTLs so that the plants with the target genes can be identified early before flowering, especially in segregating populations. Molecular markers are also employed for recovery

of the recurrent parent in the backcrosses. Gene pyramiding can be done under speed breeding protocols to fast-track the development of resistant genotypes. The simplest way for gene pyramiding is by sequential transfer of different resistance genes to a well-established cultivar targeted for improvement. The established cultivar is crossed to an identified source of resistance gene to incorporate the gene into the cultivar followed by backcrosses to the established cultivar, as the recurrent parent. Backcrossing is done to the recurrent parent for two or three generations with selection done using the appropriate molecular maker. The derived backcross progeny having the marker, hence the target gene/QTL, is then crossed to another donor with another target gene/QTL, also tagged with a molecular marker(s). The derived cross is crossed back to the recurrent parent, followed by one more or two backcrosses to the recurrent parent. Once the transfer is achieved with the traits of the established cultivar reinstated another resistance gene from another source for resistance is identified and the resistance is similarly incorporated to the established cultivar which already has the resistance genes previously incorporated. If there is another donor parent with another target gene/QTL the progenies derived from the last backcross are be used in a similar procedure followed by crossing and backcrossing to the recurrent parent.

Pyramiding of genes is achieved through multiple cross approach before all the available resistance genes are brought together in one genetic background. Different donor parents, J, G, H, I, etc., having different molecular marker-tagged resistance genes/$QTLs$ are separately crossed to the elite but resistance deficient cultivar E, used as the recurrent parent in all backcrossing. The derived F_1 progenies are independently backcrossed to the recurrent parent E. The number of donor parents to be crossed to E depends on the number of different resistance genes and the sources. The obtained backcross progenies are backcrossed further to the recurrent parent for the number of generations adequate to recover the traits of the recurrent parent then the desired lines with resistance and the E genome are selected in each case. The selected lines having different resistance genes are hybridised in a multiple-way cross method (Fig. 18.1) to obtain hybrids combining the target resistance genes e.g., $[(JE_{Bn} \times GE_{Bn})] \times [(HE_{Bn} \times IE_{Bn})]$ where $_{Bn}$ indicates the number of backcross generations to the recurrent parent E involving the progenies resulting from a cross between cultivar E and a donor parent (e.g., JE cross, etc). In this pyramiding procedure the lines selected from the backcross progenies are crossed to stack the genes in one genotype and the derived F_1 are used to obtain pure lines or are crossed further to give a higher-level multiple-way cross. This procedure was described by Jiang (2013) as convergent backcrossing. If there are more sources of the resistance genes the four-way cross progenies may be crossed to other hybrids for further multiple gene pyramiding.

Each of the selected donor parents, J, G, H and I having different resistance genes/QTLs appropriately tagged using molecular markers can also be crossed separately to the elite but disease susceptible recurrent parent E. The derived F_1's, E × J, E × G, E × H and E × I, are crossed to give double cross hybrids,

(E × J) × (E × G) and (E × H) × (E × I) which are used to get pure lines with at least four resistance genes each through some generations of backcrossing to E. The double crosses can also be crossed further to give higher level of multiple parent hybrid progenies, [(E × J) × (E × G)] × [(E × H) × (E × I)]. The progenies of the higher-level multiple-way crosses are backcrossed to the recurrent parent E for enough generations to give the desired lines with multiple gene resistance and the genome of the recurrent parent (Fig. 18.2). This method is similar to the approach described by Jiang (2013) as synchronised backcrossing or simultaneous backcrossing The relevant pathogen races are used in screening to select the progenies having the target resistance genes. Gene pyramiding for disease resistance has been realised for several crops like resistance of flax resistance to flax rust, wheat stem and leaf rust resistance, rice bacterial blight and blast resistance, barley stripe rust resistance and resistance of other crops to various diseases.

(ii) *Regional gene deployment*

Regional gene deployment entails using different vertical resistance genes in different areas so as to break epidemic development of a particular disease in a given region where the pathogen spreads in sequence from one part of the region to the next in different seasons. The survival and aggressiveness of the pathogen can be broken by growing varieties with different resistance genes in different regions, i.e., resistance genes used in one region should be different from resistance genes used in other regions. A well-established classic example is the Puccinia Path in central North America described by Frey et al. (1973) in which the Puccinia pathogen survives in the south during winter and moves to the north in spring and summer. Frey et al. (1973) proposed the subdivision of central North America into three regions, namely, most southern region, central region and far northern region, so as to help control crown rust disease of oats. In the Puccinia Path *Puccinia coronate avenae* over-winters in the southern region while oat is a summer season crop in the far north where the pathogen epidemics occur every year during growing season. In the central region oats are also grown and can serve as pathogen link between north and south. The northern and southern subregions supplement each other as regards rust development across the *Puccinia* Path extending from Mexico in the south to the Canadian prairies in the north in that the uredial stage of the *Puccinia* rust does not survive the southern summer in Mexico and Texas and also does not survive during winter in the north making the north and south interdependent as regards rust pathogen inoculum survival. The survival and aggressiveness of the pathogen can be broken by growing varieties with different resistance genes in the north, south and central regions, i.e., resistance genes used in the north should be different from resistance genes used in the central and south regions.

Horizontal resistance breeding

Where resistance components are of quantitative nature selection procedure following hybridisation employ breeding methods for development of resistant pure line varieties in autogamous crops and appropriately designed recurrent selection methods for development of resistant inbred parents and open pollinated populations in allogamous crops. Recombination and recurrent selection strategies for autogamous crops are employed to accumulate quantitative resistance genes from different sources to develop populations and pure lines with highly elevated resistance levels. Plants with low levels of disease severity and moderate levels of resistance are selected for use as parents in the recombination methods accompanied by selection for a number of generations in the populations developed. Quantitative trait loci mapping helps to identify molecular markers associated with polygenic resistance to facilitate accumulation of resistance polygenes in desired cultivars and development of pure line varieties with high levels of resistance. The relevant QTLs are selected as early as possible, particular starting in F_2 or F_3, to ensure that they are retained.

Similarly, in allogamous crops accumulation of polygenes using recurrent selection is the most effective way of developing cultivars with stable polygenic resistance. Genotypes with moderate to high levels of resistance are identified for use as parents in recombination and selection to accumulate resistance polygenes in breeding populations for development of improved open pollinated populations, inbred lines and hybrid varieties with elevated stable resistance. Recurrent selection methods have been successfully applied to improve horizontal or partial resistance to various diseases in maize, potato, barley, soybean, sorghum, etc.

Molecular breeding and transgenesis for disease resistance

Molecular breeding

Quantitative resistance is responsible for more durable resistance because of multiple small effects of polygenes but requires several cycles of conventional selection to develop such resistance in one variety. Lack of durability for qualitative resistance, the problem of identifying quantitative resistance, inconsistent QTL segregation in mapping populations and QTL interaction with environment and epistasis makes genomic selection preferable for development of quantitative resistance. Next-generation sequencing tools and DNA polymorphism detection methods facilitate the identification and precise location of the resistance gene/QTLs and their markers. The DNA markers e.g., SSRs/microsatellites, SNPs, small insertion deletions, AFLPs, RAPDs, etc, can be employed to identify genes and QTLs associated with desirable disease resistance levels for conducting indirect resistance selection. Availability of whole genome sequencing information and molecular markers distributed in the whole genome and tightly linked to their respective genes/QTLs facilitates the use of genomic selection for accumulation of the genes/QTLs controlling resistance to a given pathogen. Mapping-by-sequencing or genotype-by-sequencing and techniques like genome-wide association studies assist in precise identification of the R genes and resistance QTLs, their respective markers and cloning of genes for use in genome assisted breeding and transgenesis. The genomics and phenomics data are important in constructing models and genomic estimated breeding values

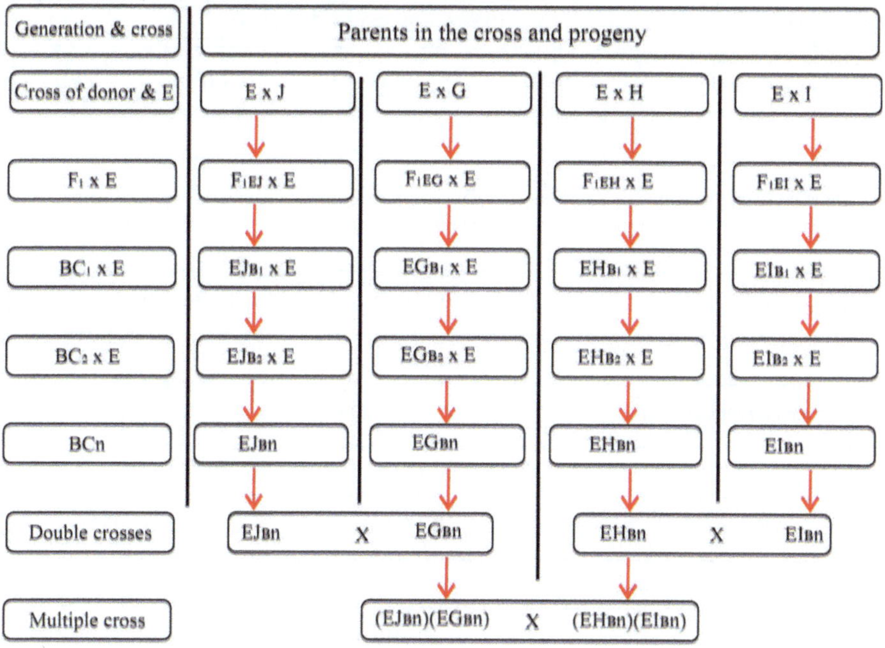

Fig. 18.1 Gene pyramiding using multiple crossing 1

proposed by Meuwissen et al. (2001) for application in genomic selection. Gene expression for plant response to the various pathogens is understood in more detail through genomics, proteomics, metabolomics, transcriptomics molecular analyses because these analyses help in detecting proteins produced/accumulated as a result of pathogen-host interaction, particularly specific response proteins for pathogenicity and resistance, e.g., effectors and receptors for both pathogens and hosts.

Transgenesis

In addition to conventional breeding methods and interspecific hybridisation, transfer of genes between plant species can be done through transgenesis, especially for the major gene resistance. Transgenesis helps in management of problematic plant diseases by introducing alien resistance genes into genomes of desired cultivars and achieving overexpression of genes that encode proteins which interfere with the pathogen and disease development based on gene-for-gene resistance. Genetic transformation methods fast-track gene transfer process and reduce the time spent on classical breeding for resistance. Different resistance genes are isolated and spliced and transferred to genomes of target cultivars making them have durable multigenic resistance. Molecular approach for gene stacking is also achievable through genetic engineering by assembling several NLR genes for transfer to target crop genomes using various DNA transfer methods. Genetic engineering for combination

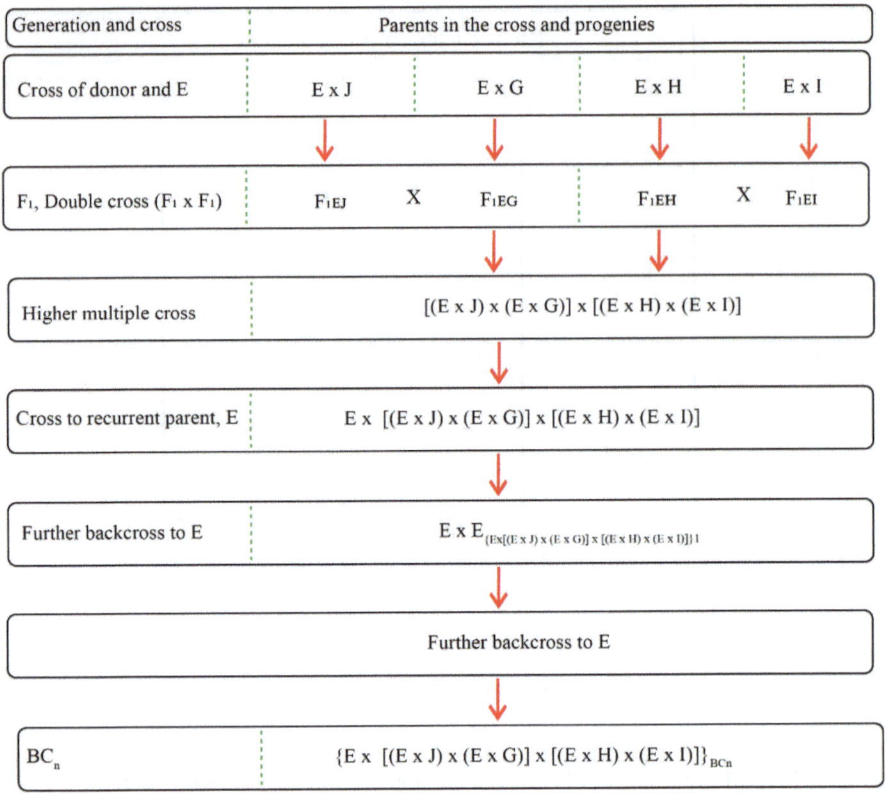

Fig. 18.2 Gene pyramiding using multiple crossing 2

of PRR and NLR genes strengthens and stabilises resistance R genes and makes them long lasting. Through transgenesis several alien genes have been incorporated into genomes of cultivated crops.

Genome editing and gene silencing

Plants can be transformed by inserting into their genomes genes that cause resistance by encoding biochemicals which degrade mycotoxins, or interfere with pathogen cell wall structure and degrading viral RNA. Through genome editing and transgenesis receptor genes can be modified to encode receptors that recognise effectors of different pathogen races. For better protection against virus pathogens the plants are engineered with genes which cause overexpression of viral RNA as regards double stranded RNA that activates the plant RNAi system to cleave the viral RNA and terminate the pathogenicity process (Hussain, 2015; Zamore, 2004). Genome editing is used to achieve similar activity by creating changes in the plant genes that are concerned with RNAi system. The host plant RNAi gene silencing system can be subjected to modification through transgenesis to develop disease resistant cultivars by equipping transgenic plants with siRNA or miRNA to silence the pathogen genome (Simon-Mateo & Garcia, 2011). It is possible to strengthen crop plant resistance through genome editing using Clustered Regularly Interspaced Palindromic Repeats – Associated Protein 9 (CRISPR-Cas9) technology or other genome editing methods to create new resistance alleles in high yielding susceptible cultivars by editing R genes and susceptibility (S) genes. The PRR and NLR gene pyramids can be construct through gene editing to develop strong and stable resistance against multiple pathogens. The NLR gene pyramids are possible to construct through gene editing within same NLR cluster to strengthen resistance while de novo domestication of resistance alien genes transferred to the crops can be achieved through step-by-step genome-editing to develop stable resistance against multiple pathogen races (Deng et al., 2020).

18.2 Breeding for Resistance to Insect Pests

Crops often suffer infestation and colonisation by various insect pests, the herbivore insects, causing significant damage, yield reductions and poor product quality, leading to low returns to the farmers and food insecurity, especially in developing countries. Insect pests cause injury to the crops through defoliation, stunted growth, boring holes on leaves, stems, flowers and fruits, dead heart, death of whole plant, feeding on fruits and seeds, damage to roots, disease transmission and destruction of stored produce as well. To address pest menace the farmers have the alternatives of chemical control or use of resistant varieties where available. Chemical control which is often unaffordable by small scale farmers has the problem of high chemical costs and technology of application, insect resistance to the insecticides, harmful chemical effects to the farmer, the general public, useful insects and other organisms and accumulation of undesirable residues causing environmental pollution. These undesirable features along with broad-spectrum property of most insecticides necessitate the use of resistant varieties and integrated pest management as the cheapest, safest and most environment friendly insect pest control measures.

18.2.1 Mechanisms of Resistance to Insect Pests

A crop variety resistance to insect pests is reflected by the inherent ability of the genotype to produce a larger crop yield of good quality than the susceptible genotype under similar pest infestation/same level of damage (Sandhu & Kang, 2017). Resistance comprises of genetically inherited characteristics that make a crop cultivar less suitable as a host to a pest and is less infested, repellent to the pest, suffers less damage or recovers from damage by the pest better than a susceptible cultivar lacking such characteristics under comparable conditions of attack by the pest. A resistant cultivar has features which reduce the extent of utilisation of the plant as a host and source of food by an insect pest. Plant resistance against insect pests may be constitutively present and displayed in the plant without any external stimulation or indirect and induced by insect attack on the host plant leading to production of anti-feeding proteins, insecticidal secondary metabolites, volatile compounds, compensatory growth and increased trichome density (Belete, 2018).

Types of resistance, morphological and phytochemical basis of resistance to insect pests

Types of resistance

According to Painter (1941) there are three categories of plant insect pest resistance mechanisms, namely, avoidance or non-preference (antixenosis), antibiosis and tolerance. The other mechanism is escape though it is not resistance per se. The plant host may have one or a combination of these resistance mechanisms. Both antibiosis and antixenosis disrupt and interfere with various stages and activities of insect pest life cycle thereby disrupting insect establishment on the plant.

Non-preference also known as avoidance or antixenosis, prevents contact between the pest and the host plant such that the pest does not parasitize the host plant (Painter, 1941). Non-preference refers to effects of the plant on insect's parasitizing activities like feeding, shelter and oviposition making the plant less attractive and be avoided by the pest because it is inappropriate and an undesirable. Antixenosis is used for genotypes with features which lead to pest differential oviposition and colonisation of the plant host such that the plant is avoided by the pest since it is an undesirable host (Kogan & Ortman, 1978). The plant may have features which prevent contact between the host and the parasite like pubescence leading to avoidance while for non-preference there are mechanisms which reduce the extent of contact between the host and the parasite. Varieties with non-preference or avoidance or antixenosis type of resistance have lower insect populations, lower insect pest reproduction rates on them and have less pest damage or not damaged at all as compared to the susceptible genotypes. Plants that have or lack certain morphological, physical and phytochemical features which interfere with colonisation activities by the insect pest leading to reduced insect feeding, reproduction and mobility have non-preference resistance characteristic. Non-preference/antixenosis or avoidance may be due to presence of features like pubescence, waxy coatings on organ surfaces, thickened and tough tissues or presence of phytochemicals that repel or deter the pests from feeding and oviposition or lower the nutritive value of the plant due to lack of certain nutrients needed by the pest or absence of chemicals that stimulate the pest oviposition (Belete, 2018).

Antibiosis comprises of features in a resistant plant that make the plant have undesirable and deleterious effects in the insect life cycle when the insect feeds on such plants (Painter, 1941). Plants with antibiosis resistance mechanism have physical and biochemical features which are unfavourable to the survival, growth, development, oviposition and reproduction of the insect. The antibiotic host plant imparts temporary or permanent adverse effects on the insect leading to insect mortality in various stages, reduced fertility and slow growth rate and smaller size, failure to convert ingested food, failure to pupate and failure of adult emergence from pupal stage, restlessness, physiological, metabolic and morphogenic irregularities and other abnormalities in the pest (Koch et al., 2016). Some resistant cultivars lack the required nutritive quality for pest survival and reproduction or have feeding deterrents, antimetabolites or growth inhibitors interfering with early stages of pest growth and development. Resistance characteristics responsible for non-preference or antixenosis also impart antibiosis resistance and vice versa.

Antibiosis is usually the most effective resistance, especially in short term. However, it breaks down very fast because of high selection pressure on insect pest population. Antibiosis and non-preference are useful where the pest has other alternative hosts but may not be helpful for prolonged long resistance under monoculture because the insect has no alternative.

Tolerance is a situation where the plant is capable of growing and reproducing or repairing the injury caused by the insect despite supporting similar insect population that causes damage to a susceptible host (Painter, 1941). Tolerance lowers the negative effects of insect pests on plant performance following attack by insect and a tolerant plant has normal growth and production despite presence of pest population that will cause serious damage and yield loss of a susceptible cultivar. Tolerance allows for continued support of the pest population, minimises yield reduction caused by pest infestation and does not interfere with survival, growth, reproduction of the insect pest and tolerant cultivars have vigour and the genetic capability to withstand pest attack and damage by outgrowing the infestation, regrowth of damaged tissue or recovery and compensatory growth with production of more organs, ending up with near normal yields that is better than non-tolerant cultivars under similar infestation conditions (Koch et al., 2016). Tolerant plants also show increased photosynthetic rate, higher growth rate, branching or tillering, relocation of pre-existing photosynthates and other metabolites from the roots to the shoots following pest attack (Koch et al., 2016). Some sorghum genotypes are known to tolerate chinch bug (*Blissus leucopterus*) and spotted stalk borer (*Chilo partellus*) while attack of tolerant young sorghum plants by sorghum shoot fly (*Atherigona soccata*) is usually followed by synchronous tillering and rapid growth leading to near normal yield. Tolerant varieties promote pest population build-up because they do not inhibit pest growth, development and reproduction.

The plants may escape attack if there is asynchrony between the insect pest parasitic stage and the host plant susceptibility stage. Escape resistance is usually manifested by earliness in a number of crops. In cotton earliness helps some cultivars escape cotton boll weevil (*Anthonomus grandis*) attack. Early flowering and short anthesis duration also make sorghum escape midge (*Stenodiplosis sorghicola*) attack since sorghum genotypes with shorth anthesis periods will be exposed for short periods only to oviposition by the midge insect.

Morphological and phytochemical basis of resistance to insect pests

Plants have morphological, biochemical and colour features making them resist insect pests by adversely affecting infestation and establishment of the pests on the host plants and interfering with pest activities like feeding and food ingestion, mating, oviposition, survival and movement. Some features may be lethal at certain pest life cycle stages. Factors which determine plant resistance to pests are genotype, growth stage, biochemical and morphological features of the pest and the host and environmental factors. Plant resistance against insect pests may be constitutively present and formed in the plant without any external stimulation like trichomes, cell wall lignification, silica deposits, organ pigmentation, repellents and antinutrients or induced by insect attack on the host plant leading to morphological changes like increased trichome density, increased lignification, rapid compensatory growth, production of anti-feeding proteins and insecticidal secondary metabolites. Plant resistance features may directly affect the herbivore insect or may act indirectly to cause resistance. Indirect resistance does not affect the insect directly but the attacked host plant attracts the natural enemies of the insect pest through production of volatile compounds and extrafloral nectar.

Morphological and physical characteristics like trichomes, thorns, setae, wax on leaf and stem surfaces, cell wall thickness, lignification of plant tissues, suberisation and hardening of plant organs and tissues and silica deposit and glandular minerals are direct resistance features acting like barriers to insect infestation and colonisation of the plant. Hardening of leaves and other organs leads to resistance to pests since it lowers palatability and digestibility of such leaves by the insect herbivores that devour the foliage or other organs (Belete, 2018). Thickness of some tissues and organs is deterrent to certain insect pests preventing them from infesting and colonising potential hosts as known in some sugarcane genotypes resistant to shoot borer. Organ and tissue toughness, cell wall composition and pubescence interfere with some insect pest feeding and swallowing. Tough plant tissues having cell walls hardened with lignin, cellulose, suberin, callose and silica particles are harmful to the mouth parts of chewing insect pests making feeding difficult. In some sorghum cultivars resistance to sorghum shoot fly (*Atherigona varia socata*) is attributed to lignified cell walls surrounding the vascular bundle sheaths (Blum, 1968) while wheat resistance to stem sawfly (*Cephus cinctus*) is due to solid stems (Wallace et al., 1973) and hardness of stem rind. Trichomes are responsible for plant surface pubescence, interference with landing, movement, settling, feeding, mating, oviposition, feeding and food ingestion by insect pests. Trichomes are of varied shapes and length, e.g., straight, hooked, spiral, stellate and glandular. Pubescence on leaves, stems and other organs of crop plants is, in several cases, a non-preference mechanism, providing protection against the pests. Pubescent crop cultivars are resistant to insects with piercing mouth parts like jassids, aphids and leafhoppers because trichomes interfere with the feeding of these pests while hooked trichomes or those that release exudates that stick to the bodies of the insects impair insect mobility or kill them (Alhousari & Greger, 2018; Mitchell et al., 2016). The hooked trichomes often trap and kill soft bodied insect pests like aphids while glandular trichomes release secondary metabolites, e.g., flavonoids, terpenoids, alkaloids and other phytochemicals which are toxic or repellent to the insects or trap them. In certain *Solanum* species, especially *S. berthaultii,* there are trichomes with distal glands that produce adhesive exudates which stick onto bodies of Colorado potato beetle (*Leptinotarsa decemlineata*), green peach aphid (*Myzus persicae*), potato leafhopper (*Empoasca fabae*) and potato flea beetle (*Epitrix cucumeris*) incapacitating these pests and leading to their death (Tingey & Laubengayer, 1981). Insect attack also leads to increased trichome density in some plant species as in black mustard attacked by cabbage white butterfly, *Pieris rapae* (War et al., 2012). Pubescence in soybean interferes with feeding of potato leafhopper while pubescent wheat also reduces feeding and food ingestion by larvae of cereal leaf beetle (*Oulema melanopus*) leading to some mortality (Webster et al., 1973). Pubescence is also known to impart cotton resistance to a number of insect pests like jassids (*Empoasca facialis*), spider mite (*Tetranychus* spp.), pink bollworm (*Pectinophora gossypiella*), cotton aphid (*Aphis gossypii*), cotton leafworm (*Spodoptera littoralis*) and tobacco budworm (*Heliothis virescens*) while resistance of sorghum to shoot fly is also attributable to the presence of trichomes and silica deposits in the epidermis of leaves. In cotton the frego mutant with narrow and twisted bracts are known to be resistant to bollworm (*Helicoverpa armigera*) and boll weevil (*Anthonomus grandis*).

Silica and wax deposits on plant surfaces prevent insect pest attack by damaging insect pest mouth parts and negatively affecting digestion and insect pest movement (Alhousari & Greger, 2018; Mitchell et al., 2016). Heavy wax deposits on leaves and stems make sorghum resist stem borer (*C. partellus*) by interfering with feeding and movement of the larvae on the plant because as the larvae move on the plant wax accumulation impedes their mobility (Belete, 2018). Some sorghum genotypes also exhibit tolerance to tunnelling by the boring larvae of this pest. Waxy coating on sorghum is known to deter migratory locust (*Locusta migratoria migratoides*) from feeding on the sorghum plant (Atkin & Hamilton, 1982).

A wide range of biochemicals produced by plants are poisonous or repellent to certain insect pests and are responsible for resistance to the pests. Phytochemicals like alkaloids, phenolics, benzoxazinoids, terpenoids, flavonoids, silica, quinones, tannins, lignin, lectins, glucosinolates, cyanogenic glycosides, certain enzymes and hormones which interfere with digestion and utilisation of food, proteinase inhibitors and terpenoids which are produced constitutively or induced due to herbivore insect attack provide resistance against the insect pests. Biochemical-based resistance may affect the insects through toxicity and lethality, reduced metabolism, retarded growth and development, reduced period of insect reproduction, feeding deterrence and repellence to the pests (Belete, 2018; Mitchell et al., 2016). Proteinase inhibitors in plant storage organs like tubers and seeds interfere with herbivore insect gut enzymes resulting in indigestion of proteins in the insect gut, starvation, poor development, reduced growth and even mortality. In some *Nicotiana* species trypsin proteinase inhibitors and nicotine create defence against certain Spodoptera species (War et al., 2012). In maize resistance to European corn borer (*Ostrinia nubilalis*) is attributable to MBOA (6-methoxybenzoxaline) that is produced in the tissues of young maize plants when the larvae of this pest damage the plant tissue making DIMBOA (2-4-dihydroxy-7-methoxy-1,4-benzoxazin-3-one) present in the plant tissue be converted to MBOA which interferes with insect reproduction, larval feeding, development, growth and reduces the number of larvae on the plant due to larval mortality (Furstenberg-Hagg et al., 2013; Klun et al., 1967). MBOA acts as a feeding deterrent to European corn borer in maize. Resistance of young sorghum plants to certain corn borers is attributed to hydrogen cyanide produced from cyanogenic glucoside in sorghum tissues (Furstenberg-Hagg et al., 2013; Klun et al., 1967). In certain Solanaceae species, including some genotypes of *Lycopersicon esculentum* glandular trichomes produce biochemicals which repel or trap aphids and whitefly (*Bemisia tabaci*) making them resistant to these pests (Belete, 2018). In maize some varieties are resistant to corn earworm (*Heliothis zea*) due to presence of C-glycosyl flavone maysin in the silk and tassels which interferes with development and growth of corn earworm larvae while in some sorghum genotypes resistance to sorghum shoot fly is due 4,4-dimethyl cyclooctene (Belete, 2018). Hormones like salicylic acid are involved in plant resistance to herbivore insects as they are known to induce resistance against piercing and sucking insects and some chewing insects like resistance to *Heliothis armigera* in tomato and resistance to spider mite (*Tetranychus* spp.) in lima bean (War et al., 2012). Pyrethrin esters in pyrethrum (*Chrysanthemum cinerariifolium)* cause reduced oviposition, poor egg development, low egg hatching and are deterrents to feeding of western flower thrips (*Frankliniella occidentalis*). Western flower thrips are found in flowers that do not have the pyrethin esters but not on leaves in which the natural concentrations of pyrethin are strongly deterrent and lethal to the thrips (Yang et al., 2012).

In a number of cases, insect feeding and oviposition on plant hosts induce the plants to produce herbivore induced plant volatile (HIPV) and extrafloral nectar phytocompounds which may be toxic, antinutritional and repellent to the pests or attractive to biological enemies of the pests like predators and parasitoids (Belete, 2018; Stenberg et al., 2015). There are over 1000 of these herbivore-induced plant volatile compounds which include aldehydes, alcohols, esters, terpenoids, green leaf volatiles, methyl salicylate, monoterpenes, etc. and extrafloral nectar compounds released from plant tissues and organs (Furstenberg-Hagg et al., 2013). Over 70 plant species are known to produce extrafloral compounds which may be secreted by leaves and other plant organs as known in some cotton genotypes, cashew, passion flowers and castor oil plant (Furstenberg-Hagg et al., 2013). The volatile organic and extrafloral nectar compounds attract natural enemies of insect pests like big-eyed bug (*Geocoris* spp.), lady bird beetles and other insect pest predators. Jasmonic acid is also known to be involved in upregulation of genes involved in biosynthesis pathways for some pest resistance phytochemicals.

Organ surface pigmentation protects some crop cultivars against certain insect pests because colours and shapes of plant organs give wrong perception to the pests leading to resistance. In sorghum leaf sheath pigmentation has protective effects against sorghum shoot fly while in cotton red leaves combined with pubescence and absence of stem glands is responsible for boll weevil resistance. Plant colouration and repellent odour are also non-preference resistance features against some insects as known for leaf sheath pigmentation which repels shoot fly in sorghum (War et al., 2012).

18.2.2 Genetics of Resistance to Insect Pests

Genetic classification of plant resistance to pests is similar to that for plant resistance to diseases and can be categorised as vertical resistance attributable to major (Mendelian) gene effects with discontinuous distribution in segregating populations and horizontal resistance controlled by polygenes or minor genes each with small effects and influenced by environment leading to continuous variation. In case of major gene resistance different cultivars with different resistance genotypes show clear cut distinction between susceptible and resistant cultivars when they are infected with the same pest biotypes. The F_1 and segregating populations, F_2, F_3, etc., and their parents can be used to determine the mode of resistance. In addition to being controlled by only one or a few genes, vertical resistance is biotype specific and is often unstable and short-lived because it exerts high selection pressure on insect populations. Major gene resistance is mostly dominant though there may be a few cases of recessive genes being responsible for resistance. Both major and minor gene resistance genes may also have epistasis and additive effects and in case of major gene control there may be more than one locus involved. In some cases, major gene resistance is influenced by modifying genes which behave as polygenes so that major and minor genes may be combined in a cultivar to exert resistance jointly. Polygenic horizontal resistance is more stable and durable than major gene resistance, is non-biotype specific and does not exhibit differential interaction when different cultivars are infested with different biotypes of the same insect pest. To overcome the resistance of a cultivar having polygenic resistance the pest needs several corresponding virulence alleles with minor effects to break the polygenic resistance which may also be attributed to unrelated resistance features. Several characters associated with resistance like morphological, biochemical, colour and phenological features have patterns which may be of Mendelian or polygenic nature.

Gene-for-gene concept of host-parasite interaction

Though gene-for-gene relationship is not adequately clear for most host-pest relationship, the gene-for-gene concept proposed by Flor (1956) to explain the plant host-pathogen interaction is also applicable to interpret the host plant-pest interaction, especially for the piercing and sucking insects. For each gene for resistance (R gene) in the host there is a corresponding locus for virulence in the pest and that a plant genotype shows resistance phenotype if it is homozygous or heterozygous for the dominant resistance gene and the pest has the dominant avirulence (Avr) gene in homozygous or heterozygous condition. On the other hand, a host plant homozygous for the recessive susceptibility allele will be successfully colonised by any pest biotype. Also, if the pest has the recessive virulence gene at the corresponding locus in homozygous state the host resistance will be broken. Dominant R genes are known to confer pest resistance in most crops.

Like R genes for resistance against pathogens in plants, dominant R genes for insect resistance have a gene-for-gene operation with the pest genes and encode the nucleotide-binding site leucine-rich repeats proteins and they share structural similarities with R genes against pathogens and it is also believed that recognition of insect pest attack by such proteins takes place through direct or indirect binding of insect effector proteins (Sandhu & Kang, 2017). In some cases, insect attack leads to hypersensitivity due to cell death at points of attack.

Due to gene-for-gene relationship several cases of the susceptible and resistant plant genotypes differ sharply and are distinct in regard to resistance features attributable to major R genes. A good example of gene-for-gene relationship in host plant and insect pest is the Hessian fly (*Mayetiola destructor*) in bread wheat and durum wheat parasitisation where virulence and resistance in Hessian fly and wheat plant are conditioned by major recessive genes and dominant genes respectively. The Hessian fly overcomes resistance of a particular wheat cultivar if at the corresponding locus the Hessian fly biotype is homozygous recessive for virulence allele. In United States of America there are several known biotypes of Hessian fly some of which are GP, A, B, C, D, E, F, G, J, L, M, N, O, D-H and L-O and a minimum of twelve wheat resistance genes among which are H7H8, H3, H6, H5, H9, H10, H11, H12, H13, H14H15, H17, H18 and H19 (Ratcliffe et al., 2000). The development and cultivation of resistant wheat varieties has put a lot of pressure on the Hessian fly populations leading to the development of several biotypes through genetic recombination which overcome wheat resistance. In rice there are several major genes conferring resistance to brown planthopper (*Nilaparvata lugens*) which also has at least four biotypes. Similar host resistance gene-pest virulence gene relationships apply to other situations where there is vertical resistance and insect pest diversity in form of biotypes like in aphids (e.g., corn leaf aphid, cabbage aphid, spotted alfalfa aphid), leafhoppers, planthoppers, greenbugs, jassids and thrips.

18.2.3 Breeding Strategies for Resistance to Insect Pests

Development of insect pest resistant cultivars with high yield, desired quality and adaptability necessitates identification of sources of resistance genes through screening of the available germplasm resources. Apart from reliable screening techniques and facilities, capital resources, knowledge on type and inheritance mode of resistance and breeding strategies, successful resistance breeding is conditioned on presence of diverse germplasm which include cultivars in production, obsolete cultivars and landraces, weedy and wild relatives of the crop in question, products of mutation breeding, tissue culture and genetic modification. Availability of diverse germplasm is necessary to cope with the dynamic development of new insect biotypes which often overcome resistance already available, especially if resistance in use is of narrow genetic base. Identification of germplasm materials having appropriate resistance features and resistance genes starts with the cultivars in production followed by obsolete and abandoned cultivars, landraces, resources in the secondary gene pool and tertiary gene pool in that order. Plants with constitutively or induced high levels of expression for various resistance morphological features and insecticidal phytochemicals can be selected for breeding and the responsible genes transferred to the desired cultivars through conventional breeding, interspecific hybridisation, molecular assisted breeding or transgenesis. Focus on use of combined resistance involving horizontal resistance and tolerance along with vertical resistance broadens the genetic base of resistance, strengthens resistance and makes resistance more durable and stable. Antibiosis, non-preference and tolerance resistance features can be combined for effective control of various pests and long resistance duration.

The choice of conventional breeding methods for development of resistant varieties depends on whether resistance to be developed is vertical or horizontal, whether the crop is self-pollinated, cross-pollinated or clonally propagated. Other breeding objectives are also be taken into consideration when deciding on breeding strategies. In self-pollinated crops the aim is to develop pure line varieties with the required resistance, high yield, adaptability and the right quality while in cross-pollinated crops the focus is to develop heterozygous varieties which may be hybrid varieties and synthetic varieties for exploitation of heterosis or other forms of open pollinated varieties.

Where resistance is of vertical type with clear expression mass selection and pure line selection for one or a few generations can improve a variable cultivar with respect to pest resistance. Major gene resistance is easy to transfer from the donor to the recipient cultivar though it has the challenge of being short-lived and pest biotype specific. Vertical resistance is transferred using hybridisation between the resistance gene donor and the susceptible established popular high yielding cultivar or other elite breeding material followed by the required generations of backcrossing to the established cultivar/elite breeding material with selection for resistance and characters of the recurrent parent until a desired resistant pure line is obtained. In cross pollinated crops identified resistance genes are similarly transferred to inbred lines through crossing followed by backcrossing or other methods of improving inbred lines. Flanking molecular markers tightly linked to the target genes are employed to tag the resistance genes, fast-track the transfer of resistance genes, eliminate the undesired traits from the donor parent and recovery of the genome of the recurrent parent efficiently and more cheaply. Other forms of parents like open pollinated populations may be improved through appropriately designed crosses to transfer vertical resistance. A number of methods are employed to maintain and prolong major gene resistance. Sequential development of resistant varieties is commonly employed such that every time a resistant variety becomes susceptible because of new pest biotypes fresh varieties with other resistance genes are developed and released. Pyramiding of genes to combine two or more non-allelic resistance genes in one variety prolongs resistance life because the pest has to overcome all the genes, hence it will take longer for new virulent pest genotypes to arise and establish themselves.

To stack genes in one variety through gene pyramiding sources of resistance genes are identified followed by multiple parent hybridisations and backcrossing. Molecular markers are particularly convenient and useful in gene pyramiding to stack genes for multigenic and polygenic resistance cheaply and fast in various crops through molecular marker assisted backcrossing. Gene pyramiding can be integrated with speed breeding protocols if facilities are available.

If a region has a collection of varieties with different resistance genes against a particular pest rotational cultivation of varieties with different resistance genes helps to interrupt the development of new virulent biotypes for longer durability of resistance. Though it is a rare practice, use of multiline varieties and varietal mixtures also prolongs the durability of vertical resistance since the multiline varieties and varietal mixtures exert low selection pressure on pest populations. Cultivar diversification whereby varieties having resistance to different biotypes of a particular pest are grown on a farm also makes it difficult for any pest biotype to spread within the farm to epidemic levels due different resistance genes in different varieties.

Breeding for horizontal resistance has the challenge of being difficult to breed for and it takes a long time to achieve the desired level of resistance due to prolonged periods needed to bring together adequate polygenes in one cultivar to raise resistance to a desired level. In self-pollinated crops horizontal resistance development follows the path for improvement of quantitative characters. After hybridisation in self-pollinated crops the segregating generations for horizontal resistance are handled using pedigree breeding, single seed descent or bulk breeding for resistant pure line variety development. Pedigree breeding has been used to improve horizontal resistance in various crops like resistances to leafhopper, brown planthopper and gall midge in rice and sorghum resistance to shoot fly and midge. Recombination and selection strategies for autogamous crops in the presence or absence of male sterility are applied for assembling quantitative resistance genes, recombination, selection and increasing the frequencies of favourable genotypes with improved horizontal resistance levels to develop populations and pure lines with desired resistance levels using plants with low levels of disease severity and moderate levels of resistance as parents in the recombination process. In cross-pollinated crops improvement of horizontal resistance in populations to be used as sources of inbred lines for hybrid varieties development or as parents in other varieties is also done through various recurrent selection schemes to bring together resistance polygenes into one genetic background and raise resistance to the desired levels.

Molecular marker assisted breeding (molecular marker assisted selection, molecular marker assisted recurrent selection and molecular marker assisted backcrossing) improve the efficiency and speed of developing the elite materials with the target resistance genes and QTLs in both self-pollinated and cross-pollinated crops. Use of molecular marker assisted recurrent selection helps to identify progenies with target genes whose effects are difficult to detect directly in the phenotype, facilitates the integration of favourable multiple resistance genes in one population, has higher genetic gain than the conventional recurrent selection and saves on time since parents with molecular markers linked to the desired resistance and other traits are selected and inter-crossed in the same season. Flanking molecular markers or markers located within the respective genes or QTLs eliminate linkage drag and improve the speed, precision, efficiency and cost of transfer of major genes for resistance while QTL markers facilitate fast and cheap transfer of chromosome segments having resistance polygenes to develop vigorous cultivars with desired levels of resistance and tolerance. Marker assisted breeding helps in full recovery of recurrent parent genome in three to four backcrosses in marker assisted backcrossing instead of six to eight generations of backcross generations needed in conventional backcrossing method. Marker assisted backcrossing is convenient in pyramiding of pest resistance genes for multigenic and polygenic resistance more easily and faster, especially if the breeder has speed breeding facilities. Quantitative trait loci associated with tolerance, non-preference/antixenosis and antibiosis resistance to various pests have been identified in several crops like wheat resistance to aphid, sorghum resistance to midge and resistance to other pests in rice, soybean, sorghum, maize, etc.

Molecular/genomic assisted breeding

Genomics technologies are employed in identification of individual plants or lines possessing genes or QTLs for use in genomic assisted breeding which integrates high through-put genotyping and whole genome sequencing for multiple trait-resistance. Genomic selection has the potential of resolving the problems associated with conventional breeding methods which are always laborious, time consuming with high costs, especially if resistance genes are obtained from wild and weedy crop relatives. Genomic selection using whole genome molecular markers and high-throughput phenotyping is helpfull in the development of high yielding varieties with desired levels of resistance to pests, especially where multigenic and polygenic resistance is involved. With the help of next-generation sequencing techniques resistance genes and QTLs can be identified and subjected to fine mapping and cloning for use in genomic selection, marker assisted breeding, gene pyramiding and transgenesis. Phenomics and whole-genome molecular marker data are essential for genomic selection since they are required for constructing models and genomic estimated breeding values proposed by Meuwissen et al. (2001) for use in genomic selection.

Use of biotechnology and other genetic technologies for resistance to insect breeding

Apart from mutation breeding which is often accompanied by backcrossing and other conventional methods, there are a number of other strategies including selection among somaclonal variants arising from various forms of tissue culture, genomic editing, genetic transformation and gene silencing which can be employed to improve pest resistance in crop plants. Through transgenesis resistance genes from other cultivars, crop relatives and other unrelated organisms are incorporated into pest susceptible elite breeding materials. The Bt genes obtained from *Bacillus thuringiensis* is an excellent example of genetic transformation with non-plant genes for resistance to insect pests where several transgenic crop varieties under commercial production have these genes which makes them resistant to Lepidopteran insect pests like European corn borer (*Ostrinia nubilalis*), tomato fruit worm *(Helicoverpa zea*), tomato pinworm (*Keiferia lycopersicella*) and cotton bollworm. If Bt genes are engineered into plants their expression leads to production of the insecticidal crystalline proteins (ICP) referred to as Bt crystals in plant tissues that are toxic to Lepidopteran larvae (*Cry*-I type-genes), Lepidopteran and Dipteran larvae (*Cry*-II-type genes), Coleopteran larvae (*Cry*-III-type genes), Dipteran larvae (*Cry*-IV-type genes) and Lepidopteran and Coleopteran larvae (*Cry*-V-type genes), resulting in mortality of the larvae making the plants that carry such genes have resistance to these insects (Sharma et al., 2004). This is because the crystalline proteins are converted into toxic proteins which bind to the midgut epithelial cells of the larvae, rupturing the cell membranes, disintegrating the cells and ultimately killing the larvae (Convents et al., 1990; Hussain, 2015). The Bt genes can also be modified and engineered into crops for protection against Hemipteran pests like aphids, *Lygus* species (Gowda et al., 2016). Several transgenic Bt cultivars of cotton, maize, rice, tobacco, tomato, soybean have been developed and are being cultivated around the globe (Clive, 2015; Emani & Hunter, 2013; ISAAA, 2018). Apart from transgenesis another technology for improving plant pest resistance to insect pests is genome editing by use of CRISPR-Cas9, Transcription-like Effector Nucleases (TALENS), Zinc Finger Nucleases (ZFNs) and Meganucleases (MN). Because resistance and susceptibility alleles at a given locus differ in only a few nucleotides genome editing technology can be applied to edit the susceptibility alleles changing them to resistance alleles and also edit specific DNA sequences in resistance genes/QTLs to improve pest resistance (Razzaq et al., 2021).

Proteinase inhibitor genes

Another control mechanism against insect pests is use of proteinase inhibitors which protect plants producing them in adequate amounts against certain insect pest. Proteinase inhibitors interfere with various specific enzymes in the insect gut by binding to the enzymes, interfering with normal functions of the enzymes and thereby inhibiting protein digestion, overall performance and survival of the pest (Belete, 2018). Genes encoding various proteinases and other metabolites toxic to insect pests can be transferred to the desired cultivars through conventional hybridisation, interspecific hybridisation, molecular marker assisted breeding, genomic selection and transgenesis. Crop plants can be stacked with genes from microbes and other organisms through transgenesis to confer resistance to various pests. Trypsin inhibitor gene from cowpea and soybean and amylase inhibitor or ribosome inactivating protein genes provide protection to various crop plants by suppressing the survival of certain insect pests, especially Coleoptera and Lepidopteran insects, if they are inserted into the genomes of susceptible cultivars (Sharma et al., 2004).

Screening: infestation and evaluation of resistance

Effective screening of germplasm for resistance genes requires an adequate supply of insects to infest the plants at the right stage when plants are most vulnerable to discriminate between susceptible and resistant and select the resistant genotypes with the desired resistance genes. The right insect biotypes with the needed parasitizing capability and virulence in form of eggs, larvae or adults are collected from the field or reared in the laboratory or greenhouse for use in screening. The reared insect populations must be adequate and representative of the natural populations and have the appropriate biotype and should be generally healthy. In breeding for resistance locations which have the pest infestation as regular production problem are desirable to use for screening of the cultivar genotype and other germplasm. A well-designed experiment with good supply of eggs or larvae or adult insects is necessary for both nursery and laboratory screening. The reared insect populations are released onto the materials to be screened at appropriate time, at the right crop and pest stages. If natural field infestation is adopted nurse crop or spreader crop can be used whereby a known susceptible crop or plant species is used to help raise a large population of the insect pest in the field when planted side by side with the crop genotypes under screening so that the pest spreads to infest the adjacent crop being evaluated. Resistance is quantified in terms of percentage damage to leaves, fruiting parts, shoot, roots, reduced crop stand, reduced plant vigour and lowered yield, fruit drop, dead hearts, number of dead plants, tissue necrosis and activities of the insect, namely, number of insects on the plant, number of eggs laid, insect aggregation, food preference, growth rate, utilisation of the plant by the pest as food and insect mortality. Longevity of resistance to the insect and important distinct resistance features have to be taken into consideration.

18.2.4 Insect Pest Biotypes and Durability of Resistance

Insect pests vary in their parasitizing ability which is a feature that varies among different biotypes. Insect biotype refers to variation within an insect pest species that is equivalent to a race within a pathogen species and relates to differences among populations of a pest for virulence and ability to exploit a plant species as a host. Different biotypes are distinguished by infestation, adaptation and development on a particular host, pheromone response, insecticide resistance, response to light and temperature and seasonal changes, dispersal or migration tendency, capacity to cause damage to the crop, disease transmission capacity and host preference for feeding and feeding behaviour, size and differential survival rates on different cultivars, oviposition, morphological and colour differences. Parasitizing ability of an insect is its capability to infect, colonise the host plant and take advantage of the host as a source of its food, site for its reproduction and general survival. Parasitizing ability is determined by the presence of virulence or avirulence genes in the pest such that a virulent pest biotype succeeds in parasitising a host plant while biotype having avirulence genes and lacking virulence genes fails to feed and multiply on a resistant host. An insect pest having the relevant virulence genes succeeds in colonising the host plant, uses the host for its reproduction and nutrition. Existence of insect biotypes was first documented by Painter (1930) who noted the presence of different biological strains of Hessian fly *(M. destructor)* with varied parasitizing abilities on different wheat cultivars. Existence of biotypes is known in several insect pests, and at least

14 different insect pest species are known to have variable biotypes, particularly those insects that are obligate parasites, those that reproduce parthenogenically and those having fast reproduction rates with several generations per year. Restricted host range, monocropping without crop rotation, use of pesticides and major gene antibiosis/antixenosis resistance also promote fast evolution of insect pest biotypes. Aphids (e.g., spotted alfalfa aphid, corn leaf aphid and other various aphids), leafhoppers, planthoppers, Hessian fly, rice gall midge, rice brown planthopper and green bugs are known to develop biotypes at high rates, especially if antibiosis is the resistance mechanism. Hessian fly has at least 15 known biotypes, e.g., G, GP, A, B, C, D, E, F, J, L, M, N, O, D-H and L-O parasitising wheat (Ratcliffe et al., 2000).

Variability in pest parasitizing ability due to presence of different pest biotypes in the insect population affects durability of host cultivar resistance because when new biotypes arise, they overcome the resistance of existent crop cultivars, especially those with major gene antibiosis and antixenosis vertical resistance which exerts high selection pressure on insect pest populations. Resistant varieties will sustain major gene resistance as long as new insect biotypes do not arise. Vertical resistance conferred by only one major gene is the most unstable and is very easy to overcome because the pest needs only one genetic change at the avirulence locus to break the resistance. When an existing resistant variety loses its resistance fresh resistance genes are identified for development of new resistant varieties, a process which leads to as many varieties as the pest biotypes in a given region since resistance of any variety may last for just a few years before a new resistance gene is again sought for.

Since single major gene antibiosis and antixenosis resistance leads to fast biotype development and short resistance durability a combination of resistance mechanisms or horizontal resistance which prolongs durability of resistance is the most preferred strategy for crop plant protection against insect pests. Horizontal resistance is more stable, lasts longer, non-biotype specific, does not encourage biotype development because it exerts no selection pressure on the pest population. Antibiosis alone leads to faster biotype development than antixenosis because it puts higher selection pressure on insect pest populations. A combination of antibiosis resistance and antixenosis resistance in one variety creates a better and more stable resistance which suppresses the rate of biotype development substantially. Antixenosis and tolerance have less selection pressure on the insect pest populations and last longer than antibiosis resistance. Furthermore, resistance attributed to tolerance is more stable and lasts longer than antixenosis resistance and antibiosis resistance because it exerts much lower selection pressure on insect pest populations than these two resistance types and tolerant varieties support insect pest populations without promoting production of new biotypes (Koch et al., 2016). Use of cultivars with multiple major gene resistance developed by gene pyramiding and use of multiline varieties, cultivar mixtures and also sequential release of varieties with different resistance genes also prolong durability of resistance.

18.3 Breeding for Resistance to Parasitic Weeds

18.3.1 Striga

Striga and broomrapes are two categories of major plant parasitic weeds belonging to the Orobanchaceae family which cause serious yield reductions in the crops that they parasitize. *Striga*, formerly classified in Scrophulariaceae family and commonly known as witchweed, is the most widely distributed and common parasitic weed affecting food production in tropical and subtropical Africa, India, Arabian Peninsula, parts of Australia and some isolated areas of USA. *Striga* is a root hemiparasite of maize, sorghum, sugarcane, rice and millets and some legumes and is highly dependent on its hosts for necessary nutrients, water and metabolites and signals for its development. It is believed to have about 42 species, with at least 33 species in Africa where it is the most important biotic stress factor affecting cereal cultivation with large tracks of farmland being *Striga*-infested in many countries. The species that have the most serious impact on crop production are *Striga hermonthica, S. asiatica, S. aspera, S. densiflora, S. euphrasioides, S. forbesii*, all being parasitic on cereals and some other grass crops like sugarcane and *S. gesnerioides* which is parasitic on certain dicotyledons, particularly legumes like cowpeas. *S. hermonthica,* which is the greatest threat to maize and sorghum production in Africa, is widely distributed in Africa, from Egypt in the north to Malagasy in the south, from eastern regions to the west coast and central parts of Africa and is the most significant *Striga* species in East Africa. Its distribution extends into western Arabia as well. *Striga* infestation in farmers' fields of sorghum, maize and millets in Sub-Saharan Africa leads to serious yield losses which can reach 100% for highly susceptible cultivars, leading to total crop failure and famine situations under heavy infestation. The amount of yield loss depends on the level of infestation, cultivar resistance, soil nitrogen and moisture contents and agronomic practices, with susceptible cultivars grown under soil nutrient and moisture deficiencies suffering the greatest impact.

Striga-host relations and host resistance

The *Striga* plant is an obligatory parasite which is totally dependent on its host for survival and requires stimulation for germination before it can parasitize the host plant. Host plant-*Striga* interaction as regards host infection and *Striga* establishment in the host tissue and subsequent emergence are influenced by intimate interactions between *Striga* and the host and prevailing environmental conditions like temperature, soil type, soil moisture and nutrient levels. Starting from seed germination stimulation, there are complex host-parasite interactions between the *Striga* plant and its host plant in relation to host resistance and *Striga* virulence reactions.

A single host plant can support up to 300 *Striga* plants, each producing between 25,000 and 500,000 small sized seeds (about 0.2 mm long) which can remain viable in the soil for a long time, ranging between two years to about 15 years (Csurhes et al., 2016; Yoneyama et al., 2010). Freshly produced *Striga* seeds are dormant and undergo after-ripening period during the dry season when their moisture content is lowered and dormancy broken before the dry seeds are preconditioned through moistening by rain and stimulated for germination (Mohamed et al., 1998). Seed preconditioning takes about ten days after onset of rains followed by germination if the seed is exposed to germination stimulants exuded by the root hairs of its hosts namely, sorghum, maize, millets, sugarcane, rice, cowpea and some non-host crops, e.g., cotton, sunflower and beans. The established *Striga* seed germination stimulants exuded by its hosts' root hairs are structurally related sesquiterpene lactones, collectively referred to as strigolactones, some of which are strigol, sorgolactones and alectrol (Hauck et al., 1992; Soliman et al., 2022; Yoneyama et al., 2010). The strigolactones known to be exuded by sorghum roots are strigol, sorgolactone, 5-deoxystrigol, sorgomol and non-*Striga* stimulating orobanchol (Gobena et al., 2017) while maize and some millets roots produce mainly strigol. Ethylene is also a strong *Striga* seed germination stimulant. Strigolactones are normally produced in high levels by crop plants under soil nutrient deficient conditions for attracting arbuscular mycorrhizal fungi for symbiotic relationships to improve plant nutrient uptake (Yoneyama et al., 2010). The signals of the germination stimulants usually reach and stimulate only those *Striga* seeds that are very close to the host root hairs and any un-germinated seeds remain in the soil until another season when they are stimulated to germinate. The germinated seed gives out a radicle which grows towards the host root producing the germination stimulant and attaches to it on contact. After seed germination and radicle formation, further development processes of host plant parasitisation by the *Striga* plant like seedling attachment, radicle penetration of the root tissue, haustoria formation, parasite establishment in root tissue and other development processes need continued signals, sugar, water, cytokinin and the necessary metabolites from the host plant. When the *Striga* radicle gets into contact with the host root it produces enzymes like polygalacturonase and pectin-esterase which decompose the host cell wall and gains entrance into the host root tissue to form a haustorium on induction by haustorium inducing factors, namely, strigolactones, flavonoids, quinones and cytokinin (Kokla and Melnyk, 2018; Yoder, 2001). The parasite uses the haustorium which is attached to the host root vascular system, establishing direct xylem-xylem connection, to inject phytotoxins in to the host tissue and draws its requirements from the host, namely, water, minerals, sugars, amino acids, photosynthates and other metabolites, thereby impairing the development of the host plant and causing hormonal imbalance in host tissues, leaf rolling, chlorotic lesions, disruption of host photosynthesis, stunted growth, reduced panicle or ear size, wilting and the damage leads to reduced or no grain production by the crop (Fig. 18.3). *Striga* does most of the damage to the host before it emerges when it is still holoparasitic and totally dependent on the host. The emerged *Striga* plant has green leaves and photosynthesises but remains a hemiparasite and withdraws part of its water, nutrients and other requirements from the host plant.

Substantial ecotypic variability exists within and between *Striga* populations and among species for preconditioning and germination requirements and interaction with its hosts. There is existence of genetic variability for virulence in this parasitic weed since it is known to have strains which have specific interactions with different

Fig. 18.3 Maize field badly damaged by *Striga* infestation in western Kenya

crops, especially as regards interaction with cereal crop hosts. It has been shown that *S. hermonthica* and *S. asiatica* strains parasitising rice are genetically distinct from those parasitising sorghum and maize (Unachukwu et al., 2016). Strains of *S. hermonthica* or *S. asiatica* which parasitize sorghum and maize do not necessarily attack rice and vice versa. There is also intra-crop specificity as regards host plant-*Striga* ecotype interaction. In studies conducted in Uganda, Kenya and Tanzania (Rodenburg et al., 2017) observed that some rice genotypes had broad spectrum resistance to *Striga* ecotypes while a number of rice cultivars exhibited ecotype-specific resistance to various *Striga* ecotypes, suggesting that the level of rice resistance in the field varied with *Striga* species and ecotypes. Rice genotypes that were resistant to *S. hermonthica* were not necessarily resistant *S. asiatica*. That is, the interactions between *Striga* and its various host species are such that if resistance exists against *S. asiatica* in a particular crop cultivar it is not a guarantee that the cultivar is protected against *S. hermonthica* and vice versa. Existence of intra-crop *Striga* specific strains is also known where strains of a certain *Striga* species parasitizing a particular crop cultivar from one region are not necessarily virulent on the same crop cultivars from other regions. Specificity is related to the response of the *Striga* seed and seedlings to the phytochemical signals from the host roots. At the molecular level host plant-*Striga* interaction is comparable to host plant-pathogen interaction system where infection triggers PTI (PAMP-triggered immunity) through PAMPS (pathogen-associated molecular patterns) leading to mechanical or biochemical defence reaction by the host (Mutinda et al., 2023). The *Striga* attempts to overcome the PTI may be followed by ETI (effector triggered immunity) host response which causes programmed cell death and hypersensitivity known in sorghum cultivars like N13, SRN 39 and Framida (Mutinda et al., 2023).

Host species and cultivars vary in levels of production of chemical signals which affect the various developmental stages in the life cycle of the *Striga* plant. Host resistance based on the absence or inadequate levels of the chemical stimuli puts pressure on the *Striga* populations leading to development of various physiological *Striga* strains. *S. hermonthica* is cross pollinating and therefore has wide variability among and within its populations which helps it to overcome the resistance of various host cultivars and species due to gradual build-up of new virulent strains. On the other hand, the self-pollinating *S. asiatica* and *S. gesnerioides* exhibit a lot of variability among populations and uniformity within populations.

Mechanisms of plant resistance to Striga and breeding for resistance to Striga

Breeding of crop plants for *Striga* resistance is based on plant resistance mechanisms which are related to *Striga*-hosts interactions. There are various known resistance mechanisms that operate before or after *Striga* attaches to the host root tissue which include inadequate or lack of *Striga* seed germination stimulants, *Striga* seed germination inhibitors, inadequate or lack of haustoria inducers, mechanical barriers, inhibitors/suppressors of *Striga* radicle exoenzymes, hypersensitivity and incompatibility reactions, phytoalexins, avoidance and tolerance. At different stages of its life cycle *Striga* takes advantage of metabolites secreted by the host plants, particularly the strigolactones, for its successful invasion of the host and establishment in the host tissue.

Resistance due to low or lack of germination stimulant

Low or lack of *Striga* seed germination stimulant production by the host roots is an established and effective resistance mechanism because cultivars which do not produce or produce only low amounts of the stimulant will not be attacked as there will be no *Striga* seed germination or only a small number of seeds will germinate. On the other hand, high stimulant producing cultivars are susceptible unless other resistance mechanisms are operational. Low stimulant production has been the main approach of breeding for *Striga* resistance, though this resistance does not last for long, especially against the variable *S. hermonthica*. Host crop varieties like sorghum, maize, sugarcane, rice and millets vary genetically in regard to the extent of *Striga* seed germination stimulant production and type of germination signals. Low stimulant production in sorghum is inherited as a recessive trait controlled by a single gene, lgs1 (Gobena et al., 2017; Ramaiah et al., 1990), and can be transferred to agronomically desirable susceptible varieties through crossing and backcrossing. There may be minor genes with additive effects as well. Mutation in lgs1 gene is known to lead to production of non-*Striga* stimulating exudate, namely, orobanchol, which still makes the host plant resistant to *Striga* (Gobena et al., 2017). A number of sorghum cultivars are low stimulant producers and known to be resistant to *Striga* in the field, namely, Framida, SRN 39, Bonganhilo, the East African landrace N13, sorghum cultivar IS9830 and wild *Sorghum drummondii* line P78 (Gobena et al., 2017; Mohamed et al., 2024).

Resistance due to inhibitory exudates and low or lack of haustoria initiation

In some host genotypes roots exude substances which inhibit *Striga* seed germination or prevent the *Striga* seedling radicle from breaking the cell wall to gain entry into the root tissue. This is known in some sorghum varieties and certain sorghum wild relatives in which roots produce exudates that are inhibitory to *Striga* exoenzymes and that *Striga* plants parasitizing such cultivars may not emerge or have delayed emergence. After *Striga* seed germination other phytohormones and secondary metabolites, especially phenolic compounds, quinones, flavonoids and cytokinin, are required for haustoria induction and the radicle attachment to the host plant root (Aoki et al., 2022; Kokla and Melnyk, 2018). The germinated seed has to be close enough to the host root hair which produces the compounds needed to initiate the development of the haustoria for attachment to the plant root tissue. Host varieties differ in their ability to induce haustoria formation because of variation in haustoria initiator production. Those varieties which are known to produce low or no haustoria inducing factors like Framida and SRN 39 will have no haustoria developing in their root tissues despite germination of *Striga*. In sorghum low haustoria inducer production is controlled by a major gene whose presence leads to *Striga*-free crop and can be transferred by simple crossing followed by backcrossing.

Resistance due to mechanical barriers

Presence of mechanical barriers and anti-haustoria resistance mechanisms in some host plants, especially certain sorghum cultivars and some sorghum wild relatives, prevent attachment and entrance of the haustoria into the host root tissues because of thickened walls of the endodermal cells and hardened vascular cylinders of the host roots through lignification, silica deposits and other forms of host cell wall modifications (Fernandez-Aparicio et al., 2020; Mbuvi et al., 2017). This leads to very low number of haustoria that successfully penetrate and establish in the host root tissues. Some of the sorghum populations known to have mechanical barrier-based resistance due to cell wall thickening are cultivars Framida, Radar and N13 and some wild sorghums of *arundinaceum* and *aethipicum* races are also known to have mechanical resistance which can be used in developing resistant superior sorghum varieties (Mbuvi et al., 2017).

Resistance through hypersensitivity and incompatibility

In some host crop cultivars necrotic lesions appear at *Striga* infection points due to secondary metabolites and other phytochemical processes like suberisation, callose deposition, lignification and release of reactive oxygen species, activation of programmed cell death of cells surrounding the haustorium evidenced by hypersensitive reaction creating physiological barriers between the *Striga* seedling and the host root tissue, thereby retarding the parasite and protecting the host plant (Fernandez-Aparicio et al., 2020). The hypersensitive necrotic lesions isolate the *Striga* seedling at the infection sites preventing the seedling from making xylem-xylem connections with host root tissue, accompanied by suppression of parasite growth and development, stunting and eventual death of the *Striga* seedling in the host root tissue. This resistance mechanism exists in sorghum varieties Dobbs, Serena, and Framida and in some wild sorghum relatives. There is also another post-infection resistance reaction similar to hypersensitivity response but is suggestive of incompatibility or antibiosis which also impedes the development of *Striga* seedling after attachment to the host tissue without necrosis. In another mechanism reported by (Kavuluko et al., 2021) the parasite fails to identify the host vascular tissue and does not form vascular connection with the host xylem. The parasite haustorium grows past the cortex and emerges at the side opposite to infection point without linking with host vascular system.

Avoidance and tolerance

Rooting pattern in host crops also determine the extent of *Striga* attack if there are no other resistance mechanisms operating. Normally the *Striga* seeds are in the top 10 cm of soil. Host varieties that have reduced root-length-density in the upper 10–15 cm of the soil but have higher root density lower than this soil depth are exposed to less *Striga* attack (Amusan et al., 2008). Therefore, deep rooting cultivars with low root branching density within the top 15 cm of the soil profile avoid *Striga* attack and are deemed resistant. This mechanism which simply takes advantage of soil distribution pattern of the *Striga* seeds and is known in the vertical rooting Framida cultivar can be incorporated in promising susceptible genotypes through crossing and backcrossing.

Some sorghum landraces, particularly certain African sorghum landraces, are known to be tolerant to *Striga* and can withstand high levels of *Striga* without significant growth and yield reduction as compared to the other genotypes when infested by *Striga.* Seredo is an established East African sorghum variety known to be tolerant to *Striga* and gives fairly high yields in *Striga* infested fields (Ayiecho & Nyabundi, 1994). *Striga* tolerance is a disadvantage in that it leads to increased *Striga* infestation under African peasant production. Tolerance has quantitative nature of inheritance.

Screening methods for resistance to Striga

Striga agronomic control methods like weeding, pulling and soil fertility amendments have failed to give long lasting solutions to the farmers, especially for the resource poor peasants whose fields are often heavily *Striga*-infested in regions of the world where *Striga* is endemic. Chemical control has rarely been advocated for use by peasant farmers because of lack of appropriate chemicals, costs, pollution and technical problems associated with it. Furthermore, no effective and practical chemical control method has been developed that can clear *Striga* seeds in the soil. Host plant resistance remains the only practical solution that can save the situation for the small-scale cereal farmers. Resistance breeding needs a diverse germplasm base on which to conduct screening. Apart from germplasm within the crop gene pool like landraces, abandoned and modern cultivars, wild and weedy relatives of various crops, especially sorghum, maize, rice and millets are also screened as possible sources of resistance genes. Breeding materials obtained through induced mutagenesis, biotechnology and transgenesis are useful in development of *Striga* resistant cultivars as well.

Development of cultivars with resistance requires reliable and precise screening techniques for identification of resistance gene sources and determination of mode of inheritance. Gene action for host-*Striga* interaction in cereals may take the form of dominance, partial dominance, overdominance and additive effects for major gene or polygene control of various resistance mechanisms. Resistance in some sorghum genotypes like SRN39, N13, Framida, and Dobbs may be due to single recessive nuclear genes (Hess & Ejeta, 1992; Ramaiah et al., 1990) or partial dominance as known in Dobbs and multigenic resistance effects known in several sorghum cultivars. A number of methods including field and pot tests, greenhouse and laboratory screening methods employed for *Striga* resistance screening are based on Striga emergence, stimulant production and haustoria attachment.

Field and pot screening methods

Field screening tests which may comprise of preliminary observations and advanced screening tests are designed based on the resources and the number of germplasm materials to be screened and usually conduced in *Striga* hot spots which are areas where *Striga* is endemic and in which *Striga* sick fields can be established. *Striga* sick plot is created by tilling a low fertility plot having good drainage and adequate even distribution of *Striga* seed to a depth of 10–15 cm. Alternatively, infestation is done by putting *Striga* seed in the hole in which the crop seed is to be planted. *Striga* seed should be at least one year old. Assessment of crop resistance to *Striga* in the field gives field resistance of the host plant which is a product of interactions between *Striga* and the host plant and environmental conditions at the various stages of infection leading to expression of resistance or susceptibility as a net result of different resistance mechanisms. Field resistance is often inherited as a quantitative character with variable magnitudes of additive, dominance and epistatic gene effects and is best improved through recurrent selection.

In the field host susceptibility is assessed based on counts of emerged *Striga* plants per host plant, damages to the host plant in form of chlorosis, reduction in host plant vigour and growth, reduced ear or head size, reduced head exertion or panicle sterility, host plant stunting, wilting and seed yield of host plant. *Striga* virulence is scored in terms of *Striga* plant performance traits, e.g., field *Striga* emergence counts, *Striga* height, number of branches, flowers and seed capsules. The number of emerged *Striga* plants does not precisely imply that the host plants with fewer emerged plants are more resistant than those with higher number of emerged *Striga* plants since there may be many more un-emerged plants underground that are already attached and parasitizing other plants.

Field resistance screening has the weaknesses of being unreliable, inefficient and slow. Field screening has shortcomings of wide variation in *Striga* resistance field scores that are highly influenced by both genotype and environmental effects, heterogeneity of *Striga* seed distribution and infestation, soil variability and its effect on the host and the parasite, effects of soil moisture stress and other environmental factors (Ayiecho & Nyabundi, 1992, 1994). Despite its challenges field screening remains one of the most important screening methods because it assesses plant growth, yield, quality and other agronomic characteristics under natural field conditions. Field screening is also an important pointer to the genotypes which may have or lack resistance and provides useful information for further screening of the germplasm. Since screening of host plants for resistance to *Striga* in the field suffers from *Striga* infestation heterogeneity in the field, variable *Striga* emergence, soil conditions heterogeneity and other environmental factors appropriately designed replicated field screening experiments in the *Striga*-infested fields are necessary to identify resistant genotypes. Such screening experiments have to include at least one known susceptible genotype and/or resistant genotype planted adjacent to the test genotypes for use as checks and yardstick for comparison.

Pot screening experiments in the field or greenhouse supplement field and laboratory experiments. Pots vary in shapes and size and should have holes at the bottom for water drainage and adequate size for growth of the host plants to be screened. The pots are filled with appropriate soil, infested with *Striga* seed and watered to precondition the *Striga* seeds and are then sowed with host seeds and regularly watered. Usually, one plant is grown per pot to avoid competition.

Another screening method that is conducted in the field or in pots is use of a cloth bag commonly referred to as Eplee bag developed by Eplee (1975). In this screening method about 100 preconditioned *Striga* seeds or any convenient seed amount are placed in a porous nylon screen cloth, tying up the cloth at the point where the seeds are placed and the cloth is buried next to host plant in the field or pot. The bag should be porous enough to allow in germination stimulant, moisture and gases but not allow *Striga* seed to pass through to avoid loss of the seeds into the soil during the study. The bag is later removed, opened and the germinated *Striga* seeds counted under a microscope and expressed as percentage of total seeds originally placed in the bag.

Laboratory and in vitro methods

The most commonly used laboratory screening method for Striga resistance is agar gel assay described by Hess et al. (1992) and Ochiel et al. (2021). In this technique preconditioned *Striga* seeds are dispersed in 0.7% water agar-gel in a petri dish before the agar solidifies and once the agar has solidified radicles of host test genotype seedlings are carefully inserted into slits in the agar in the petri dish so that each petri dish contains a single host plant seedling. The germinated *Striga* seeds are counted after five days of incubation and distance from the host radicle is measured. Identification and counting of germinated *Striga* seeds are achieved by examining the underside of the plate using a dissecting microscope. Genotypes with maximum *Striga* germination distance less than 10 mm are considered as low stimulant producers.

Mohamed et al. (2024) described a modified assay, the extended agar gel assay (EAGA), based on the agar gel assay of Hess et al. (1992). The extended agar gel assay (EAGA) helps to distinguish specific resistance mechanisms. The EAGA method uses large petri dishes which can hold deep agar gel layers for support of host plant seedling growth. Preconditioned sterilised *Striga* seeds are evenly spread in the dish followed by pouring of agar solution into the dish. Seed of pre-germinated sorghum or other host crop under screening are placed on the agar gel and the petri dish is covered and incubated under appropriated conditions for about three days. After three days data are taken on *Striga* seed germination and *Striga* radicle attachment on the host roots. Ethylene is then applied to nullify any cultivar difference among the host genotypes in regard to *Striga* seed germination. Ethylene induces the germination of the remaining ungerminated viable *Striga* seeds. After two days excess ethylene is removed and the host roots are observed for *Striga* seedling attachments and signs of necrosis around the parasite haustorium using a dissecting microscope. Where the host genotype has inhibitory effects on the germination of *Striga* seed there may be no germination despite the presence of ethylene around the roots with inhibitory effects (Mohamed et al., 2024). Mbuvi et al. (2017) described a screening method in which host seeds are germinated using cotton wool lined with filter paper then transferred to root observation chamber Perspex having vermiculite, covered with aluminium foil and provided with appropriate nutrients daily for 11 days. The host roots are infested with pre-germinated *Striga* seeds and after a few days the host roots are observed for *Striga* seedling infestation using a stereo microscope. The *Striga* seedlings are detached and the host roots dissected and sectioned for microscopic determination of the extent of parasite development in the root cortex.

***Striga* resistance breeding strategies**

Though some resistance mechanisms are controlled by major genes the overall resistance to *Striga*, especially field resistance, has complex inheritance, not adequately understood and is under the influence of multiple genes, polygenes and environmental factors. In sorghum, some *Striga* resistant cultivars have general agronomic undesirability with poor quality and low yields. Incorporation of resistance into good genetic backgrounds of elite breeding materials that lack resistance is achieved through hybridisation with resistance sources which may be known *Striga* resistant crop cultivars and wild crop relatives followed by appropriate segregating population handling strategies. The strategies may be fast-tracked through molecular marker assisted methods. Major gene resistance mechanisms like low strigolactones production and low haustoria initiation in sorghum are simply transferred to the acceptable varieties from the resistant cultivars followed by backcrossing and selection until the desired resistant lines are obtained. A good resistance breeding strategy assembles resistance features in one genetic background through pyramiding of genes for different pre-attachment and post-attachment resistance mechanisms, namely, low stimulant production, germination inhibition, mechanical barrier presence, poor haustorium initiation and hypersensitivity. In case of partial dominance or resistance mechanisms that are of quantitative nature pure line and pedigree breeding methods are applicable for sorghum, millets, rice and cowpea. Selection of agronomically accepted lines is conducted as from F_3 onwards employing pedigree breeding approach. Population improvement involving various forms of recurrent selection are applicable for maize, pearl millet, sorghum and even for rice, other millets and cowpea, especially if there is male sterility in the autogamous crops. Recurrent selection has been applied in maize to develop *Striga* resistant open-pollinated populations, inbred lines and hybrids leading to some reduction in *Striga*-induced damage and elevated qualitative and quantitative *Striga* resistance levels. In maize major problems associated with *Striga* resistance breeding is limited genetic variability for *Striga* resistance, unreliable resistance sources from maize populations and lack of strong or high-level major gene resistance as compared to sorghum and other crops which are native to *Striga* endemic regions and evolved with *Striga* side-by-side. However, promising levels of resistance and tolerance are known to exist in some Africa landrace maize populations for exploitation to improve *Striga* resistance and a number of maize genotypes have been developed with promise for *Striga* resistance and tolerance. Developments in genetic engineering make it feasible to transfer the various resistance mechanisms controlled by major genes like lgs1 gene for low strigolactones production in SRN39 sorghum cultivar (Gobena et al., 2017; Ramaiah et al., 1990) and resistance QTLs in crops like rice, millets, sorghum and sorghum relatives to maize genome through transgenesis and also engage genome editing to improve resistance levels in maize populations.

The unstable nature and short duration of single major gene resistance calls for pyramiding of both major gene resistance mechanisms and QTLs associated with polygenic resistance by employing molecular marker assisted breeding and genomic technologies. Advances in genomics, proteomics, metabolomics and transcriptomics, sequencing technology and high-throughput phenotyping make it possible to ascertain genotypes with resistance to *Striga*, characterise and map individual resistance genes and resistance QTLs. The advances help in the understanding of ecotypic variation in *Striga*, molecular basis of resistance mechanisms and susceptibility, elucidation of transcription factors, regulatory roles in the genome in regard to host plant-*Striga* interaction, adaptive pathways linked to synthesis and activation of parasitism, resistance related proteins and metabolites. Such information facilitates and fast-tracks efficient breeding of resistant cultivars through marker assisted selection, marker assisted backcrossing, marker assisted pyramiding of resistance traits, genomic selection, transgenesis and genome editing. Molecular genomic technologies overcome challenges of field screening for *Striga* resistance as relates to failure to distinguish clearly the resistant genotypes, inadequate information on genetics of resistance, *Striga* ecotypic variation, non-uniform *Striga* infestation and poor field screening techniques. Genotype sequencing techniques like genome-wide association studies (GWAS), molecular markers, e.g., single nucleotide polymorphisms (SNPs), amplified fragment length polymorphisms (AFLPs), single sequence repeats (SSRs) and restriction fragment length polymorphisms (RFLPs) and quantitative trait loci (QTLs) associated with resistance traits have improved the use of molecular breeding for *Striga* resistance in sorghum, millets, maize, rice, cowpea and sugarcane by identifying and fast-tracking the transfer of resistance genes and QTLs.

Sequencing and mapping of genes encoding low stimulant production, low haustoria inducing factors, phytochemicals blocking haustorium access into the host root vascular system, hypersensitivity, secondary metabolites involved in countering parasitism and other resistance mechanisms facilitate crop breeding for resistance to *Striga*. Genetic transformation of the crops parasitised by *Striga* through introduction of RANi system that silences the *Striga* parasitism genes when the parasite invades the host tissues also has the potential of elevating host plant resistance.

Apart from some known resistant sorghum, finger millet, pearl millet and bulrush millet landrace cultivars, work undertaken by various national research and breeding institutions and International Crops Research Institute for Semi-Arid Tropics (ICRISAT) led to identification and development of several resistant or tolerant genotypes like Seredo, Framida, Serena, SAR13, SAR19, SAR35, SRN39, Bonganhilo, N13, IS9830 sorghum varieties. Resistant cultivars like Radar and Dobbs had been developed much earlier in South Africa and East Africa respectively. Some resistant sorghum cultivars have undesirable agronomic traits and therefore unpopular since they may be low grain yielders and of poor quality. For example, while cultivar Framida has stable *Striga* resistance, resistance to grain mould and birds with good yields, its high seed tannin content and brown seed colour make it undesirable. Resistant sorghum cultivars SRN39 and N13 are also known to be low yielding under field conditions (Ayiecho & Nyabundi, 1994). However, resistance in these cultivars can be transferred to other desirable cultivars. Cultivars with varying degrees of resistance or tolerance have also been developed in millets and rice.

18.3.2 Broomrape and Other Parasitic Weeds

Broomrapes (*Orobanche and Phelipanche* spp.), belonging to Orobanchaceae family, are chlorophyll-lacking holoparasites of dicotyledons (Fig. 18.4) like tomato, oilseed rape, carrot, tobacco, clover, cucumber, sunflower, Faba bean, lentils, chickpea, peas and forage legumes and are commonly found in western Asian countries of the Mediterranean region, northern Asia, central and eastern Europe and northern African countries where they cause serious crop damage (Fernandez-Aparicio et al., 2016). Among the known broomrapes that parasitize crop plants are *Orobanche minor, O. foetida, O. crenata, O. cumana, O. cernua, Phelipanche aegyptiaca* and *P. ramose*. The parasite attaches to the roots of its hosts and causes heavy yield losses and sometimes forces the farmers to abandon cultivation of the affected crops.

Like *Striga,* a single *Orobanche* plant produces large amounts of seed reaching about 500,000 which undergo a period of dormancy and need environmental conditioning before they can germinate to infect the host (Fernandez-Aparicio et al., 2016; Yoneyama et al., 2010). The *Orobanche* seed germinates on stimulation by strigolactones exuded by the host roots and when its radicle gets in contact with the host root it forms an appressorium and a haustorium which exerts mechanical pressure and enzymatic impairment of host cell wall for its penetration and establishment in the root cortex for withdrawal of its water and metabolite requirements

Fig. 18.4 Broomrape plant parasitising an herbaceous plant

through the root vascular system (Fernandez-Aparicio et al., 2016; Yoneyama et al., 2010). The seedling also develops into a tubercle which later emerges from the soil as a chlorophyl-less flowering plant (Fig. 18.4). Because of lack of chlorophyl broomrapes are totally dependent on their hosts for their water, nutrient, photosynthates and metabolite requirements thereby causing damage by reducing total biomass, yields and impart negative effects on host photosynthetic activities and hormonal balance.

Since no economically effective mechanical or chemical method is available for controlling broomrapes, crop resistance is the most viable, reliable and cheapest control approach. Resistance to *Orobanche* is mostly complex and of quantitative nature, with low heritability and difficult to breed for. There may be a few cases of unstable short-lived major gene resistance as known in sunflower. Identification of QTLs associated with resistance is a reliable way of mapping resistance genes for breeding purposes. Genomic tools and gene expression analyses assist in understanding of molecular basis of host response to *Orobanche* infection and adaptive pathways related to synthesis and activation of parasitism/resistance proteins and metabolites and the analyses are also employed in identification of resistance genes and QTLs for use in marker assisted selection, marker assisted backcrossing and genomic selection for the purpose of accumulating resistance genes in elite breeding materials.

Like the case with *Striga*, some effective resistance mechanisms to broomrapes are low germination stimulant and low haustoria inducer production, *Orobanche* seed germination suppressor and mechanical resistance which are known to occur in some hosts of these parasitic plants. The most reliable and stable resistance combines different resistance mechanisms, bringing together low or absence of germination stimulant production, low or lack of haustoria initiation factors, barriers to infection and haustoria penetration and establishment in one genotype. Cultivars with the best resistance also combine both quantitative and qualitative resistance making them have long lasting and stable resistance for withstanding new races of the weed. This can be achieved through pyramiding of resistance genes/QTLs using conventional breeding combined with marker assisted recurrent selection, marker assisted backcrossing, genomic selection and transgenesis.

Dodder (*Cuscuta* species), a shoot parasite belonging to the Convolvulaceae or Cuscutaceae family, is an aggressive chlorophyll-less holoparasitic weed that has currently hit parts of Kenya, parasitizing and killing a wide variety of dicotyledonous tree and herbaceous plants some of which being crop species like tobacco, tomato, eggplant, pepper, potatoes, fruit tree species, legumes like pigeon pea, alfalfa, clovers and species of agroforestry importance. Since it is a serious threat to production of dicotyledonous crops and plant species of agroforestry importance there is need to study its host range, spread and parasitizing mechanisms and control methods so as to manage it through resistance breeding. It identifies it host through the host's volatile organic compounds and intimately winds around the young tender shoots of its host (Fig. 18.5), gaining access into the vascular system using its haustoria to penetrate and establish itself in the shoot tissue. It draws all its requirements from the host, e.g., water, minerals, sugars, amino acids, photosynthates and metabolites for growth, development and reproduction, and in most cases, it impairs its host which may end up wilting and dying. Sometimes, its luxuriant growth forms a carpet totally covering the host's shoot hindering the host plant from getting adequate solar radiation for photosynthesis (Fig. 18.6).

Problems

18.1 Explain how gene-for-gene host-pathogen interaction works in regard to host resistance and pathogen virulence

Fig. 18.5 The intimate attachment of the tendril of *Cuscuta* to the young shoot of its host plant

18.2 What are the roles of the following in plant-pathogen interaction?

(i) Effectors
(ii) Receptors
(iii) PAMPS
(iv) PTI
(v) ETI

18.3 Distinguish between antixenosis and antibiosis.

18.4 Why is major gene resistance to Hessian fly not maintained for long?

18.5 Give an account of *Striga* plant response to environment and host signals from its seed maturity to emergence of a new *Striga* plant.

Fig. 18.6 Luxuriant *Cuscuta* growth totally covering the lower foliage of a host tree plant

18.6 What are the weaknesses of field screening for *Striga* resistance in sorghum and how can they be addressed?

18.7 Which morphological features prevent aphids and other suckers from successfully parasitising crop plants like cotton and tobacco?

References

Alhousari, F., & Greger, M. (2018). Silicon and mechanisms of plant resistance to insect pests. *Plants, 7*, 33. https://doi.org/10.3390/plants7020033. Creative Commons Attribution International License (CC BY).

Amusan, I. O., Rich, P. J., Menkir, A., Housley, T., & Ejeta, G. (2008). Resistance to *Striga hermonthica* in a maize inbred line derived from *Zea diploperennis. New Phytology, 178*, 157–166. https://doi.org/10.1111/j.1469-8137.2007.02355x. Wiley, No. 5770261419717.

Andersen, E. J., Ali, Byamukama, E., Yen, Y., & Nepal, M. P. (2018). Disease resistance mechanisms in plants. *Genes, 9*, 339. https://doi.org/10.3390/genes9070339. Creative Commons Attribution International License (CC BY).

Aoki, N., Cui, S., Ito, C., Kumaishi, K., Kobori, S., Ichihashi, Y., & Yoshida, S. (2022). Phenolic signals for prehaustorium formation in Striga hermonthica. *Frontiers in Plant Science, 13*, 1077996. https://doi.org/10.3389/fpls.2022.1077996. Creative Commons Attribution International License (CC BY).

Atkin, D. S. J., & Hamilton, R. J. (1982). The effects of plant waxes on insects. *Journal of Natural Products, 45*, 694–696. https://doi.org/10.1021/np50024a007. Permission from American Chemical Society (ACS).

Ayiecho, P. O., & Nyabundi, J. O. (1992). Sorghum *Striga* resistance screening. In A. Blum, P. O. Ayiecho, J. O. Nyabundi, H. H. Geiger, & B. Haussmann (1992). Sorghum improvement for semi-arid areas of Kenya. Annual report for second year, 1992 GIARA Project No. 91/4. Project Report with no Copyright.

Ayiecho, P. O., & Nyabundi, J. O. (1994). Screening of sorghum cultivars for *Striga* resistance. In A. Blum, P. O. Ayiecho, J. O. Nyabundi, H. H. Geiger, & B. Haussmann (Eds.), Sorghum improvement for semi-arid areas of Kenya. Annual report for third year, 1993. GIARA Project No. 91/4. Project Report with no Copyright.

Belete, T. (2018). Defense mechanisms of plants to insect pests: from morphological to biochemical approach. *Trends in Technical and Scientific Research, 2*(2), 555584. https://doi.org/10.19080/TTSR.2018.02.555584. Creative Commons Attribution International License (CC BY).

Blum, A. (1968). Anatomical phenomenon in seedlings of sorghum varieties resistant to the sorghum shoot fly *(Atherigona varia socata)*. *Crop Science, 8*, 388–390. https://doi.org/10.2135/cropsci1968.0011183X000800030037x. Wiley, No. 5757220265555.

Bonardi, V., Cherkis, K., Nishimura, M. T., & Dangl, J. L. (2012). A new eye on NLR proteins: focused on clarity or diffused by complexity. *Current Opinion in Immunology, 24*, 41–50. https://doi.org/10.1016/j.coi.2011.12.006. Elsevier, No. 5755530066164.

Clive, J. (2015). Global status of commercialised Biotech/GM crops: 2015. *ISAAA Brief No. 51. ISAAA.* Ithaca, New York. ISAAA Director Permission.

Convents, D., Houssier, C., Lasters, I., & Lauwereys, M. (1990). The *Bacillus thuringiensis* delta-endotoxin. Evidence of a two-domain structure of the minimal toxic fragment. *Journal of Biological Chemistry, 265*(3), 1369–1375. https://doi.org/10.1016/S0021-9258(19)40023-9. Creative Commons Attribution International License (CC BY).

Csurhes, S., Markula, A., & Zhou, Y. (2016). Invasive plant risk assessment: Witchweeds—*Striga* spp. Department of Agriculture and Fisheries, Queensland Government, State of Queensland. Creative Commons Attribution International 3.0 Australia (CC BY) License.

Deng, Y., Ning, Y., Yang, D. L., Zhai, K., Wang, G. L., & He, Z. (2020). Molecular basis of disease resistance and perspectives on breeding strategies for resistance improvement in crops. *Molecular Plant, 13*, 1402–1419. https://doi.org/10.1016/j.molp.2020.09.018. Elsevier, No. 5626441167719. Also, Creative Commons Attribution International License (CC BY).

Emani, C., & Hunter, W. (2013). Insect resistance. In C. Kole (Ed.), *Genomics and breeding for climate-resilient crops* (Vol. 2) https://doi.org/10.1007/978-3-642-37048-9_9. Springer. Springer Nature, No. 5758351004877.

Eplee, R. E. (1975). Ethylene: A witchweed seed germination stimulant. *Weed Science, 23*, 433–436. https://doi.org/10.1017/S0043174500062822. Cambridge University Press, No. 5681950184973.

Fernandez-Aparicio, M., Delavault, P., & Timko, M. P. (2020). Management of infection by parasitic weeds: a review. *Plants, 2020*(9), 1184. https://doi.org/10.3390/plants9091184. Creative Commons Attribution International License (CC BY).

Fernandez-Aparicio, M., Reboud, X., & Gibot-Leclerc, S. (2016). Broomrape weeds. Underground mechanisms of parasitism and associated strategies for their control: A review. *Frontiers in Plant Science, 7*, Article 135. https://doi.org/10.3389/fpls.2016.00135. Creative Commons Attribution International License (CC BY).

Flor, H. H. (1956). The complimentary genic systems in flax and flax rust. *Advances in Genetics, 8*, 29–54. https://doi.org/10.1016/S0065-2660(08)60498-8. Elsevier, No. 5625520215135.

Frey, K J., Browning, J. A., & Simons, M. D. (1973). Management of host resistance genes to control diseases. *Z. Pflanzenzuechtr., 80*, 160–180. https://doi.org/10.1111/j.1749-6632.1977.tb34245.x. Wiley, No. 5804930449202.

Furstenberg-Hagg, J., Zagrobelny, M., & Bak, S. (2013). Plant defense against insect herbivores. *International Journal of Molecular Science, 15*, 10242–10297. https://doi.org/10.3390/ijms/140510242. Creative Commons Attribution International License (CC BY).

Gobena, D., Shimels, M., Rich, P. J., Ruyte-Sptra, C., Bouwmeerster, H., Kanuganti, S., Mengiste, T., & Ejeta, G. (2017). Mutation in sorghum low germination stimulant 1 alters strigolactones and causes Striga resistance. *Proceedings of the National Academy of Sciences (PNAS), 114*(17), 4471–4476. https://doi.org/10.1073/pnas.1618965114. Creative Commons Attribution International License (CC BY).

Gowda, A., Rydel, T. J., Wollacott, A. M., Brown, R. S., Akbar, W., *et al.* (2016). A transgenic approach for controlling *Lygus* in cotton. *Nature Communications, 7*, 12213. https://doi.org/10.1038/ncomms12213. Creative Commons Attribution International License (CC BY).

Hauck, C., Muller, S., & Schildknecht, H. (1992). A germination stimulant for parasitic flowering plants from Sorghum bicolor, a genuine host plant. *Journal of Plant Physiology, 139*, 474–478. https://doi.org/10.1016/S0176-1617(11)80497-9. Elsevier, No. 5767731378712

Hess, D. E., & Ejeta, G. (1992). Inheritance of resistance to Striga in sorghum genotype SRN39. *Plant Breeding, 109*, 233–241. https://doi.org/10.1111/j.1439-0523.1992.tb00178.x. Wiley, No. 5807631505000.

Hess, D. E., Ejeta, G., & Butler, L. G. (1992). Selecting sorghum genotypes expressing a quantitative biosynthetic trait that confers resistance to Striga. *Phytochemistry, 31*, 493–497. https://doi.org/10.1016/0031-9422(92)90023-J. Elsevier, No. 5676640161381.

Hussain, B. (2015). Modernization in plant breeding approaches for improving biotic stress resistance in crop plants. *Turkish Journal of Agriculture and Forestry, 39*, 515–530. https://doi.org/10.3906/tar-1406-176. Creative Commons Attribution International License (CC BY).

ISAAA. (2018). Global status of commercialised Biotech/GM crops in 2018. Biotech crops continue to help meet the challenges of increased population and climate change. *ISAAA Briefs No. 54.* ISAAA, Ithaca, New York. ISAAA Director Permission.

Jiang, G. (2013). Molecular markers and marker-assisted breeding in plants. *Plant Breeding from Laboratories to the Field Chapter, 3*, 45–83. https://doi.org/10.5772/52583. Creative Commons Attribution International License (CC BY).

Jones, J. D., & Dangl, J. L. (2006). The plant immune system. *Nature, 444*, 323–329. https://doi.org/10.1038/nature05286. Springer Nature, No. 5755520130045.

Kavuluko, J., Kibe, M., Sugut, I., Masanga, J., Mutinda, Wamalwa, M., Magomere, T., Odeny, D., & Runo, S. (2021). GWAS provides biological insights into mechanisms of the parasitic plant (*Striga)* resistance in sorghum. *BMC Plant Biology, 21*, Article Number 392 (2021). https://doi.org/10.1186/s12870-012-03155-7.Creative Commons Attribution International License (CC BY).

Kim, S. K., & Brewbaker, J. L. (1977). Inheritance of general resistance in maize to *Puccinia sorghi*. *Crop Science, 17*, 450–461. https://doi.org/10.2135/cropsci1977.0011183X001700030029x. Wiley, No. 5755470175604.

Klun, J. A., Tipton, C. L., & Brindley, T. A. (1967). 2,4-Dihydroxy-7-methoxy-1,4-benzoxazin-3-one (DIMBOA), an active agent in the resistance of maize to European corn borer. *Journal of Economic Entomology, 60*, 1529–1533. https://doi.org/10.1093/jee/60.1529. Oxford University Press, No. 5626430773603.

Koch K. G., Chapman, K., Louis, J., Heng-Moss, T., & Sarath, G. (2016). Plant tolerance: A unique approach to control of Hemipteran pests. *Frontiers in Plant Science, 7*, Article 1363. https://doi.org/10.3389/fpls.2016.01363. Creative Commons Attribution International License (CC BY).

Kogan, M., & Ortman, E. E. (1978). Antixenosis—A new term proposed to replace Painter's nonpreference modality of resistance. *Bulletin of the Entomological Society of America, 24*, 175–176. https://doi.org/10.1093/besa/24.2.175. Oxford University Press, No. 5625540048842.

Kokla, A., & Melnyk, C. W. (2018). Developing a thief: Haustoria formation in parasitic plants. *Developmental Biology, 442*, 53–59. https://doi.org/10.1016/j.ydbio.2018.06.013. Elsevier, No. 5767741215628.

Leach, J. E., Leung, H., & Tisserat, N. A. (2014). Plant disease and resistance. In *Encyclopaedia of agriculture and food systems* (Vol. 4). Elsevier Inc. https://doi.org/10.1016/B978-0-444-52512-3.00165-0. Elsevier, No. 5754940796984.

Mbuvi, D. A., Masiga, C. W., Kuria, E., Masanga, J., Wamalwa, M., Mohamed, A., Odeny, D. A., Hamza, N., Timko, P., & Runo, S. (2017). Novel sources of witchweed (*Striga*) resistance from wild sorghum accessions. *Frontiers in Plant Science, 8*, Article 116. https://doi.org/10.3389/fpls.2017.0116. Creative Commons Attribution International License (CC BY).

Meuwissen, T. H. E., Hayes, B. J., & Goddard, M. E. (2001). Prediction of total genetic value using genome-wide dense marker maps. *Genetics, 157*, 1819–1829. https://doi.org/10.1093/genetics/157.4.1819. Oxford University Press, No. 5626070526316.

Mitchell, C., Brennan, R. M., Graham, J., & Karley, A. J. (2016). Plant defense against herbivorous pests: Exploiting resistance and tolerance traits for sustainable crop protection. *Frontiers and Plant Science, 7*, Article 1132. https://doi.org/10.3389/fpls.2016.01132. Creative Commons Attribution International License (CC BY).

Mohamed, A. H., Ejeta, G., Butler, L. G., & Housley, T. L. (1998). Moisture content and dormancy in *Striga asiatica* seeds. *Weed Research, 30*, 257–265. https://doi.org/10.1046/j.1365-3180.1998.00102.x. Wiley, No. 5767581065976.

Mohamed, A. H., Housley, T. L., & Ejeta, G. An *in vitro* technique for studying specific *Striga* resistance mechanisms in sorghum. *African Journal of Agricultural Research, 5*, 1868–1875. http://www.academicjournals.org/AJAR. Creative Commons Attribution International License (CC BY).

Mutinda, S., Mogeni, F. M., Hale, B., Dayou, O., Ateka, E., Wijeratne, A., Wicke, S., Bellis, E. S., & Runo, S. (2023). Resolving intergenotypic Striga resistance in sorghum. *Journal of Experimental Botany, 74*(17), 5294–5306. https://doi.org/10.1093/jxb/erad201. Creative Commons Attribution International License (CC BY).

Nelson, R. R. (1984). Strategy of breeding for disease resistance. In P. B. Vose & S. G. Blixt (Eds.), *Crop breeding: A contemporary basis* (pp. 32–49). Pergamon Press. https://doi.org/10.1016/B978-0-025505-7.50005-1. Elsevier, No. 5754501779776.

Ochiel, D. C., Dida, M. M., Ouma, E., Olweny, P. A., & Gudu, S. (2021). Response of selected sorghum (Sorghum bicolor (L.) Moench) genotypes to Striga hermonthica (Del.) Benth in western Kenya. *Journal of Plant Breeding and Crop Science, 13*, 103–114. https://doi.org/10.5897/JPBCS2020.0933. Creative Commons Attribution International License (CC BY).

Painter, R. H. (1930). Biological strains of Hessian fly. *Journal of Economic Entomology, 23*, 322–326. https://doi.org/10.1093/jee/23.2.322. Oxford University Press, No. 5761811398302.

Painter, R. H. (1941). The economic value and biological significance of insect resistance in plants. *Journal of Economic Entomology, 34*, 358–367. https://doi.org/10.1093/jee/34.3.358. Oxford University Press, No. 5758360023178.

Parlevliet, J. E., & Zadoks, J. C. (1977). The integrated concept of disease resistance: A new view including horizontal and vertical resistance in plants. *Euphytica., 26*, 5–21. https://doi.org/10.1007/BF00032062. Springer Nature, No. 575400023444.

Piquerez, S. J. M., Harvey, S. E., Beynon, J. L., & Ntoukakis, V. (2014). Improving crop disease resistance: Lessons from research on Arabidopsis and tomato. *Frontiers in Plant Science, 5*, Article 671. https://doi.org/10.3389/fpls.2014.00671. Creative Commons Attribution International License (CC BY).

Ramaiah, K. V., Chidley, V. L., & House, L. R. (1990). Inheritance of Striga seed-germination stimulant in sorghum. *Euphytica, 45*, 33–38. https://doi.org/10.1007/BF00032147. Springer Nature, No. 5807650984340.

Ratcliffe, R. H., Cambron, S. E., Flanders, K. L., Bosque-Perez, N. A., Clement, S. L., & Ohm, H. W. (2000). Biotype composition of Hessian fly (Diptera: Cecidomyiidae) populations from the southeastern, midwestern and northwestern United States and virulence to resistance genes in wheat. *Journal of Economic Entomology, 93*, 1319–1328. https://doi.org/10.1603/0022-0493-93.4.1319. PMID:10985049. Oxford University Press, No. 5627240882268.

Razzaq, A., Kaur, P., Akhter, N., Wani, S. S., & Saleem, F. (2021). Next-generation breeding strategies for climate-ready crops. *Frontiers in Plant Science, 12*, 620420. https://doi.org/10.3389/fpls.2021.620420. Creative Commons Attribution International License (CC BY).

Ribeiro do Vale, F. X., Parlevliet, J., & Zambolin, L. (2001). Concepts in plant disease resistance. *Fitopatologia Brasileira, 26*, 577–589. https://doi.org/10.1590/S100-41582001000300001. Creative Commons Attribution International License (CC BY).

Rodenburg, J., Cissoko, M., Kayongo, N., Dieng, I., Bisikwa, J., Irakiza, R., Masoka, I., Midega, C. A. O., & Scholes, J. D. (2017). Genetic variation and host-parasite specificity of *Striga* resistance and tolerance in rice: The need for predictive breeding. *The New Phytologist, 214*(3), 1267–1280. https://www.jstr.org/stable/90004232. https://doi.org/10.1111/nph.14451. Creative Commons Attribution International License (CC BY).

Sandhu, S., & Kang, M. S. (2017). Advances in breeding for resistance to insects. In *Breeding insect resistant crops for sustainable agriculture* (pp. 67–99). Springer Nature. https://doi.org/10.1007/978-981-10-6056-4_3. Nos. 5760700632463 and 5805420622704.

Sharma, H. C., Sharma, K. K., & Crouch, J. H. (2004). Genetic transformation of crops for insect resistance: Potential and limitations. *Critical Reviews in Plant Sciences, 23*, 47–72. https://doi.org/10.1080/07352680490273400. Taylor and Francis, Nos. 5760791305216 and 5760851238359.

Simon-Mateo, C., & Garcia, J. A. (2011). Antiviral strategies in plants based on RNA silencing. *Biochimica et Biophysica Acta, 1809*, 722–731. https://doi.org/10.1016/j.bbagrm.2011.05.011. Elsevier, No. 5626450264018.

Soliman, S., Wang, Y., Ha, Z., Pervaiz, T., & El-kereamy, A. (2022). Strigolactones in plants and their interaction with the ecological microbiome in response to abiotic stress. *Plants, 2022*(11), 3499. https://doi.org/10.3390/plants114399. Creative Commons Attribution International License (CC BY).

Stenberg, J. A., Heil, M., Ahman, L., & Bjorkman, C. (2015). Optimising crops for biocontrol of pests and diseases. *Trends in Plant Science, 20*, 698–712. https://doi.org/10.1016/j.tplants.2015.08.007. Creative Commons Attribution International License (CC BY).

Tingey, W. M., & Lubengayer, J. E. (1981). Defence against the green peach aphid and potato leafhopper by glandular trichomes of Solanum berhaulii. *Journal of Economic Entomology, 74*, 721–725. https://doi.org/10.1093/JEE/74.6.721. Oxford University Press, No. 5760061074844.

Unachukwu, N. N., Menkir, A., Rabbi, I. Y., Oluoch, M., Munaraka, S., Elzein, A., Odhiambo, G., Farombi, E. O., & Gedi, M. (2016). Genetic diversity and population structure of Striga hermonthica populations from Kenya and Nigeria. *Weed Research*. https://doi.org/10.1111/wre.12260. Creative Commons Attribution International License (CC BY).

Wallace, L. E., McNeal, F. H., & Berg, M. A. (1973). Minimum stem solidness required in wheat for resistance to wheat stem fly. *Journal of Economic Entomology, 66*, 1121–1123. https://doi.org/10.1093/jee/66.5.1121. Oxford University Press, No. 5757211181674.

War, A. R., Paulraj, M. G., Ahmad, T., Buhroo, A. A., Hussain, B., Ignacimuthu, S., & Sharma, H. C. (2012). Mechanisms of plant defense against herbivores. *Plant Signal Behaviour, 7*, 1306–1320. https://doi.org/10.4161/psb.21663. Taylor and Francis, No. 5626031353148.

Webster, J. A., Gage, S. H., & Smith, D. H., Jr. (1973). Suppression of the cereal leaf beetle with resistant wheat. *Environmental Entomology, 2*, 1089–1091. https://doi.org/10.1093/ee/2.6.1089. Oxford University Press, No. 5757710897328.

Yang, T., Stoopen, G., Wiegers, G., Mao, J., Wick, W., Dicke, M., & Jongsma, M. A. (2012). Pyrethrins protect pyrethrum leaves against attack by western flower thrips, Frankliniella occidentalis. *Journal of Chemical Ecology, 38*, 370–377. https://doi.org/10.1007/s10886-012-0097-7. Creative Commons Attribution International License (CC BY).

Yoder, J. I. (2001). Host plant recognition by parasitic Scrophulariaceae. *Current Opinion in Plant Biology, 4*, 359–365. https://doi.org/10.1016/S1369-5266(00)00185-0. Elsevier, No. 5767780603987.

Yoneyama, K., Awad, A. A., Xie, X., Yoneyama, K., & Takeuchi, Y. (2010). Strigolactones as germination stimulants for root parasitic plants. plant *Cell Physiology, 51*, 1095–1103. https://doi.org/10.1093.pcp/pcq055. Creative Commons Attribution International License (CC BY).

Zamore, P. D. (2004). Plant RNAi; How a viral silencing suppressor inactivates siRNA. *Current Biology, 14*, R198–R200. https://doi.org/10.1016/j.cub.2004.02.021. Elsevier, No. 5804700933211.

Chapter 19
Breeding for Resistance to Abiotic Stress Factors

Crops are often subjected to unfavourable abiotic conditions, especially in marginal lands, due to different soil types, varying soil fertility levels, drought stress, water logging and submergence, heat, chilling and frost temperatures, pollution, cultural practices, mineral toxicity or deficiency, acidity, salinity, etc., leading to undesirable impacts on crop production and accounting for substantial yield reduction, low produce quality and food insecurity in various parts of the world. Drastic reduction in crop yield due to exposure to these factors is a result of agro-morphological, physiological, molecular and biochemical modifications which have negative impacts on all phases of plant growth, development and ultimately, yield and produce quality. These factors are often interrelated leading to serious damage to crop plants.

Plant response to various abiotic stress factors, drought, heat stress and mineral stress, are interlinked through physical, physiological, biochemical and molecular changes that arise due to the stresses. Occurrence of stress leads to signal transduction pathways that alter gene expression and activation of genes encoding various enzymes and proteins for protection against stress. The consequence of the changes and gene activation is the synthesis of certain mRNAs and proteins for enhancing tolerance to stress. Manipulation of the environmental conditions to resolve agricultural production problems arising from abiotic stress factors is difficult and expensive and, in some cases, impractical or impossible. The most practical and cheapest strategy to address environmental abiotic challenges is development of crop cultivars that have resistance or tolerance genes which control resistance/tolerance component mechanisms.

19.1 Breeding for Drought Resistance

Drought occurrence and high temperatures are the key abiotic factors responsible for water stress reducing accumulation of biomass in the shoot and root with consequent low crop yields, especially in dryland and semi-arid lands where rainfall is always

P. O. Ayiecho and J. O. Nyabundi, *Conventional and Contemporary Practices of Plant Breeding*, https://doi.org/10.1007/978-3-031-74998-8_19

limiting and droughts are prolonged. Water stress always acts in tandem with heat to limit crop yields with serious negative impacts on produce quality, particularly under conditions of terminal drought. Elevated temperatures due to global climate change have aggravated the levels of stress arising from droughts and environmental degradation resulting in heavy agricultural production losses and food insecurity, with the heaviest impact being felt in the dry lowlands within the tropics.

Water stress affects overall morphological, physiological, biochemical and molecular features of the plant and impairs cell division, cell growth, cell elongation and enlargement leading to suppression of germination, seedling growth, crop stand, overall plant growth and development, plant size, branching, tillering, spikes per plant, flowering, grains/seeds per plant or fruits per plant, grain or seed filling duration, grain/seed weight, and ultimate crop performance. Drought stress reduces relative water content in plant tissues, leaf water potential, turgor potential, stomatal conductance, transpiration rate and increases leaf and plant canopy temperatures. Under normal circumstances transpiration dissipates excess heat in the plant canopy and helps to lower leaf and canopy temperatures below the immediate surrounding environment. The stress causes stomatal closure thereby limiting CO_2 uptake by the leaves and reduces nutrient uptake and translocation, photosynthesis, assimilate flux for seed or grain and fruit development resulting in abortion of flowers, the developing fruits and seeds. Moisture deficit causes scorching and abscission of plant organs, general chlorosis and necrosis, reduced pollination and fertilisation rate, poor seed and fruit development and lowered plant fresh and dry weight and seed yield. There is reduced assimilation and photosynthetic rates, inactivation of enzymes, production and accumulation of toxic phytochemicals, particularly reactive oxygen species (ROS) like peroxides and superoxide responsible for oxidative damage accompanied by interference with cell metabolism, cell organelles, integrity of cell membrane lipids, nucleic acids, protein synthesis and protein integrity with the consequence of protein denaturation. The damage attributed to prolonged moisture stress may lead to death of the cells, tissues and organs, precocious leaf senescence and abscission, leaf firing, schorching and the whole plant may wilt and die (Fig. 19.1), particularly for highly susceptible genotypes.

Farmers faced with the problems of drought who do not use irrigation get lower than expected crop yields. For the resource-poor small-scale farmers drought stress is a problem that they can rarely address through irrigation because of the costs and infrastructure involved unless the governments or other institutions provide support. The cheapest way to address drought problem for the benefit of farmers is development of drought and heat resistant varieties with high and stable yields. Drought resistance in plants is a complex trait with various mechanisms, influenced by both environmental factors and pant morphological, physiological, molecular and biochemical features controlled by the plant genotype. The many genes associated with plant response to water stress are also under complex control systems involving several processes and expression leading to synthesis of proteins that signal other activities, functional proteins protecting cell membranes and other metabolites associated with water uptake and transport under moisture deficit situations.

Fig. 19.1 A young sugarcane crop damaged by a prolonged spell of dry and hot weather in western Kenya

19.1.1 Drought Resistance Mechanisms

Plants have various morphological, physiological, anatomical, molecular and biochemical features which help them adapt to various water deficit and heat prone conditions. Some of the important features of drought resistant genotypes are acceptable fresh and dry biomass yield, survival rate, seed-setting rate, spikelet fertility, grain or seed yield and stay-green capacity. Drought resistance is a multi-trait and multigenic character that is mostly of quantitative nature, involving both major genes and minor genes. The several component traits of plant resistance mechanisms to water stress are categorised as escape, avoidance, tolerance and recovery.

Drought escape and Recovery

Drought escape is a feature of annual crop plants and is associated with the plant life cycle, growth period and development plasticity which make it possible for a plant to complete its life cycle in a short period when moisture is still available and escapes drought and high temperatures, especially situations of terminal drought. Through escape mechanism the plant undergoes all its developmental stages and matures before drought sets in by being fast growing, early flowering, fast grain or seed filling and early maturing. Earliness is usually negatively associated with yield and biomass but is a useful trait for dealing with terminal drought challenges which do not suit long maturing genotypes. Late maturing varieties are higher yielding under adequate rain than early varieties but under drought prone conditions earliness leads to a crop yield where there would be a total crop failure for late maturing varieties. Slow growing and late maturing varieties are better adapted to areas with more rain and wetter conditions. Therefore, under situations of terminal drought selection for earliness is the recommended breeding strategy. With genetic recombination, molecular marker assisted breeding, genome editing, and transgenesis, it is possible to combine earliness and high yield genes. The problem of low yields in early varieties can be addressed through increased planting rate since early varieties are shorter and smaller in size, have smaller leaves and lower leaf area indices and can withstand high planting densities which compensates for low individual plant yields to give high yields per unit area (Blum, 1970). Crop varieties that are adapted to short growing periods and escape terminal drought usually have early seedling growth and fast crop stand establishment. Early maturity is associated with reduced water use because early plants have lower leaf area indices, lower evapotranspiration rates and maintain favourable leaf water potential during their growth and development as compared to leafier late maturing genotypes. Early maturity is a highly heritable and reliably selectable trait in nearly all crop plants and is an important breeding objective in maize, beans, rice, sorghum, barley, wheat, pigeon peas, cowpeas, cotton, sunflower, chick pea, sesame and many other annual crops so as to suit the crop to short rainy seasons in regions with terminal droughts.

In some crops there are genotypes which can recover from water stress damage when rain returns or when water is supplied. Such genotypes recover well and still give a good crop yield. Capacity for recovery is variable among crops and crop cultivars and can be exploited for improving drought resistance in crop varieties.

Drought avoidance

Drought avoidance is ability of a plant to maintain high leaf water potential when subjected to water stress either through water uptake or maintenance of its tissue water status, thereby maintaining high levels of tissue hydration despite drought stress. Drought avoidance is usually associated with certain plant morphological, anatomical, physiological, molecular and biochemical features which improve plant water status, water uptake and sustenance of physiological processes. Morphological features which play roles in water stress avoidance include leaf rolling, increased pubescence on aerial parts, thick epicuticular wax, stomatal density, leaf shape and leaf orientation, increased reflectance of solar radiation, increased heat dissipation by aerial organs, glaucousness, smaller leaves and reduced leaf area, high root/shoot ratio, deep and extensive rooting system (Tardieu, 2013; Fang & Xiong, 2015). Physiological plant features that contribute to drought avoidance are photosynthetic rate and efficiency, reduced transpiration, abscisic acid-mediated stomatal closure, osmotic adjustment, high leaf water retention, relatively higher leaf water potential under soil water deficits, well developed vascular bundle sheath and canopy temperature, leaf firing, leaf rolling, desiccation tolerance, plant recovery following dehydration and stability of cellular membranes (Tardieu 2013; Fang & Xiong, 2015) . Water stress to plants leads to altered expression of genes associated with response to drought as regards transcription and translation, accumulation of phytochemicals concerned with molecular control of drought resistance, reduction and repair of damage due to water stress, drought resistance associated hormones, organic acids, amino acids, antioxidants and sugars in the tissues (Luo et al., 2019; Martignago et al., 2020).

Plants capable of photosynthesising under water deficit and those that close their stomates are both considered resistant. Control of stomatal aperture is an important mechanism of controlling water loss by plants and increased stomatal resistance is a common feature of drought resistance making stomatal sensitivity and conductance effective for control of transpiration under water stress and useful criteria for selecting promising drought resistant genotypes. Crop cultivars vary in stomatal regulation to control transpiration rates under varying levels of water deficit and in some cases stomatal closure for internal moisture conservation and reduced water loss under drought stress may have negative effects on photosynthesis. However, since stomatal conductance is believed to be positively associated with yield under stress and other drought resistance traits it can be used as a selection criterion for improved performance under drought conditions. Crop cultivars adapted to dryland conditions usually exhibit higher photosynthetic rates, higher stomatal conductance, deeper rooting tendency and high osmotic adjustment than cultivars adapted to high rainfall environments. In studies with sorghum (Nyabundi & Ayiecho, 1994) observed that sorghum cultivars adapted to dryland conditions exhibited higher photosynthetic rates, higher stomatal conductance, deeper rooting tendency and high osmotic adjustment than cultivars adapted to high rainfall environments.

When plants are exposed to drought stress, accompanied by osmotic and oxidative stress, the osmotic and homeostasis integrity is be disrupted and proteins and plasma membrane are damaged because of accumulation of harmful compounds in the plant tissues. A number of genes are induced for synthesis of chaperons, reactive oxygen species scavengers and osmoprotectants, antioxidants and metabolites responsible for plasma membrane integrity and protein protection for the promotion of drought stress tolerance. There is also induction of genes concerned with pathways for signalling, perception, transduction, accumulation and activation of enzymes, proteins and other metabolites related to stress response, transcription factors and heat shock proteins.

As the plant experiences water deficit under drought there are adverse effects on cell membranes, organelles, hydration and structure of macromolecules like proteins and nucleic acids, plant hormone function, cell expansion, cell division and photosynthesis which arise resulting in reduced plant growth, development and yield. In some genotypes, osmotic adjustment or osmo-regulation helps in the maintenance of high turgor potential needed for normal cell function, growth and development. Osmotic adjustment helps to sustain leaf water potential through closure of stomata and its influence on turgor maintenance has desirable effects on growth and yield under water stress (Khan et al., 2018). Osmotic adjustment in cells within plant tissues is achieved through osmoregulation as the plant responds to water deficit by synthesis and accumulation of various phytochemicals which include plant hormones, organic acids, amino acids, antioxidants and sugars like abscisic acid (ABA), brassinosteroids, jasmonic acid, proline, glycinebetaine, fructans, trehalose, osmoprotectants, mannitol and other phytochemicals to maintain turgor as water potential falls, protect cell membranes, growth and other physiological processes and suppress dehydrative damage to plant tissues (Farooq et al., 2009; Khan et al., 2018). Accumulation of such phytochemicals in drought stressed plants vary according to plant genotypes and growth stage. As water stress develops within the plant tissues due to limited water supply leaf water potential becomes a good parameter for measuring the plant water status because the leaf is directly involved in transpiration and assimilate production for development, growth and yield. Selection for osmotic adjustment leads to improved dehydration avoidance and maintenance of high yields since there is ample genetic variation for this mechanism in many crops.

In some plant genotypes there are early warning systems regarding response to drought stress through phytohormone abscisic acid (ABA) production in root tips as the soil dries up and ABA thus produced is transferred to the shoots for accumulation in the leaves where it triggers stomatal closure resulting in reduced dehydration in plant tissues and also promotes synthesis and accumulation of proline (Choi et al.,2000; Khan et al., 2018). Therefore, in water saving genotypes the stomata are influenced by ABA accumulated in the leaves such that they close as water stress progresses. As ABA increases in tissues of water stressed plants it improves drought tolerance by triggering the expression of drought related genes and helps in maintaining high relative water content and chlorophyll (Choi et al., 2000). Water stress leads to accumulation of ROS which are harmful to DNA, proteins and plasma membrane, react with proteins, lipids and other cellular components thereby interfering with cell functions (Reddy et al., 2004; Langridge et al., 2006) . Reactive oxygen species are natural biochemical by-products arising from some cell metabolic activities. Within the tissues of plants subjected to water stress the expression of some genes are altered as regards protein synthesis, water transport, oxidative damage, damage repair mechanism, osmotic balance and other plant physiological functions to enhance plant survival under stress. Though ROS may be beneficial to the plant cells they are also toxic to the cell making it necessary to remove excess amounts, their synthesis suppressed or the cell should have a process of repairing the damage they cause. As a defence response against the ensuing water deficit stress the plant synthesises and accumulates antioxidants, enzymes and osmolytes like glycosylates, heat shock proteins, ascorbic acid, glutathione, α-tocopherol, superoxide dismutase, peroxidase, proline, glycinebetaine, late embryogenesis abundant proteins (LEA), mannitol, trehalose, fructan, inositol and ectoine which scavenge the ROS, improve the cell membrane integrity, protect the proteins and help in adjustment of osmotic potential within the plant cells and tissues (Langridge et al., 2006; Khan et al., 2018). Genes encoding these drought resistance phytochemicals exist in crop plants and can be used to improve drought resistance in drought susceptible elite cultivars by applying conventional selection strategies, molecular marker assisted breeding, genomic selection, transgenesis and genomic editing.

Fructans promotes root branching and increased root surface area for improved water uptake leading to elevated drought tolerance and plants with high levels of fructans tend to be drought tolerant. Proline accumulation in plant tissues under water deficit conditions is considered to be a symptom of water stress in plants because moisture stress causes a blockage in the synthesis of certain amino acids resulting in its accumulation in plant tissues (Delauney & Verma, 1993). Glycinebetaine, proline, mannitol, fructans, trehalose, late embryogenesis abundant proteins and sucrose are among the phytochemicals that maintain membrane stability and other biological cellular structures by suppressing molecular reactions that cause damage to cell membranes under water stress and genotypes that accumulate them under drought can be selected for drought tolerant cultivar development (Farooq et al., 2009). Brassinosteroids belong to another group of phytohormones that signal different responses during root growth, root development and stress response and genetic manipulation of their pathways can lead to drought resistance through changes in expression of root related traits (Planas-Riverola et al., 2019). Jasmonic acid is another phytohormone which helps in better rooting, fruit ripening and good general plant performance under water stress conditions and combating various forms of abiotic stress (Khan et al., 2018) and genotypes which accumulate it can be identified for drought tolerant cultivar development. Genetic variation in production of ABA, fructans, glycinebetaine, proline, jasmonic acid and other phytochemicals associated with response to drought stress among plant genotypes is important for drought resistance selection and incorporation of the relevant genes into the target cultivars.

Among morpho-physiological traits used to assess plant drought avoidance are absence or presence of wilting, leaf rolling, leaf firing, ability to recover turgor overnight (i.e., absence of early morning wilting), water use efficiency, use of infrared thermometer to assess canopy temperature under drought conditions. Leaf rolling is a drought susceptibility symptom and is related to leaf water potential such that it reflects loss of turgor, especially in cereals. It is delayed by osmotic adjustment and genotypes with delayed leaf rolling can be selected to obtain cultivars with dehydration avoidance. Genotypes with low osmotic potential have high leaf rolling and leaf tip and margin drying (leaf firing) while the genotypes which maintain turgor status exhibit delayed leaf firing under water stress resulting in better performance (Nyabundi & Ayiecho, 1994; Haussmann et al., 1999). Leaf rolling also reduces leaf area exposed to the atmosphere and solar radiation thereby reducing transpiration substantially in plants under drought stress. The extent and onset of leaf rolling and its elasticity in terms of unrolling can be incorporated in the selection process for drought resistance breeding. Plant canopy temperature is another parameter that is used to assess water stress resistance and select resistant genotypes. High canopy temperature is an undesirable character since genotypes exhibiting this trait have reduced crop yields. Genotypes with low midday canopy temperatures are drought resistant and suffer less yield reduction under water stress conditions. Low canopy temperature under water deficit is a selectable trait which is used for crop improvement and identification of desirable genotypes for improved performance under water deficit conditions. Low canopy temperatures and plant water status are known to be associated with good rooting characteristics and drought resistance.

Glaucousness (pale yellow-green colouration) in plants reduces radiation and leaf temperatures by increasing leaf reflectance. Glaucous appearance in plants is attributed to scattering and reflection of light by wax crystal deposits on plant organ surfaces (Wurschum et al., 2020). Crop plants with glaucous leaves reflect more light than those with dark green leaves, hence keeping the canopy cooler, leading to drought and heat resistance and relatively high yields under such stress conditions (Wurschum et al., 2020). Some plant genotypes have delayed senescence feature often referred to as stay-green or green leaf area at maturity (GLAM) which helps in extended photosynthesis under terminal drought situations (Martignago et al., 2020). Under drought stress post-flowering stay-green feature maximises yield since it maintains transpiration and photosynthesis. Stay-green genes have been identified in some cultivars for a number of crops and in some of them the feature is believed to be of quantitative nature and associated with some identified QTLs.

Internal water deficit in plant tissues (lower water potential) occurs when water uptake by the roots is not sufficient to replace the water lost by the plant through transpiration. This happens when there is water deficit in the soil or the roots are not adequately developed to extract sufficient water from the soil for plant needs. Maintenance of leaf water potential under soil water deficit depends on the supply of water to the leaves and control of water loss from the leaves. Rooting characteristics are related to leaf water potential since deep and dense roots with high volumes are helpful in drought avoidance through water uptake from the deeper soil horizons to meet the transpiration demands when the upper horizons are depleted of water. Extensive lateral root development also helps in keeping the plant tissues and leaves hydrated as water stress sets in. Roots have the greatest effects on leaf water potential under water stress, and as root depth, width and branching increase negative impact of soil water deficit on the plant is reduced. Deep and branched rooting system with large numbers of slender lateral roots give large surface for tapping of water and nutrients in deep soil horizons and desirable adaptation to water stress conditions. Most crops have selectable genetic variability among their cultivars and related species for rooting characteristics which can be mobilised for improving adaptability to water deficit situations, taking into consideration that deep rooting and high root number traits tend towards dominant gene control and therefore selectable and transferable in breeding programmes. Leaf water potential is also a selectable useful selection criterion which can be used to improve crop cultivars with respect to water stress resistance.

Reduced leaf area increases plant survival under water deficit because of low surface area for evapotranspiration. Plants with smaller leaves are usually more drought resistant, have reduced leaf rolling and higher midday water potential than cultivars with larger leaves. In cereal crops like sorghum there are some genotypes with epicuticular wax layer on some aerial parts including the peduncle, leaf sheaths and basal parts of abaxial leaf surfaces which help to reduce water loss through reduced transpiration, lowered radiation, increased reflectance and protection of leaf epidermis against UV light, thereby promoting drought resistance (Martignago et al., 2020). High pubescence on leaves and other organs also increases leaf reflectance under high radiation helping to lower temperatures and evapotranspiration and consequently leads to improved water use efficiency by the plant. Leaf orientation as a drought avoidance trait works by affecting radiation received by the plant. Maximum radiation is received on leaf surfaces when leaves are in perpendicular position while erect leaves receive less radiation, have less transpiration loss and are at a better leaf water status, especially under water stress (Ludlow & Bjorkman, 1984). Genotypes with erect leaves yield better under drought than genotypes with perpendicularly placed leaves and can be selected to develop varieties with improved adaptation to drought conditions.

Tolerance

Plants tolerant to drought withstand tissue dehydration as a result of osmotic adjustment and activities of osmo-protectants making them sustain high cellular turgor potential, retain water in the tissues, undergo all their developmental and growth stages while under drought stress and ultimately give good yields in spite of stress conditions. Plants can exhibit tolerance to dehydration at various stages of their growth and development, namely, seedling growth, adult plant growth to flowering, flowering and grain filling or seed development and survival to give a good yield. While early seedling vigour, especially good seedling root growth and development, is associated with seedling drought stress tolerance organs like the flag leaves and awns help in increasing photosynthetic area for improved photosynthesis and better survival of the adult cereal crop plants under waters stress conditions. Phytochemicals like proline, mannitol, glycinebetaine, trehalose, fructan, LEA, inositol, alkaloids, polyols, jasmonic acid, dehydrin and ABA accumulated in plant tissues under water stress help in osmotic adjustment for water stress tolerance by contributing to cytoplasm osmotic balance, cell turgor maintenance, protection to various enzymes and chlorophyll under dehydration and prevention of cell membrane disruption due to water and heat stress (Fang & Xiong, 2015; Khan, 2018). Proline is known to bind to both proteins and water, enhancing protein solubility in water and at the same time preventing protein denaturation (Ashraf & Foolad, 2007). While trehalose also helps in stabilising structure of proteins, nucleic acids and other biological molecules under drought, glycinebetaine stabilises structure and activities of photosynthetic apparatus and enzymes and protects cell membranes from damage by water stress (Ashraf & Foolad, 2007). Dehydrin, whose genes are induced by stress and ABA, is another molecular protectant that prevents denaturation of enzymes and other proteins due to drought and heat stress. Genotypes may vary with respect to heat tolerance, accumulation of the various phytochemicals and desiccation survival making it possible to use such attributes as selection criteria for improved drought tolerance.

19.1.2 Drought Resistance Breeding Strategies

Breeding for drought resistance requires knowledge about the genetic control of the resistance features, appropriate source of resistance genes and information about the environment as regards soil characteristics, weather data, especially rainfall and temperature, other abiotic, biotic and agronomic factors pertaining to the region for which drought resistant crops are to be developed. It is necessary to understand effects of such factors on the crop plant, physiological and biochemical changes associated with drought stress and the stages of the plant most affected by water deficit situations.

Plant response to drought is complex because it is determined by interacting component traits which differ in the way they respond to drought stress and other environmental factors and are controlled by several major genes and minor genes with additive, dominance and epistasis effects. Plants usually employ a combination of different resistance mechanisms and most drought resistance traits are interrelated such that selecting for one or a few of them will alter others. For example, selection for earliness may alter the plant architecture and size affecting inflorescence, fruit or grain size number, tillering or branching pattern, height, determinate or indeterminate nature, size of organs like leaves and overall yield. Selection for root drought resistance features is an effective morphological drought resistance breeding strategy since in certain crops some root characteristics may be major gene controlled or have high heritability (Luo et al., 2019). In most crops there is ample genetic variation for morphological and physiological characteristics, e.g., rooting characteristics, leaf water potential, canopy temperature, leaf rolling, leaf firing, carbon exchange rate, osmo-regulation and stomatal conductance, osmotic adjustment, stomatal control of transpiration, osmotic adjustment, chlorophyll content, improved photosynthesis, epicuticular wax load, carbon fixation at low water potential, cellular membrane stability and seedling recovery from severe water stress and yield under drought stress conditions. Genetic variation for these traits and several other drought resistance features like escape features, production of phytochemicals associated with drought response and resistance, namely, proline, glycinebetaine, trehalose, ABA, fructans, mannitol, brassinosteroids, antioxidants, inositol, alkaloids, polyols, jasmonic acid, dehydrin and others have also been documented in many crops and non-crop organisms as well by various crop scientists. The identified drought resistance genes can be transferred to the target crop cultivars through conventional methods of crossing and backcrossing, gene pyramiding for multiple resistance mechanisms and interspecific hybridisation using molecular marker assisted selection, molecular marker assisted backcrossing and transgenesis for better yield performance under drought conditions. Incorporation or upregulation of genes linked to accumulation of ABA, trehalose, glycinebetaine, LEA, proline, jasmonic acid, brassinosteroids, antioxidant enzymes and other related phytochemicals in plant tissues and genetic modulation of signalling pathways associated with them improves plant survival and yield performance under drought conditions as is known for brassinosteroids in *Arabidopsis* and certain crop plants (Planas-Riverola et al., 2019; Martignago et al., 2020). Since ABA is the main phytohormone responsible for stomatal closure under water stress incorporation or upregulation of genes linked to its synthesis in the roots and accumulation in leaves and genetic modulation of its signalling system elevates stomatal response to water deficits and plant survival under drought conditions (Bi et al., 2017; Martignago et al., 2020).

In heterogeneous cultivars belonging to autogamous crop species fast selection for drought resistance is achievable through mass selection and pure line selection by conducting selection in the relevant environments if the cultivars have acceptable agronomic traits and yield. Where characters of importance are controlled by major genes like earliness hybridisation of selected genotypes is followed by backcrossing facilitated by molecular markers accompanied with selection for pure line development.

Breeding for escape mechanism as a drought resistance strategy entails identifying genes for fast growth, early flowering and maturity. Breeding for early maturity to escape terminal drought in annual crops is easy and straight forward since in several cases earliness is a major gene trait and is more simply inherited than many other drought resistance traits and is incorporated by hybridisation and backcrossing. However, days to flowering is also under the influence of quantitative polygenes and environmental factors. Any of the conventional breeding methods is applicable for modifying characters associated with escape mechanism, depending on which one is appropriate, the germplasm materials available, other breeding objectives, genetics of the trait, other resistance mechanisms to be combined, the crop in question and its reproductive system and the type of varieties to be developed. Other genes are also known to influence flowering pathways and may need to be addressed, especially in temperate crops in which some genes associated with flowering have to be induced by vernalisation and day length like wheat and barley which are long day plants (Cockram et al., 2007). In wheat, barley and other temperate crops the varieties planted in autumn require vernalisation while the cultivars grown in spring season do not require vernalization.

Incorporation of drought avoidance and tolerance mechanisms depend on the drought resistance mechanism components in the germplasm at hand. Most of dehydration avoidance mechanism traits are selectable because of presence of genetic variability, and the genotypes that wilt, roll leaves at different stages, show leaf firing and other symptoms attributed to water stress can be eliminated in a breeding programme. Selection is done for genotypes with low canopy temperatures, those having cuticular wax, pubescence, high leaf water retention, desirable rooting characteristics, seedling survival under stress, cellular membrane stability and other resistance traits. Epicuticular wax and some rooting characteristics are known to be under major gene control and can be incorporated into the target cultivars by conventional hybridisation and interspecific hybridisation facilitated by use of molecular markers and transgenesis where necessary (Martinagno et al., 2020). Leaf rolling is also easily selected for, especially in crops like wheat and rice where it is believed to be influenced by major genes.

Hybridisation of desired resistance gene sources with genotypes targeted for improvement is followed by the selection approach as determined by genetics of the traits to be transferred and the crop reproductive system so as to develop water stress resistant pure lines, inbred lines, hybrid varieties and open pollinated populations with the desired agronomic traits and yield. In autogamous crops like sorghum, millets and rainfed rice cultivars, wheat, beans, barley, soybean and cowpea selection methods following hybridisation, namely mass selection, pure line, pedigree and backcrossing, bulk breeding and single seed descent/single pod descent are determined by the nature of inheritance of the resistance. In major gene resistance traits simple hybridisation followed by backcrossing for a few generations facilitated by molecular markers is usually adequate in stacking of the resistance genes in the elite breeding materials. In cases of resistance mechanisms that are of quantitative nature like stay-green, osmotic adjustment and low canopy temperature pure line selection, pedigree breeding and single seed descent/single pod descent are often applied. In pedigree breeding selection can be initiated in F_3 followed by testing of F_4 and F_5 selected lines and the subsequent genotypes evaluated further under stress conditions to identify the best pure lines. Traits of quantitative nature are also improved through recurrent selection and recombination methods designed for autogamous crops to accumulate resistance polygenes for elevating resistance to desired levels before drought resistant superior and high yielding pure lines are obtained from improved populations. In allogamous crops identified major genes for drought resistance traits are transferred to inbred lines through crossing followed by backcrossing or other methods of improving inbred lines. Mass selection and population improvement methods involving various forms of recurrent selection are often employed to increase the frequencies of resistance polygenes to improve the various resistance features to the desired levels with respect to quantitative drought resistance mechanisms.

Molecular breeding for drought tolerance

Drought tolerance genes are often induced to encode various metabolites when plants are exposed to waters stress so as to enhance cellular and tissue tolerance to dehydration and protect cellular functions, plasma membrane, proteins and lipids from damage. Trehalose, brassinosteroids, superoxide dismutase, peroxidase, ABA, LEA, trehalose-6-synthetase, fructans, proline, glycinebetaine, other signalling pathway components and osmoprotectants are encoded by dehydration responsive genes which can be manipulated and transferred to desired cultivars (Planas-Riverola et al., 2019; Martignago et al., 2020). Transcription factors, which are also accumulated and activated as the plant responds to drought stress, enhance or suppress the rate of expression of certain drought tolerance genes by binding to specific promoter regions of such genes upon activation or deactivation of upstream signalling steps and are involved in regulating the expression of the genes (Choi et al., 2000; Bi et al., 2017). Based on the fact that molecular control for resistance to drought is determined by activation and regulation of drought response multigenes involved in signalling and transducing, encoding and transcription for phytochemicals responsible for plasma membrane stability, homeostasis, scavenging of reactive oxygen species, etc., molecular breeding employing molecular markers and genomic technologies offer the best breeding approaches for developing stress tolerant cultivars. Genomics, proteomics, transcriptomics, metabolomics (omics technologies), high-throughput phenotyping assays along with bioinformatics have the potential to facilitate molecular level screening for drought resistance variation in germplasm materials and make marker assisted breeding easy through identification of molecular processes involved in gene expression, gene function and interaction as the plant responds to stress. The 'omics' assays generate the required data for application in genomic selection as well. Genes which encode the various features, proteins and metabolites for stress adaptation, whether biochemical, physiological or morphological or involved in regulation of pathways related to response to water deficit stress, can be precisely identified, mapped and tagged, using appropriate markers. The genes can be distinguished and isolated through their expression in form of transcripts and proteins which reveal qualitative and quantitative changes associated with plant response to drought stress.

Since plant tolerance to water stress is multi-trait, multigenic and polygenic, the best approach of accumulating drought resistant traits in one genetic background is associating tolerance traits to relevant genes and QTLs tagged using flanking molecular markers for identification of the chromosome segments linked to the expression of such traits. Genes and QTLs for root characteristics, leaf-based features, phytochemical and physiological traits linked to drought stress response well documented in a number of crops for various traits like stay-green trait, rooting characteristics, yield, tillering and other agronomic traits under drought stress can be mobilised to improve crop drought resistance (Harris et al., 2007; Baytar et al., 2018). Molecular marker assisted breeding, particularly marker assisted recurrent selection, increase the frequencies of tolerance associated genes and QTLs in target cultivars while marker assisted backcrossing helps to combine donor tolerance traits and the recurrent parent superiority traits in one genetic background. Molecular marker assisted backcrossing is important in gene pyramiding for developing multi-trait drought tolerant cultivars. In some cases, tolerance trait QTLs, yield and other desired traits colocalise and co-segregate, making the development of stress tolerant cultivars much easier because of such genetic association (Lafitte et al., 2004). The tagged drought tolerance genes and QTLs are introgressed into selected elite cultivars to incorporate tolerance in pure lines or inbred lines for inclusion in the mainstream programme for drought tolerant variety development.

Alien genetic variation for drought resistance in crop wild and weedy relatives can be mobilised through interspecific hybridisation while drought resistance genes identified in plants not related to crop plants can be moved to crops by way of transgenesis to create drought resistant genotypes. Though there may be challenges due to multigenic/polygenic and complex nature of drought resistance in plants the identified alien genes associated with drought resistance pathways and regulation can be transferred to the drought susceptible elite cultivars and accumulated in the respective cultivars using genetic engineering methods for purpose of gene pyramiding of different drought tolerance genes affecting various drought resistance traits at different stages of crop growth and development. Genome editing technology employing clustered regularly interspersed short palindromic repeats (CRISPR-Cas9), zinc finger nuclease (ZFN) and transcription activator-like effector nucleases (TALENs) is another molecular strategy for rapid improvement of drought resistance by altering a few nucleotides in the target genes, knock out genes, modify expression of genes by inserting transgenes at certain locations through homologous recombination or disrupting gene promoter sequences and creating new alleles (Khan et al., 2018; Martignago et al., 2020). Transgenesis and genome editing, particularly CRISPR-Cas9, are being applied in a number of crops to improve abiotic stress resistance.

Screening tests for drought resistance

Screening and tests for drought resistance are done in the field and under controlled temperature and moisture supply conditions in laboratory, growth chambers and greenhouse so as to select promising genotypes. Field tests are often unreliable since temperature and moisture conditions are unpredictable and fluctuate from location to location, time to time and even within the same field location soil moisture is never uniform. Despite lack of uniformity of field conditions in regard to water stress occurrence and effects of other stress and environmental factors, field screening is usually preferable for large populations, especially if sites that help in broad discrimination among the genotypes are used for the tests. Drought conditions are usually simulated by controlling moisture supply in plant pots or appropriate plant growth containers for more efficient screening. Hydroponic tanks and PVC tubes in the greenhouses, growth chambers and laboratories are used to distinguish genotypes in terms of root length, density and dry weight to screen for rooting characteristics. Rooting characteristics screening is also done by growing plants in solutions in plastic or glass faced boxes and sand to distinguish the rooting patterns. Use of detached plant leaves for laboratory tests to screen genotypes for transpiration, stomatal behaviour and net photosynthesis, laboratory tests at germination and seedling stages and in vitro selection under simulated water stress conditions help in easy and fast selection of drought resistant genotypes. Speed breeding and high-throughput phenotyping techniques using automated and computerised systems under greenhouse, nursery and growth chamber conditions can be applied to enhance screening efficiency and also improve throughput and accuracy of phenotyping for the advantage of development of stress tolerant cultivars much faster.

Seed germination tests to identify drought avoiding genotypes are often conducted to compare germination ability and germination rates of different genotypes in the laboratory and pots with simulated water stress conditions and in the field under moisture stress situations. The test genotypes are assessed and compared to known selected drought resistant or susceptible genotypes for germination rates and overall germination performances, early seedling growth and development. Osmoticums like D-mannitol or polyethylene glycol are also used to simulate drought stress for purpose of conducting germination tests in the laboratory and screening tissues in culture to select the variants that can be regenerated into drought resistant plants by taking advantage of somaclonal variation.

Metabolite concentration changes due to stress also imply that metabolomics assays provide important information on the mechanisms of stress response for use in drought tolerance breeding. Processes of stress response like signalling pathway components, e.g., transcription factors, heat shock proteins, chaperones, LEA proteins, reactive oxygen species scavenging, synthesis and accumulation of osmoprotectants, ion and water transporters can be assessed in the laboratory using appropriate facilities to distinguish and develop cultivars that are adequately stress tolerant.

19.2 Breeding for Heat Tolerance

Heat stress (thermo-stress) is among the factors causing yield reduction in crop plants making heat tolerance an important breeding objective. High temperatures arising from intense solar radiation, impact negatively on crop productivity and are a threat to agricultural production and food security, especially in light of climate change which causes global warming due to increased greenhouse gases. Heat stress arising from global elevated temperatures has adverse effects on crop plant growth and development making it a major crop production challenge globally, especially in underdeveloped and developing tropical and subtropical regions. Heat stress is always accompanied by drought stress so that the ultimate combined effects of the two stress factors turn out to be extremely disastrous on the plant.

19.2.1 Morphological, Physiological and Biochemical Changes in Plants Due to High Temperatures

Rising temperatures due to climate change affect crop performance through negative influence on various stages of plant growth and development with reproductive stage being the most vulnerable. Various plant species have critical cardinal temperatures below which or above which the temperatures are too low or too high for the plant, affecting morphological, physiological, biochemical and molecular features of the plant and ultimately leading to poor growth, poor development and low yields. The impact of high temperatures depends on the intensity, timing and duration of stress, plant species, genotype and growth stage of the plant. Heat stress causes varied morpho-anatomical, physiological and biochemical changes in plants which are detrimental to carbon assimilation, plant growth, development, fruit and seed yield and produce quality. This situation can be mitigated by breeding crop genotypes with better levels of thermo-tolerance.

Increasing temperatures beyond the critical cardinal thresholds causes heat stress that leads to irreversible harmful effects and damage to growth and development through undesirable changes on agronomic, morphological physiological and biochemical features. High temperatures much above the ambient temperatures cause heat shock or heat stress to the plant interfering with seed germination, seedling growth and development, crop stand, flowering and seed or fruit production. Heat stress may make the plant have precocious leaf senescence and abscission, firing of leaves, scorching and abscission of stems and branches and supressed shoot growth and development, shoot chlorosis and necrosis, reduced pollen viability and fertility, reduced pollination and fertilisation rate, poor seed and fruit development, poor seed germination, reduced assimilation and photosynthetic rate and general reduction in seed and biomass yield (Bita & Gerats, 2013; Driedonks et al., 2016). The whole plant may wilt and die in heat susceptible genotypes. Reproductive stage is the most vulnerable to elevated temperatures, especially during panicle/inflorescence formation and development with consequences of reduced fertility and flower abortion. Since plant's reproductive organs and tissues, mainly male gametophytes, are highly sensitive to heat stress reproductive processes and features such as meiosis, pollen production, anther number, anther dehiscence, pollen germination and stigma receptivity, pollen tube growth and fertilisation, fruit set and development are suppressed by high temperatures and there is reduced protein and micronutrient contents in cereal and legume seeds and inadequate starch synthesis during grain development in heat sensitive genotypes (Driedonks et al., 2016; Janni, 2020). There is reduced pollen viability and the female organs and female gametophyte are also affected by high temperatures. After pollination, pollen tube, endosperm and embryo development are affected as well, leading to embryo abortion and malformation. Because heat stress affects reproductive and seed filling processes it causes serious reduction in crop yield by reducing seed number, seed or grain weight and grain and seed quality. Temperatures above optimal are detrimental to growth, development and metabolism of root system resulting in reduced water and nutrient uptake while transpiration is increased.

At the cellular and physiological levels heat stress has undesirable effects on plasma membrane integrity, proteins and enzymatic functions and stability, hormonal and transcriptional changes and integrity of nucleic acids (Janni et al., 2020). Due to heat stress reactive oxygen species (ROS), namely, superoxide and peroxides, which are destructive to cell membrane, protein integrity, DNA and lipids, are synthesised and accumulated in the plant tissues as heat stress progresses (Shanmugavel et al., 2020). Photosynthesis process and sucrose metabolism are also affected because ROS cause damage to the chloroplast structure and interfere with photosynthetic process. Heat stress causes damage and death to the cells, tissues and organs, interferes with cell metabolism, protein integrity with the consequences of protein denaturation, fluidity of cell membrane lipids, interference with protein synthesis and production of toxic compounds in addition to reactive oxygen species.

19.2.2 Morphological, Physiological, Genetic and Molecular Basis of Heat Resistance in Plants

Heat tolerant or thermo-tolerant plants have the ability to resist heat by undergoing growth, development and producing economic yield despite heat stress. Plants overcome heat stress by avoidance, escape and tolerance. By escape the plant accomplishes most its development and growth including reproduction before the onset of heat stress through early flowering and early maturity. Plants also survive heat stress by avoidance through cooling effect arising from transpiration, leaf orientation, other morpho-physiological features and phytochemical processes. Tolerance is when the plant endures stress with alteration of growth habit so that the plant continues with its growth to give an appreciable yield despite the stress. Plant heat resistance mechanisms are controlled by several major and minor genes which are responsible for morpho-physiological changes like leaf orientation, cooling effect arising from transpiration, canopy temperature depression and changes at the cellular and phytochemical levels involving signalling, gene regulation and expression. Heat resistant genotypes often have morphological, physiological and phytochemical features which make them grow and undergo development under stress with less negative effects of heat stress on reproduction, photosynthesis, physiological and metabolic processes and can be selected for use in development of heat tolerant cultivars with desirable agronomic traits and high yields. Characters which reflect heat tolerance include good performance in germination, coleoptile growth, seedling growth, canopy temperature depression, plant shoot growth and height, proper primary and secondary root growth, root hairs and root tip condition, general root architecture, spatial and temporal configuration of root system, root length and density, sufficient number of lateral roots and high root growth rate. Plant heat tolerance is influenced by high harvest indices, respiration rate, delayed senescence and canopy architecture, root morphological features like root number, length, basal root number, lateral and adventitious root number, tap root length and root dry weight and some other traits considered associated with heat stress resistance (Chaudhary et al., 2020). Most heat tolerance features have correlations with yield under stress. Plant tolerance to heat may be in form of sustained metabolic activities, e.g., high photosynthetic rate, stay-green feature, water and nutrient use efficiency, cell membrane thermostability, low canopy temperature, high chlorophyl content, leaf conductance, sustenance of growth and development, sustained fertility, spike number, biomass and seed yield. Canopy temperature depression, often measured using infra-red thermometer, is a heat stress tolerance trait which has high heritability and genetic correlation with stay-green, leaf area, earliness, yield and is of value in selection for thermo-tolerance and high yield. Sustained leaf stomata conductance, which is a good heat tolerance selection trait, is measured using leaf porometer. Chlorophyl fluorescence, leaf chlorophyll content and stay-green feature which is a sign of sustained chlorophyll content under heat stress, are thermo-tolerance traits and useful for enhanced heat resistance selection.

When plants are exposed to heat stress, the genes related to heat stress response are induced for expression or upregulated for synthesis and accumulation of heat stress resistance related transcripts (Driedonks et al., 2016). This leads to accumulation of various protective phytochemicals including free radical scavengers/reactive oxygen species scavengers, membrane lipids, phytochromes, antioxidants like salicylic acid, nitrous oxide, ABA, proline, peroxidase, superoxide dismutase, glutathione, glutaredoxin, brassinosteroids, trehalose, sugar alcohols, glycinebetaine, LEA proteins and other osmoprotectants to assist in tolerance to heat stress through osmotic adjustment, detoxification of reactive oxygen species, stabilisation of cell membrane and enzymes and protection of the cells and tissues against stress (Bita & Gerats, 2013; Driedonks et al., 2016; Chaudhary et al., 2020). Genes that encode enzymes like catalase, peroxidases, superoxide dismutase and other compounds are also activated for survival of the plant under stress. Heat shock proteins encoding genes are induced or their expression modified under heat stress for synthesis and accumulation of such proteins for preventing formation of deleterious compounds (Driedonks et al., 2016). Heat shock proteins act as molecular chaperons to protect the proteins and maintain protein integrity, proper aggregation, prevent undesirable harmful protein conformations and damage to the photosynthetic apparatus (Bita & Gerats, 2013; Chaudhary et al., 2020). Over-expression of heat shock transcription factors which control the expression of heat shock proteins leads to improved level of heat tolerance (Driedonks et al., 2016). High levels of IAA and brassinosteroids also reduce the levels of reactive oxygen species. Genes that are responsive to heat stress are also influenced by a variety of metabolites accumulated as a response to heat stress.

19.2.3 Screening and Breeding Strategies for Heat Stress Resistance in Plants

To harness and exploit variability existent in the available genetic resources for addressing heat stress challenges appropriate and effective screening methods to identify tolerance genes and QTLs have to be designed accompanied by use of breeding strategies that employ efficient gene/QTL transfer techniques and, where possible, apply speed breeding protocols to fast-track the breeding process for higher genetic gain and fast varietal development. The sources of resistance genes are landraces and obsolete cultivars, varieties in cultivation, crop wild and weedy relatives, products of biotechnology, induced mutagenesis, transgenesis and genome editing. Field screening for heat tolerance is usually done in high temperature environments to identify the genotypes and breeding materials with heat resistance genes. Screening germplasm for heat tolerance is also conducted in greenhouses and growth chambers in which high temperature conditions are created. Where resources and facilities permit speed breeding protocols and high-throughput phenotyping are integrated to improve efficiency and fast-track the screening process. The germplasm materials are screened for the various tolerance traits including higher and stable photosynthetic rates and stay-green trait, enhanced membrane thermostability, abundant flower production, stable pod set or grain production under high temperatures, canopy temperature depression, protein thermostability and accumulation of metabolites associated with heat stress tolerance. Screening for membrane and protein thermostability and accumulation of heat tolerance metabolites can be done in the laboratory to select the genotypes with high cell membrane and protein thermostability. In case of canopy temperature depression, it is better to take measurements on cloudless days. Plants are also screened for stomatal conductance which assesses stomatal behaviour, leaf chlorophyll content, low leaf firing, stay-green feature and low senescence under high temperature conditions to isolate heat stress tolerant cultivars. Screening for pollen traits related to heat tolerance among crop plant species and within crop species helps to identify genotypes with stable pollen viability under high temperature conditions. Some fertility traits associated with reproductive success under heat stress that can serve as selection criteria include days to flowering, sufficient pollen production, pollen viability and germination, stigma receptivity, pollen tube growth and successful fertilisation, seed set percentage, grain filling and grain weight. Since heat stress causes poor pollination and fertilisation, seeds/spikelet or seeds/pod or grains/ear can also be used as selection criteria for heat tolerance. Seed weight and grain weight which are also directly related to seed yield or grain yield are other useful parameters for identifying genotypes with heat tolerance.

For the purpose of identifying genotypes with the required level of phytochemicals that promote heat tolerance the available germplasm materials are screened with respect to heat tolerance metabolite accumulation. Metabolomics, proteomics and transcriptomics assays on plant tissues and screening for altered metabolite regulation identify the genotypes which accumulate ROS scavengers, protective enzymes, heat shock proteins and other protective metabolites in response to heat stress. The assays reveal the stress responsive proteins and other metabolites, their functions under stress, complex molecular mechanisms underlying plant growth and development, their interactions with the environment, the pattern of plant response to stress and stress response gene expression and regulation under stress (Driedonks et al., 2016; Janni et al., 2020) to help in identification of genotypes having desired responses. The genotypes with appropriate heat shock protein genes expression and expression of other genes important in plant heat tolerance are also identified for selection.

Because heat tolerance is multigenic and controlled by polygenes as well, field screening for tolerance always has the challenge of large environmental effects reducing the precision and efficiency of screening exercise in identifying the target genes and QTLs. Use of high-throughput phenotyping facilities, especially under speed breeding protocols, improve the efficiency and fast-track screening of germplasm to identify resistance sources. Genome sequencing employing high-throughput next generation sequencing, genome-wide association studies (GWAS), genotyping and molecular analyses identify and map genes/QTLs with their associated molecular markers that are tightly linked to heat tolerance traits like days to flowering, pollen viability and germination, pollen tube growth, fertilisation, seed/fruit set, grain weight, grain filling, stay-green, chlorophyl fluorescence, photosynthesis, lower canopy temperature, water uptake and other features (Chaudhary et al., 2020; Janni et al., 2020) . High-throughput next-generation sequencing, genotyping-by-sequencing, high-throughput phenotyping and GWAS facilitate efficient and fast sequencing of genomes leading to generation of a large number of markers with genome-wide distribution, particularly single nucleotide polymorphisms (NSPs) for narrowing down on genomic regions having the resistance genes/QTLs (Janni et al., 2020). Genome-wide analysis along with 'omics' technologies assays are important in elucidating plant response to heat stress and identification of genomic regions encoding tolerance traits, especially accumulation of tolerance related metabolites, and generating data for genomic selection and other molecular marker breeding methods (Driedonks et al., 2016; Chaudhary et al., 2020). That is, DNA sequencing and mapping and 'omics' analyses assist in better understanding of the concerned genes and QTLs in regard to their expression and regulation. Molecular marker-tagged genes and QTLs encoding various metabolites associated with heat tolerance, e.g., heat shock proteins, transcription factors, antioxidants, reactive oxygen species scavengers and others, are identified through sequencing techniques and isolated so that they can be transferred to the elite breeding materials using conventional breeding, molecular marker assisted breeding and also subjected to genome selection and genome editing.

Germplasm materials identified to have target tolerance traits are crossed to heat sensitive elite materials to transfer the tolerance genes/QTLs to the heat sensitive cultivars. Conventional breeding employing pure line development, pedigree breeding, bulk breeding, backcrossing, recurrent selection for population improvement is applied to develop heat tolerant cultivars which are also better adapted to other conditions. In allogamous crops such hybridisation and backcrossing is done to improve inbred lines for improved hybrid and synthetic varieties development. Population improvement methods for autogamous and allogamous crops are employed to elevate heat tolerance levels to acceptable levels before improved cultivars are used to derive desired varieties.

Strategies employing molecular markers, especially those flanking or located within the genes/QTLs, help stack the resistance genes/QTLs from the donor genotypes to the target heat sensitive cultivars followed by selection in segregating populations. Marker assisted recurrent selection and marker assisted backcrossing are conducted to accumulate and pyramid the QTLs or the genes in the recipient cultivar, leading to greater variability for the tolerance traits in the generated populations, creating opportunities for genetic recombination, increasing the frequencies of desired favourable allele combinations and giving room for incorporating exotic germplasm in the population. This leads to fast development of pure lines in autogamous and improved populations and inbred lines in cross-pollinated crops that are heat stress tolerant by manipulating and transferring morpho-physiological, reproductive and phytochemical traits like earliness, canopy temperature depression, high photosynthetic rate, chlorophyll fluorescence, chlorophyll content, cell membrane stability, stay-green trait, pollen viability and pollen related traits, high fertility rate, high seed setting, seed number and weight, desirable metabolite contents, including proteins and water soluble carbohydrates, to elite breeding materials or target superior cultivars.

Desirable tolerance genes from crop relatives and landraces may be linked to undesirable traits in some cases. This is resolved through pre-breeding and genome sequencing for precise identification and mapping of QTLs and genes as has been achieved in a number of crops like maize, rice, sorghum, etc. Molecular markers are applied in pre-breeding activities to identify germplasm materials with useful alien genes and QTLs for the various complex tolerance traits that are expressed at different plant growth and development stages for creating intermediate breeding materials that are in turn used in mainstream breeding programmes.

Genomic selection employing molecular markers, transgenesis and gene editing can also be applied for development of tolerant cultivars through transferring genes for transcription factors, induction and accumulation of heat shock protein, molecular chaperones, osmolytes like glycinebetaine, antioxidants, other proteins and metabolites, growth regulators and altering membrane properties for heat tolerance. Genomic selection is more effective and efficient than other selection methods since it employs several markers covering the whole genome and phenotyping data to develop prediction for genome estimated breeding values (Meuwissen et al., 2001) for individuals or lines under selection leading to high genetic gains with the consequence of accumulating multi-trait and multigenic tolerance in desirable cultivars. Despite the complex and multigenic nature of heat resistance in plants, transgenesis approach can be employed once the desired genes have been identified in the source species, including genes from non-plant species. Alien genes associated with heat resistance pathways and regulation may be transferred to the target crop plants by genetic engineering methods and stacked through gene pyramiding of different tolerance genes affecting various heat resistance traits at different crop growth and development stages. Plants are also engineered with specific transcription factors associated with tolerance response so that the expression of their transcriptomes improves the level of tolerance. Genome editing using CRISPR-Ca9 is also a potentially powerful tool for creating targeted mutations in crop genomes to elevate heat stress tolerance and yield in crop plants.

19.3 Mineral Stress Resistance Breeding

Mineral stress refers to too high levels and toxic concentrations or too low levels of a mineral nutrient in the soil or irrigation water for normal plant growth and development. Macro- and micro-nutrients are essential for normal plant growth and development and if they are too low or too high in the soil or irrigation water, they cause mineral deficiency or toxicity respectively leading to poor growth and development, low yields and undesired product quality.The soil or irrigation water may have undesirable levels of minerals like iron (Fe), aluminium (Al), copper (Cu), boron (B), nitrogen (N), potassium (K), manganese (Mn) and zinc (Zn) making plants show characteristic symptoms for respective nutrient mineral stress and poor performance. Crops and crop varieties differ in their response to various mineral stress conditions as known for wheat, maize, soybean, cotton, potatoes, sorghum, barley, oats, tomatoes, potatoes, forage grasses and many other crops resulting in reduced yield or no yield and death of the plant. Mineral efficient plant genotypes are tolerant to mineral deficiency and are capable of extracting sufficient and adequate amounts of the minerals available in the soil to meet their demands until maturity giving better yields than the mineral deficiency susceptible or mineral inefficient genotypes.

19.3.1 Mineral Deficiency

Mineral deficiency is a form of soil infertility and plants growing in mineral deficient soils or irrigated with mineral deficient water show symptoms characteristic of the respective minerals. The mineral nutrients are important for plant growth and development, metabolism and any inadequate uptake of the nutrients suppresses physiological and biochemical processes, cell division, development and growth leading to poor performance of the plant. Since mineral efficient genotypes tend to have good rooting characteristics, root growth and traits like fast root growth and roots with greater depth, volume, weight, root hair length, root density and high root absorptive surface are associated with tolerance to mineral deficiency and efficient nutrient uptake in the soil, especially in soils low in immobile minerals like phosphorous (P). Breeding efforts to improve mineral deficiency tolerance focuses on rooting features to increase surface area for improved mineral absorption. Different crop species and cultivars have variable capabilities to adapt to the mineral deficient conditions. In case of Fe^{2+} deficiency some crop cultivars release hydrogen ions into the rhizosphere to reduce Fe^{3+} to Fe^{2+} which is more absorbable by the roots. Some mineral deficiency tolerant genotypes have better efficiency of utilising the minerals within the tissues.

19.3.2 Mineral Toxicity

Plants react to too high and toxic levels of mineral nutrients by showing symptoms that reflect interference with growth and development and ultimate yield. Aluminium and Mn toxicities are agriculture production problems in acid soils affecting many crops like maize, sorghum, rice, etc. Toxic levels of Al and Mn are associated with pH below 5.50, low organic matter, low and deficient levels of calcium (Ca), and P and K and for every decrease in soil pH Al concentrations increases 1000-fold in the soil leading to serious harmful impacts on the plants (Carver et al., 1988). Iron toxicity is associated with soil acidity which is usually common in swampy lowlands, irrigated lands, moist forests and grasslands and heavy soils. Ferritin (found in animals and some plants), a protein that stores Fe in cells, works by enhancing high iron storage in plants. Cadmium which is also toxic to plants has high concentrations in acid soils.

Development of crops that can tolerate mineral toxicity and give acceptable high yield despite such conditions is the best and cheapest option to address mineral toxicity challenge, especially for sustained agricultural production in resource challenged developing countries. Aluminium resistant genotype roots induce increased pH in rhizosphere lowering Al solubility in the soil as is known for tolerant maize, wheat, barley, rice. Within various crop species like wheat, tea, rye, triticale and alfalfa there is selectable variation for Al toxicity resistance which can be exploited for improvement of Al toxicity resistance. Resistance to Mn toxicity involves reduced transport of Mn from the roots to the shoots as it happens in resistant maize cultivars.

19.3.3 Breeding Strategies for Mineral Stress Resistance in Plants

Crop breeding plans for mineral stress resistance necessitates identification of sources of tolerance and application of mass selection, pure line selection, hybridisation and backcrossing, pedigree, bulk and single seed descent breeding methods and population improvement recurrent selection methods. Screening for mineral stress is often done in the field with soils having toxic levels or deficient levels of the respective nutrients and in pots filled with soils having toxic or deficient levels of the respective mineral nutrients. Plants are also grown for screening in hydroponic nutrient solutions or sand culture irrigated with nutrient solution lacking or having a particular nutrient in excess and varying pH levels. Crop wild relatives and the landraces have the diversity and the genes needed for developing tolerance to mineral stress which can be mobilised for improving crop plant resistance to mineral stress. Simply inherited mineral tolerance traits as Mn and Fe efficiency in barley and Al tolerance in sorghum are transferrable by crossing and backcrossing while those that are of quantitative nature like N and P efficiency in many crops are addressed through more elaborate methods like pedigree breeding and recurrent selection.

Products of induced mutagenesis, biotechnology and transgenesis are other sources of genes for tolerance to mineral stress tolerance. Mutation breeding and in vitro culture are helpful in generating variability followed by selection for genotypes with resistance to mineral stress. Protoplast culture, cell culture and protoplast fusion methods are used to generate diversity through somaclonal variation for in vitro selection to identify mineral stress resistant variants. The biotechnology methods can be employed along with induced mutagenesis by applying mutagens to the culture. Molecular marker assisted selection and transgenesis facilitate the transfer of target specific genes or blocks of genes between crop cultivars or across species. Next-generation sequencing and the ‘omics’ technologies enhance the application of marker assisted selection, marker assisted recurrent selection, marker assisted backcrossing, transgenesis, genome editing and genomic selection to develop mineral stress tolerant crop plants. Sequencing and precise mapping and screening for root characteristics and other mineral resistance traits assist in identification of markers associated with mineral stress tolerance genes and QTLs. Root system characteristics like root growth rate, root depth, root volume and weight are quantitative traits and are known to vary among crop genotypes and can be subjected to improvement through molecular marker tagged QTLs with selection for better performing genotypes under mineral stress conditions. The QTLs can be accumulated in the desired cultivars through gene pyramiding or marker assisted recurrent and marker assisted backcrossing.

Problems

19.1 Indicate the damage reactive oxygen species (ROS) cause to plant tissues and how do plant counteract these phytochemicals?

19.2 What is the role of abscisic acid (ABA) in tissues and organs of plant subjected to drought stress?

19.3 How does rise in temperature above optimal affect the cellular and physiological processes in the plant tissues?

19.4 How is selection at the molecular level done for multi-trait and polygenically controlled heat tolerance done?

References

Ashraf, M., & Foolad, M. (2007). Roles of glycine betaine and proline in improving plant abiotic stress resistance. *Environmental and Experimental Botany, 59*, 206–216. https://doi.org/10.1016/j.envexpbot.2005.12.006. Elsevier, No. 5630771269042.

Bi, C., Ma., Y., Wang, X. F., & Zhang, D. P. (2017). Overexpression of the transcription factor NF-YC9 confers abscisic acid hypersensitivity in *Arabidopsis*. *Plant Molecular Biology, 95*, 425–435. https://doi.org/10.1007/s11103-017-0661-1. Creative Commons Attribution International License (CC BY).

Bi, C., Ma., Y., Wang, X. F., & Zhang, D. P. (2017). Overexpression of the transcription factor NF-YC9 confers abscisic acid hypersensitivity in *Arabidopsis*. *Plant Molecular Biology, 95*, 425–435. https://doi.org/10.1007/s11103-017-0661-1. Creative Commons Attribution International License (CC BY).

Bita, C. E., & Gerats, T. (2013). Plant tolerance to high temperature in a changing environment: Scientific fundamentals and production of heat stress-tolerant crops. *Frontiers in Plant Science, 4*, 273. https://doi.org/10.3389/fpls.2013.00273. Creative Commons Attribution International License (CC BY).

Blum, A. (1970). Effect of plant density and growth duration on sorghum yield under limited water supply. *Agronomy Journal, 62*, 333. https://doi.org/10.2134/agroj1970.00021962006200030007x. John Wiley and Sons, No. 5771460122840.

Carver, B. F., Inskeep, W. P., Wilso, N. P., & Westerman, R. L. (1988). Seedling tolerance to Aluminium toxicity in hard red winter wheat germplasm. *Crop Science, 28*, 463–467. https://doi.org/10.2135/cropsci1988.00118X002800030006x. John Wiley and Sons, No.5630950649262.

Chaudhary, S., Devi, P., Bhardwaj, A., Jha, U. C., Sharma, K. D., Prasad, P. V. V., Siddique, K. H. M., Bindumadhava, H., Kumar, S., & Nayyar, H. (2020). Identification and characterisation of contrasting genotypes/cultivars for developing heat tolerance in agricultural crops: Current status and prospects. *Frontiers in Plant Science, 11*, 587264. https://doi.org/10.3389/fpls.2020.587264. Creative Commons Attribution International License (CC BY).

Choi, H., Hong. J., Ha, J., Kang, J., & Kim, S. Y. (2000). ABFs, family of ABA-responsive element binding factors. *Journal of Biological Chemistry, 275*, 1723–1730. https://doi.org/10.1074/jbc.275.3.1723. Creative Commons Attribution International License (CC BY).

Cockram, J., Jones, H., Leigh, F. J., O'Sullivan, D., Powell, W., Laurie, D. A., & Greenland, A. (2007). Control of flowering time in cereals: Genes, domestication and sustainable productivity. *Journal of Experimental Botany, 58*, 1231–1244. https://doi.org/10.1093/jxb/erm042. Oxford University Press, No. 5811220689539.

Delauney, A. J., & Verma, D. P. S. (1993). Proline biosynthesis and osmoregulation in plants. *The Plant Journal, 4*, 215–223. https://doi.org/10.1046/j.1365-313X.1993.04020215.x. John Wiley and Sons, No. 5771890122258.

Driedonks, N., Rieu, I., & Vriezen, W. H. (2016). Breeding for plant heat tolerance at vegetative and reproductive stages. *Plant Reproduction, 29*, 67–79. https://doi.org/10.1007/s00497-016-0275-9. Creative Commons Attribution International License (CC BY).

Fang, Y., Xiong, L. (2015). General mechanisms of drought response and their application in drought resistance improvement in plants. *Cellular and Molecular Life Sciences, 72*, 673–689. https://doi.org/10.1007/s00018-014-1767-0. Springer Nature, No. 5771320098428, 5771510478288 and 5772370762801.

Farooq, M., Wahid, A., Kobayashi, N. Fujita, D., & Basra, S. M. A. (2009). Plant drought stress: Effects, mechanisms and management. *Agronomy for Sustainable Development, 29*, 185–212. https://doi.org/10.1051/agro:2008021. Springer Nature, No. 5631210405422

Harris, K., Subudhi, P. K., Borrell, A., Jordan, D., Rosenow, D., Nguyen, H., Klein, P., Klein, R., & Mullet, J. (2007). Sorghum stay-green QTL individually reduce post-flowering drought-induced leaf-senescence. *Journal of Experimental Botany, 58*, 327–338. https://doi.org/10.1093/jxb/erl225. Oxford University Press, No. 577380533108.

Haussmann, B. I. G., Obilana, A. B., Ayiecho, P. O., Blum, A., Shipprack, W., & Geiger, H. H. (1999). Quantitative genetic parameters of sorghum (*Sorghum bicolor* (L.) Moench) grown in semi-arid areas of Kenya. *Euphytica, 105*, 109–118. https://doi.org/10.1023/A.1003469528461. Springer Nature, No. 5771940012095.

Janni, M., Gulli, M. Maestri, E., Marmiroli, M, Valliyodan, B., Nguyen, H. T., & Marmiroli, N. (2020). Molecular and genetic bases of heat stress responses in crop plants and breeding for increased resilience and productivity. *The Journal of Experimental Biology, 71*, 3780–3802. https://doi.org/10.1093/jxb/eraa034. Creative Commons Attribution International License (CC BY).

Khan, X., Pan, X., Najeeb, U., Tan, D. K. Y., Fahad, S., Zahoor, R., & Luo, H. (2018). Coping with drought: Stress and adaptive mechanisms and management through cultural and molecular alternatives in cotton as vital constituents for plant stress resilience and fitness. *Biological Research, 51*, 47. https://doi.org/10.1186/s40659-018-0198-z. Creative Commons Attribution International License (CC BY).

Lafitte, H. R., Price, A. H., & Courtis, B. (2004). Yield response to water deficit in an upland rice mapping population: Association among traits and genetic markers. *Theoretical and Applied Genetics, 109*, 1237–1246. https://doi.org/10.1007/s00122-004-1731-8. Springer Nature, No. 5630811342867.

Langridge, P., Paltridge, N., & Fincher, G. (2006). Functional genomics of abiotic stress tolerance in cereals. *Briefings in Functional Genomics and Proteomics, 4*, 343–354. https://doi.org/10.1093/bfgp/eli005. Oxford University Press, No. 5630790788981.

Levi, A., Ovnat, L., Paterson, A. H., & Saranga, Y. (2009). Photosynthesis of cotton near-isogenic lines introgressed with QTLs for productivity and drought related traits. *Molecular Breeding, 23*, 179–195. https://doi.org/10.1016/j.plantsci.2009.04.001. Elsevier, No. 5630831327182.

Ludlow M. M., & Bjorkman, O. (1984). Paraheliotropic leaf movement in *Sirato* as a protective mechanism against drought-induced damage to primary photosynthetic reactions: Damage by excessive light and heat. *Planta, 161*, 505. https://doi.org/10.1007/BF00407082. Springer Nature, No. 5772091168804.

Luo, L., Xia, H., & Lu, B. (2019). Editorial: Crop breeding for drought resistance. *Front: Plant Science*. https://doi.org/10.3389/fpls.2019.00314. Creative Commons Attribution International License (CC BY).

Martignago, D., Ric-Medina, A., Blasco-Escamec, D., Fontanet-Manazaneque, J. B., & Cano-Delgado, A. I. (2020). Drought resistance by engineering plant tissue-specific responses. *Frontiers in Plant Science, 10*, 1676. https://doi.org/10.3389/fpls.2019.01676. Creative Commons Attribution International License (CC BY).

Meuwissen, T. H. E., Hayes, B. J., & Goddard, M. E. (2001). Prediction of total genetic value using genome-wide dense marker maps. *Genetics, 157*, 1819–1829. https://doi.org/10.1093/genetics/157.4.1819. Oxford University Press, No. 5626070526316.

Nyabundi, J. O., & Ayiecho, P. O. (1994). Comparative evaluation of sorghum for drought resistance using physiological parameters. In Blum, A., Ayiecho, P.O., Nyabundi, J. O., Geiger, H. H., & Haussmann, B. I. G. (Eds.), *Sorghum improvement for semi-arid areas of Kenya*. Third Year Annual Report for Third Year, 1993. GIARA Project No. 91/4. Project Report With no Copyright.

Planas-Riverola, A., Gupta, A., Betegon-Putze, I., Bosch, N., Ibanes, M., & Cano-Delgado, A. I. (2019). Brassinosteroid signalling in plant development and adaptation to stress. *Development, 146*(5), dev151894. https://doi.org/10.1242/dev.151894. Creative Commons Attribution International License (CC BY).

Reddy, A. R., Chaitanya, K. V., & Vivekanandan, M. (2004). Drought-induced responses of photosynthesis and antioxidant metabolism in higher plants. *Journal of Plant Physiology, 161*, 1189–1202. https://doi.org/10.1016/j.jplph.2004.01.013. Elsevier, No. 5630840464972.

Shanmugavel, P., Rajarprakasam, S., Chockalingam, V., Ramasamy, G., & Thiyagarajan, T. (2020). Breeding mechanisms for high temperature tolerance in crop plants. In Addurakhmonov (Ed.), *Plant breeding, current and future views*. https://doi.org/10.5772/intechopen.94693. Creative Commons Attribution International License (CC BY).

Tardieu, F. (2013). Plant response to environmental conditions: Assessing potential production, water demand and negative effects of water deficit. *Frontiers in Physiology, 4*, 17. https://doi.org/10.3389/fphys.2013.000. Creative Commons Attribution International License (CC BY).

Wurschum, T., Langer, S. M., Longin, C. F. H., Tucker, M. R., & Leiser, W. L. (2020). Refining the genetic architecture of flag leaf glaucousness in wheat. *Theoretical and Applied Genetics, 133*(3), 981–991. https://doi.org/10.1007/s00122-019-03522-x. Creative Commons Attribution International License (CC BY).

Chapter 20
Speed Breeding, Breeding of Climate Resilient Crops and Functional Foods

20.1 Speed Breeding

20.1.1 Need for Speed Breeding

One of the major hindrances in crop improvement is long generation times due to the fact that in conventional breeding programmes the transfer of multiple traits to identified cultivars requires several generations of evaluation and selection before a desired superior cultivar is obtained. This is particularly so if quantitative traits are targeted for transfer to the cultivars to be improved because under regular circumstances the breeder can afford only up to two generations per year for the annual field crops like wheat, barley, oat, cowpea, rice, soybean, millets, sorghum, field beans, maize, sunflower, groundnut, pigeon pea, lintel, etc. The aim of speed breeding is to reduce the duration for doing the breeding work so that the breeder can have several generations per year for the annual field crops.

Apart from the required space and resources, conventional plant breeding is expensive in terms of time needed for selection and hybridisation of selected genotypes and selection in the segregating generations because of generation time taken from planting until harvesting. From hybridisation of selected parents to the release of a new variety in annual crops the duration needed may be at least eight years but can take up to 20 years depending on the crop, resources available, variety to be released and breeding objectives. Continuously increasing world population that is growing at high rate and challenges related to climate change like rising global temperatures, frequent devastating floods and droughts accompanied by novel plant diseases and frequent pest outbreaks, call for non-conventional faster and more efficient breeding methods to address food security problems, especially in the developing and underdeveloped countries.

Methods designed to reduce plant generation time and speed up the plant breeding cycle for faster varietal development are collectively referred to as speed breeding. Speed breeding protocols aim at fast crop growth and development rates, early flowering and early maturity so as to reduce generation time in the various crops to

P. O. Ayiecho and J. O. Nyabundi, *Conventional and Contemporary Practices of Plant Breeding*, https://doi.org/10.1007/978-3-031-74998-8_20

fast-track crop improvement initiatives thereby reducing breeding cycle duration and address food security problems (Shanmugavel et al., 2022). Speed breeding restructures breeding outputs so that there are more than two generations per year for annual field crops instead of only 1–2 generations per year under conventional breeding which makes breeding cycle to develop improved varieties take too long. Speed breeding protocols resolve the problems associated with crop generation time for wheat, barley, rice, rice, canola, oats, cowpea, soybean, millets, lentil, common bean, pea, sorghum, Faba bean, chickpea and other annual field crops by improving gain per unit time and increasing the number of generations per year to between four and eight, depending on the crop and the methods used, thereby allowing breeders to fast-track the development of new varieties having desired characters more efficiently (Razzaq et al., 2021; Samantara et al., 2022). This is achieved by controlling temperature, humidity, light duration and intensity, soil moisture content, carbon dioxide levels, plant nutrition and planting density to manipulate the generation time of photo-insensitive or long day plants and a few short-day cultivars. Speed breeding imitates dusk-dawn system with an extended day length duration and controlled temperatures, humidity, influencing stomatal behaviour, gas exchange, transpiration and photosynthetic rates, making plants grow and develop faster and flower earlier with quicker seed development followed by harvesting of immature seeds to reduce generation time (Shanmugavel et al., 2022).

20.1.2 Speed Breeding Procedure

In tropical regions with bimodal rainfall pattern, the main rains and the short rains, climatic conditions allow two crops of breeding material to be grown in the same location by having one crop in the main rainy season and the second crop in the short rainy season giving two generations per year for annual crops. If there are irrigation facilities a third crop is possible to grow during the dry season. Under other situations growing of two crops within a year to speed up development of new varieties can be achieved for annual crops by the growing of the same breeding materials between different locations in consecutive seasons. The main season is grown in one place and the off-season crop in another place when the conditions at first location are not favourable. Both approaches of growing at least two crops in a year, because of the bimodal rain pattern in one place along with an irrigated dry season nursery and changing the experiments between different locations, shorten the breeding cycle since there are at least two generations per year instead of only one.

Strictly controlled environmental conditions for speed breeding protocols involve growing the plants under manipulated photoperiod and temperature regimes in the greenhouse or growth chamber. The plants are grown in extended photoperiod of 22 h of light and 2 h of dark with temperatures of 22 and 17 °C respectively and humidity of 70% (Watson et al., 2017; Ghosh et al., 2018). Carbon dioxide supplementation is also done as part of speed breeding to increase plant growth rate and yield. The light source for extended light duration is light emitting diode (LED) or sodium vapour lamps (Ghosh et al., 2018). The prolonged light period, changes in temperature regimes of 22 °C/17 °C, increased carbon dioxide levels and water stress make the plants grow and develop much faster because of longer duration of photosynthesis leading to earlier flowering and seed formation induction, accelerated seed filling, earlier maturity, reduced harvesting time and reproductive periods without adverse effects on flowering and plant vigour (Ghosh et al., 2018; Shanmugavel et al., 2022). The plants are grown at high densities for increased competition for further induction of early flowering and fast maturity. The plants grown under high densities are stressed because they experience reduced water and nutrient availability making them grow faster and flower earlier than under regular planting and growth conditions (Ghosh et al., 2018). High density planting also makes it possible to have large populations under such protocols. Under speed breeding conditions the plant matures in half of the time as compared to those grown under normal growth conditions in the field or greenhouse. Plant growth hormones like cytokinin (e.g., kinetin, zeatin and 6-benzylaminopurine) and auxins (e.g., indole-3-acetic acid, chloroindole-3-acetic acid, naphthalene acetic acid, 2, 4-dichlorophenoxyacetic acid and indole-3-butyric acid) may be used to speed up growth and development, induce floral primordia initiation and early flowering followed by harvesting of immature seeds for fast generation advancement (Mobini et al., 2015; Shanmugavel et al., 2022). The seed is harvested a few weeks before maturity for those embryos which have reached physiological maturity without adverse effects on viability and gemination or the embryo may be rescued and subjected to embryo culture to reduce generation time further, leading to high generation number per year.

Though conventional breeding methods like pedigree breeding, mass selection, pure line breeding and recurrent selection are not amenable to speed breeding, the speed breeding protocols are suitable for backcross breeding, single seed descent, single pod descent and bulk breeding and can be used to turn these methods into speed breeding strategies for fast generation advancement in the growth chambers, greenhouse and field nurseries. Use of speed breeding technology in backcross breeding, single seed descent, single pod descent and bulk breeding programmes for handling of segregating generations makes it possible to get homozygous pure lines in self-pollinated long day and day neutral annual field crops in a single year instead of a minimum of three years because of application of fast advancement of generations. Development of recombinant lines is also fast-tracked by speed breeding protocols. In all cases, fast growth, early flowering and early harvesting of immature seeds are key to reducing generation time. In single seed decent and single pod descent conducted under speed breeding conditions it is possible to raise F_2, F_3, F_4 and F_5 generations in one year for any cross followed by growing of F_6 in the field as F_5 derived families for evaluation and selection, especially in marker assisted selection for target traits. Selected F_6 families are grown for another round of family selection in F_7, even up to F_8 to identify best pure lines. Therefore, under speed breeding from the time of the making the crosses to F_7, only four years are needed for pure line development instead of eight to ten years.

Under speed breeding protocols single seed descent can be integrated with in vitro techniques through methods like embryo rescue (Forster et al., 2014; Potts et al., 2023). An embryo is rescued from each F_2 plant and cultured in vitro into a plantlet which is used to obtain an adult plant from which another embryo is rescued after anthesis and fertilisation. This procedure for fast generation advancement through single embryo rescue and culture from each plant to regenerate adult plants is done until F_5 or F_6 when mature seeds are harvested from each plant and used to raise family progenies to initiate family-based selection in the field.

The fastest method of generating homozygous lines is doubling chromosome numbers of haploids. The haploids obtained through pollen/anther culture can be diploidised by use of colchicine or other chromosome doubling agents, thereby obtaining 100% homozygous genotypes in one generation instead of prolonged periods of self-pollination which may need up to 6 or 7 generations before homozygous lines are obtained. Apart from haploids obtained through anther/pollen culture, haploids can be obtained from ovary/ovule culture for diploidisation to obtain 100% homozygous genotypes. Some interspecific hybridisations also create haploids spontaneously resulting in the development of haploid zygotes that can be diploidised. A classic case is the cross between *Hordeum vulgare* (barley) and *H. bulbosum* (bulbosum barley) in which the *H. bulbosum* chromosomes are naturally eliminated from the developing embryo leading to development of haploid embryos which can be cultured and diploidised (Kasha & Kao, 1970). Once the chromosome numbers of haploids in cultures are doubled, they can be regenerated into plantlets and adult plants having normal chromosome number and 100% homozygosity with purpose of obtaining pure line cultivars in autogamous crops or inbred lines in allogamous crops. Haploids for obtaining doubled haploids can be sourced from gametes of segregating populations derived from crosses between selected parents or from gametes of cultivars improved through recombination/recurrent selection methods, subjected to culturing, chromosome doubling and regenerated into adult homozygous plants. Doubled haploids are subjected to evaluations as lines and the most promising are selected for further breeding exercise.

Speed breeding is suitable for crossing, backcross breeding, single seed descent, bulk breeding, development of pure lines, inbred line development and early stages of other conventional breeding methods when field evaluation is not yet required. It is also suitable for all forms of molecular marker assisted breeding, including molecular marker assisted selection, molecular marker assisted recurrent selection, molecular marker assisted backcrossing and gene pyramiding to fast-track the transfer of desirable genes and QTLs into target genotypes for fast development of pure lines, inbred lines and other desired materials. It is applicable to procedure like genomic selection, genome editing, rapid gene identification, mapping populations and development of transgenic plants. Indirect selection under speed breeding condition may be more precise for predicting the performance in the target traits than if such selection is done in field conditions, especially for multi-trait selection. Evaluation and multi-trait phenotyping under speed breeding protocols is highly efficient because the traits are more distinctly expressed (Alahmad et al., 2018).

20.1.3 Advantages of Speed Breeding and Compatibility with Other Breeding Methods

Speed breeding works well for screening of multiple traits for characters like disease and pest resistance, tolerance to abiotic stress, morphological and phenological traits like height, flowering and maturity time, leaf characteristics, etc. (Alahmad et al., 2018). Under speed breeding conditions there is synchronous flowering and distinct expression of contrasting characters controlled by different alleles like disease susceptibility or resistance expression and morphological traits (Watson et al., 2017). Plant development under speed breeding is always normal and the obtained seeds are viable with high germination rate despite early harvesting before seed maturity (Ghosh et al., 2018). Because of faster growth and development speed breeding makes it possible for the breeder to do faster selection.

Speed breeding is usually done in the greenhouse, laboratory and growth chambers and requires less space, is simpler and cheaper in terms of huma resource need than the regular breeding methods. Fast advancement of generations through speed breeding protocols fast-tracks the breeding process leading to high number of generations per year and ultimately shortens the period to release a variety. Because of its impact on improving efficiency of plant breeding strategies including conventional methods and reduction of generation time and breeding cycle duration, speed breeding maximises the benefits of various breeding methods resulting in higher genetic gains as compared to working under ordinary field conditions in the process of developing improved varieties. Speed breeding can be integrated with all forms of molecular marker assisted breeding, gene pyramiding, genomic selection, genome editing, transgenesis and is easily applicable to multi-trait selection to obtain desired genotypes rapidly (Alahmad et al., 2018; Potts et al., 2023). Speed breeding makes genomic selection, genome editing, backcrossing, gene pyramiding, transgenesis, molecular marker assisted backcrossing and molecular marker assisted selection more effective, efficient and cheaper with improved selection accuracy and much higher gain per unit time and resources. In interspecific hybridisation exercise between crops and their relatives speed breeding reduces the duration needed to introgress the alien genes into the crop plants and obtain improved cultivars with such genes.

Next-generation sequencing techniques are employed to identify DNA sequences associated with traits of interest for use in breeding programmes and also to ascertain lines with desirable traits/genes. Next-generation sequencing techniques, high-throughput phenotyping, proteomics, metabolomics, transcriptomics analyses integrated with speed breeding and bioinformatics help in distinguishing markers associated with desirable genes or QTLs, gene functions, regulation and interaction and materials with desired genes. Data from such analyses provide information regarding genetic control of useful traits and new alleles or haplotypes for various characters which are useful for development of new crop varieties.

20.1.4 Challenges of Speed Breeding

Use of extended light duration in speed breeding is suitable for long day and day length neutral plants but not for short day plants which require photoperiods to be less than respective critical day length for flowering induction. Speed breeding needs sophisticated infrastructure for control of lighting period, light intensity, moisture, temperature, humidity, CO_2 levels and other factors which need well trained and skilled personnel for maintenance and repairs (Shanmugavel et al., 2022; Potts et al., 2023). Facilities used in speed breeding are usually too expensive for researchers in some countries, especially in some developing and underdeveloped countries which may also lack resources, expertise and consistent electric power supply. Speed breeding conditions are also different from field conditions, raising the question of adequate adaptability of varieties developed under speed protocols.

Excessive photoperiod control leads to limited plant growth, photo-oxidation, high starch production and elevated levels of stress hormones and there may be inadequate yield under speed breeding conditions (Cazzola et al., 2020; Potts et al., 2023). Phenotyping of agronomic traits, tolerance and resilience to stress factors can be biased by controlled environmental conditions under speed breeding protocols making it necessary to ascertain the traits by growing the materials under normal field conditions (Potts et al., 2023). Speed breeding also has the problem of drift, plant loss and inadequate opportunities for selection in early generations. The usual problem of single seed descent or single pod descent breeding of genetic drift leads to loss of some materials due to restricted population size.

20.2 Climate Change Resilient Crop Breeding

20.2.1 Significance of Climate Change and Need for Climate Change Resilient Crops

Climate change is a global phenomenon associated with increased CO_2, melting of ice, rise in sea levels, elevated global temperatures and natural calamities like prolonged devastating droughts, more unpredictable destructive stormy rains of high intensity, hailstorms, severe floods submerging agricultural fields destroying crops and infrastructure, hurricanes and increased frequency of wild fires during the dry season. These impact negatively on agricultural production and are destructive to the crops in the field, making climate change a major threat to global food security. Elevated temperatures accelerate crop insect pests' and pathogens' metabolic activities, making them more aggressive with increased consumption rate. This leads to higher incidences of severe crop disease and pest infestation, especially outbreaks of new races and biotypes of diseases and pests that were previously insignificant or non-existent. Climate change phenomenon is a major challenge to agricultural production and global food security because it causes significant global reduction in the yields of major crops the world depends on, i.e., rice, wheat, maize, potatoes, sorghum, millets, field beans, soybean, cotton, barley, etc., especially in developing and underdeveloped countries. Climate change also affects agricultural activities and production through changes in temperature which cause heat stress to crop plants. Apart from reduced agricultural production, climate change is expected to cause global biodiversity reduction, particularly in marginal environments (Bakala et al., 2020). The rapid global climate change and fast increasing world population necessitate fast-tracking the development of new crop varieties that withstand conditions associated with climate change so as to alleviate the increasing food insecurity problems, especially in the tropical developing and underdeveloped countries where the impacts of climate change are most severely felt. Agricultural production challenges attributed to climate change can be addressed by adoption of efficient breeding methods to develop climate smart crop cultivars with good quality and high yields which also withstand and recover from abiotic and biotic stress factors associated with climate change. High yielding multi-stress tolerant crop varieties with climate change resilience should have appropriate shoot and root architecture, morphological, physiological and biochemical features that make them tolerate or resist the stress factors and adapt to various soil conditions and soil types. There is need for crop scientists to develop desirable high yielding multi-stress tolerant crop varieties in quick successions and also undertake the diversification of crop range so that even those crops considered as orphan or under-utilised crops are given attention for increased cultivation and production to improve food security situations in the light of unpredictable weather patterns and new disease and pest strains.

20.2.2 Methods for Climate Change Resilient Crop Breeding

Fast varietal development requires that the breeding approaches that fast-track crop improvement be integrated with conventional breeding methods which often take long periods of between eight and twenty years to develop a new variety in an annual field crop. Some of the strategies that can be used along with the established autogamous and allogamous crop breeding methods for developing climate change resilient varieties in quick successions employ modern breeding methodologies like application of speed breeding protocols to shorten breeding cycle, interspecific hybridisation, pre-breeding, crop diversification, molecular marker assisted breeding strategies, transgenesis, genomic selection and genome editing which help to fast-track the development of crop varieties by shortening the breeding cycle (Varshney et al., 2018; Razzaq et al., 2021). Molecular breeding technologies require high-throughput phenotyping using automated and computerised equipment, genotyping using next-generation sequencing methods, other genomic techniques and application of proteomics, transcriptomics, metabolomics technologies and bioinformatics to handle massive data obtained in the analyses.

Speed breeding protocols fast-track the development of crop varieties that are tolerant to stress conditions associated with climate change by increasing generation number per year to as high as eight for annual field crops, thereby reducing the breeding cycle. The procedure for speed breeding includes application of prolonged light duration of 22 h using LED and 2 h of dark, controlled humidity and temperatures of 22 and 17 °C respectively in the greenhouse or growth chamber and high planting densities (Watson et al., 2017; Ghosh et al., 2018). The plants may be subjected to CO_2 supplementation and water and nutrient deficit as well (Ghosh et al., 2018; Shanmugavel et al., 2022). Such conditions promote faster growth and development and earlier flowering than under regular planting conditions. Harvesting is also done early before seed maturity or the embryo may be rescued and subjected to in vitro culture and regenerated plants grown in growth chambers or greenhouses nurseries. These activities lead to fast advancement of generations and higher generation number per year than the regular growth conditions. Fast advancement of generations can be done for single seed descent and other established conventional breeding methods like backcrossing, bulk breeding and inbred line development under speed breeding conditions in growth chambers or greenhouse nurseries. Doubled haploid production also fast-tracks crop breeding through production of pure lines or inbred lines that are 100% homozygous in one generation instead of prolonged periods of self-pollination taking up to 6 or 7 generations to get pure lines and inbred lines that are close to 100% homozygousity.

Pre-breeding is used in mobilising the genetic diversity in crop wild and weedy relatives and unadapted materials to broaden the genetic base of breeding materials, identify and characterise the traits for climate change resilience in the wild and weedy relatives and unadapted materials. Next-generation phenotyping, genome sequencing and precise mapping are employed in identifying molecular markers linked to the target traits to eliminate linkage drag during transfer of the identified genes/QTLs (Varshney et al., 2018; Bakala et al., 2020).

Molecular marker assisted breeding and genomic selection for transfer of desired genes and QTLs to elite breeding genotypes requires data for whole genome distributed markers, use of high-throughput next-generation sequencing, genome-wide association studies (GWAS), high-throughput phenotyping, metabolomics, transcriptomics and proteomics assays and bioinformatics to identify the target genes and QTLs and understand their functions, interaction and regulation (Varshney et al., 2018; Bakala et al., 2020; Razzaq et al., 2021). Molecular markers e.g., restriction fragment length polymorphisms (RFLPs), amplified fragment length polymorphisms (AFLPs), random amplified polymorphic DNAs (RAPDs), micro-satellite loci or simple sequence repeats (SSRs), variable number of tandem repeats (VNTR), single nucleotide polymorphisms (SNPs), etc., are used to identify and locate genes or QTLs associated with tolerance and resistance to climate change related stress factors and help facilitate genotype-based molecular marker assisted breeding and genomic selection. Molecular marker assisted recurrent selection is important where characters are multigenic and polygenic and the selected genotypes are subjected to inter-crossing to develop populations with desirable combinations of favourable alleles for another round of selection to accumulate the target genes and QTLs in the desired populations. Molecular marker assisted backcrossing is useful in pyramiding of genes and QTLs for multi-trait and multigenic climate change resilience traits in the target cultivars. Genomic selection employs genomics and phenomics information to get genome-wide molecular markers data for obtaining genomic estimated breeding values (Meuwissen et al., 2001) that are in turn used in improving the crops with respect to multigenic traits rapidly, especially if used along with speed breeding. The molecular marker-tagged genes and QTLs can also be cloned and transferred to target cultivars by transgenesis through *Agrobacterium*-mediated transformation or direct DNA transfer techniques. Genome editing enables plant breeders to change and improve genes including alien alleles from crop relatives and even convert unadapted materials to superior varieties resilient to climate change within short durations. Genome editing gives opportunities to create changes that make established cultivars more adaptable and climate change resilient. Genome editing applying Zinc Finger Nucleases (ZFNs), Transcription-like Effector Nucleases (TALENS), Meganucleases (MN) and Clustered Regular Interspaced Short Palindromic Repeats (CRISPR-Cas9) (Holme et al., 2019; Bakala et al., 2020; Razzaq et al., 2021) and transgenesis are applicable in conjunction with speed breeding to fast-track the development of the desired climate smart genotypes.

Increased cultivation and utilisation of orphan or under-utilised crops and crop relatives is one way of diversifying crops available for agricultural production, especially those crops which have good adaptability traits. Under-utilised crops can be promoted to diversify sources of food and climate change resilience. Promotion of the use of crop wild and weedy relatives through pre-breeding, gene pyramiding employing molecular markers, genome editing and chromosome manipulation by developing chromosome substitution lines, segment substitution lines and introgression lines accelerate the use of the relatives in crop improvement. Development of such materials can be fast-tracked by speed breeding and the application of genomics and phenomics tools that help generate and identify molecular markers linked to the desired traits. Materials with alien genes from wild relatives are utilised to develop backcross populations, recombinant inbred lines and multiple-way crosses for gene pyramiding used in deriving desired genotypes.

20.3 Breeding for Functional Foods

Functional foods are food items or food products having additional attributes by having certain ingredients which make them have health benefits to the consumer beyond their regular nutritional value. Functional foods are also referred to as nutraceuticals or bioprotective foods. Functional foods also apply to attributes bred into edible plants such as purple or gold potatoes or golden rice having increased anthocyanin or increased carotenoids to improve their nutritional qualities for the consumers (Nayak et al., 2021).

The crops with nutraceutical properties provide a variety of vitamins, mineral nutrients and various other metabolite compounds that are beneficial to humans. Functional foods improve the human health and prevent diseases in humans since they have ingredients like vitamins A, B_1, B_2, C, E, K, etc., polyphenolic compounds, alkaloids, flavonoids, carotenoids, β-carotenes, catechins, tannins, phytosterols, lignans, quinone, anthocyanins, β-glucans and tocopherols, and micronutrients like zinc and iron at much higher levels than regular foods (Saxena & Cramer, 2013; Nayak et al., 2021; Shvachko et al., 2021). These ingredients are usually available at required levels in edible parts of crop plants having acceptable nutraceutical properties. The other compounds synthesised in the plants that form part of the functional foods derived from the edible parts are pigments, isoflavones, antioxidants, lycopene, folic acid, health oils, flavanols, saponins and sapogenins. Most of these ingredients, other than the micronutrients, are natural secondary metabolites synthesised within the plant tissues and organs and are ingredients for food or edible parts of plants consumed by humans. They may be found in fruits, vegetables, spices, beverages, cereals, roots, tubers, stems, bulbs, pulses, and oilseeds and different ingredients are unique to certain plant families or species. Some of the fruits that are known to have functional food components are guava, black currant, blackberry, mango, banana, pineapple, kiwi, raspberry, avocado, watermelon, cherry, guava berry, papaya, strawberry, pears, peaches, apples, custard apple, soursop (guanabana), citrus, *Pepino melon,* tamarind, African horned cucumber or kiwano while some of the known nuts with functional food potential are almonds, cashew nut, macadamia and Brazil nuts. Some vegetables that have potential as sources of high content of functional food ingredients belonging to Brassicaceae, Amaranthaceae, Solanaceae, Leguminoseae, Cucurbitaceae families and include wild cabbage (*B. oleracea),* leaf mustard (*B. juncea*), scurvy grass (*Cochlearia officinalis*), garden cress (*Lepidium sativum*), *Amaranthus caudatus, A, cruentus, A. hypochondraicu, A. tricholor, A. lividus,* cowpea (*Vigna unguiculata*), ash gourd (*Benincasa hispida*) (Fotev et al., 2012). Some others are okra, yam, pumpkin, kale, spinach, cauliflower, celery, green pea, cinnamon, ginger, carrot, tomato and common chicory (*Cichrium intybus*). The crop plants whose seeds have nutraceutical properties are pumpkin, flax, chia, hemp, legumes like black beans, chickpea, lentil, navy bean, field pea and cowpea and cereals like oats and barley, sorghum and millets and golden (yellow) rice cultivars. The beverages that are known to have potential for functional ingredients are coffee, tea, cocoa and lemongrass while some root and tuber crops with nutraceutical properties are onion, carrot, potato, yellow/orange flesh sweet potato, garlic, leek, turnip, radish, vegetable beet and turmeric. Because of their health and physiological benefits to the consumer arising from their bioprotective ingredients the functional foods are used to prevent and treat some diseases like cancer, diabetes, eye disorders, obesity, cardiovascular problems and other health issues.

Iron and zinc are important mineral nutrients for human nutrition. Iron is an essential component of haemoglobin and myoglobin and is involved in oxygen transport in the blood. Zinc is an important cofactor for many enzymes and regulatory proteins and is essential for DNA and RNA synthesis (Garcia-Oliveira et al., 2018). The two nutrients, iron and zinc, are accumulated by the plants in the grains where they are availed to the humans on consumption of the grains. Flavanols and flavonoids are believed to have anti-HIV and anti-cancer properties while saponins and sapogenins reduce the cholesterol levels and prevent cancer and tumours and quinones hinder blood from clotting and help in bone repair (Saxena & Cramer, 2013). Antioxidants are known to assist in regulation of DNA repair. The β-glucans are beneficial for human gastrointestinal tract processes and reduce the risks of heart diseases, lower blood cholesterol and improve liver function while antioxidant β-carotenes protect body cells from damage by free radicals and has beneficial effects to the brain function, skin, lungs and prevent eye diseases (Shvachko et al., 2021).

20.3.1 Nutritional Deficiency

In many parts of the world there are health problems associated with nutritional mineral deficiency because of low intake of zinc and iron, especially in the developing and underdeveloped countries. In the developing and underdeveloped countries, with most affected regions being sub-Saharan Africa, southern Asia and Latin America, the main staple foods are cereals and root crops like cassava, sweet potatoes and yams which have low contents and low bioavailability of important mineral nutrients. In sub-Saharan Africa, southern Asia and Latin America regions poor quality food intake leads to malnutrition and poor human health due to a variety of diseases like blindness, stunted growth in children arising from Zn deficiency, kwashiorkor, lowered immunity to a number of diseases and anaemia caused by iron deficiency (Garcia-Oliveira et al., 2018). Such deficiency problems are addressed through medical supplements, food fortification and addition of micronutrients to processed food. Biofortification which involves increasing the levels of essential nutrients in the edible tissues and organs of crop plants is the best approach for the benefit of the consumers (Bouis & Saltzman, 2017; Garcia-Oliveira et al., 2018). Application of fertilisers is used to fortify the crops but the costs make them difficult to use, especially in developing and underdeveloped countries. The cheapest remedy is genetic biofortification through plant breeding. Genetic biofortification enhances the contents and even bioavailability of the nutrients, vitamins and other nutraceutical compounds in the food crops, hence improving the quality of the foods to be consumed by the populations and alleviating the problem of mineral and nutrient deficiency (Bouis & Saltzman, 2017).

20.3.2 Functional Foods Breeding Methods

Genetic variability in the various cultivars of the respective crops, landraces and crop weedy and wild relatives is usually available for use in biofortification of elite materials through conventional breeding, molecular marker assisted breeding, genomic selection, transgenesis and genome editing to develop crop varieties with desired levels and bioavailability of the nutrient minerals, vitamins and other nutraceutical metabolites. The landraces and the wild and weedy crop relatives tend to be more diverse and richer in Fe and Zn nutrient contents and bioactive metabolites than the improved crop cultivars. This is the case for wheat relatives like *Triticum boeoticum, T. monococcum, T. dicoccoides, Aegilops tauschii*, and *Ae. speltoides* (Rawat et al., 2009) and rice relatives like *Oryza nivara, O. rufipogon, O. latifolia* and *O. officinalis* (Anuradha et al., 2012) which have higher Fe and Zn contents as compared to cultivated wheat and cultivated rice respectively. To identify potential sources of the required genes the available germplasm resources are assessed for target nutrients and other nutraceutical metabolite contents and the genotypes with desired functional food ingredients are selected for use in hybridisation for improving the cultivars in the mainstream breeding programme or other identified materials. Management of arising segregating generations depends on the reproductive system of the crop, genetics of the trait and intended varieties which may be pure lines or hybrid varieties.

Iron, Zn and various metabolites accumulation is influenced by quantitative genes in some crops. The genes and QTLs associated with control of Fe and Zn and some metabolite accumulation and their respective molecular markers are well documented in cereal crops like wheat, barley, maize, sorghum, rice and pearl millet (Nayak et al., 2021) making it possible to conduct conventional breeding and molecular marker assisted breeding to elevate the levels of these minerals in the cereals. Molecular marker tagged QTLs and genes for phenolic, anthocyanin and flavonoids in rice and for carotenoid and polyphenol synthesis and Zn, Fe, Ca and Mg accumulation in wheat are also characterised (Jin et al., 2009) and can be used for manipulation to raise the levels of these metabolites and mineral nutrients. Similarly, in maize gene sequencing analyses have identified and located genes and QTLs for Zn and Fe micronutrient contents, bioactive compounds like β-carotene, amino acids like lysine (Shetti et al., 2020) for exploitation to elevate nutraceutical qualities of deficient maize cultivars. In addition to raising the levels of these minerals in the cereals through genetic biofortification, their bioavailability to the consumers can be improved by reducing the content of phytic acid in cereals. Phytic acid usually binds to Ca, K, Mg, Mn, Fe, and Zn making them unavailable for absorption in the human alimentary system (Nayak et al., 2021).

Barley and oats have good qualities as functional bioactive compound sources, qualifying them as functional foods with the major bioactive functional compounds being β-glucans, phenolic acids, flavones, flavonoids, coumarins, terpenes, glutathione and phytosterols, vitamins E, B_1, B_2, B_3, minerals P, K, Mg, Ca, Zn and S and dietary fibre (Shvachko et al., 2021). Oat contains significant amounts of unsaturated fatty acids, sterols, basic minerals, globular proteins and β-glucans and compounds having antioxidant properties. While cultivated oat and barley globulins have appreciable contents of essential amino acids (arginine, histidine, lysine, tryptophan, etc.) as compared to other cereals oat wild relatives, e.g., *Avena fatua, A. sterilis, A. ludoviciana,* have much higher levels of some of these compounds and oil contents than the cultivated oat (Shvachko et al., 2021). The wild oats may serve as sources of nutraceutical properties to improve the cultivated oats and even to other cereals by way of interspecific hybridisation and transgenesis. Mapping and sequencing of genes and QTLs concerned with anthocyanin synthesis and other nutraceutical compounds in barley, rice, tomato and other crops have been done and molecular markers associated with them established (Nayak et al., 2021; Shvachko et al., 2021) to assist in the incorporation of such genes and QTLs into the target cultivars for improved functional food qualities. Sorghum and millets have nutraceutical qualities by virtue of being gluten-free and having high nutritional, mineral, vitamins and dietary fibre contents which make them prevent, control and reduce diabetes (Anitha et al., 2021). Genes and $QTLs$ that are associated with Vitamin A, β-carotene, lysine, Zn and Fe grain contents for improving the qualities of these cereals have been mapped (Nayak et al., 2021). A number of sorghum breeding materials have been genetically improved for elevated lysine, β-carotene and threonine contents, improved protein digestibility and reduced phytic acid contents. Millets also have secondary metabolites like glucosinolates, isoflavones, genistein, daidzein which are believed to reduce cancer, heart diseases and osteoporosis (Nayak et al., 2021). Apart from being rich in proteins, the grain legumes have high soluble fibre contents which make them reduce cholesterol and sugar levels in the blood and also have functionally bioactive saponins (Singh et al., 2017). The levels of Zn and Fe in legume seeds can also be raised through molecular marker assisted breeding by identification of genes associated with high levels of these micronutrients. Most vegetable oil crops have the potential for improvement with respect to oil quality, especially in regard to resveratrol content and the compounds linked to colour like carotenoids. Among the vegetables, tomato has some of the well-studied bioactive functional compounds, namely, carotenoids, anthocyanins and lycopenes. Genes have been identified for carotenoids, Vitamins C and E and phenolics (Nayak et al., 2021) which can be employed to improve the targeted tomato cultivars. In other vegetables secondary metabolite synthesis genes have been identified as well, like the gene for carotenoid accumulation in carrot, the gene for anthocyanin synthesis in cabbage and the gene for α-tocopherol content in sweet potato which can be transferred to deficient elite cultivars for the respective crops.

For the purpose of molecular breeding approach, high-throughput next generation sequencing, genome-wide association studies (GWAS), high-throughput phenotyping along with metabolomics, phenomics, proteomics transcriptomics and genomics analyses are undertaken to provide data for whole genome distributed markers required for locating genomic regions and identifying molecular marker-tagged genes and QTLs associated with the synthesis and accumulation of various beneficial metabolites and mineral nutrients in various crop plant tissues. Integration of omics technologies with conventional breeding methods, speed breeding, genomic selection, molecular marker assisted breeding, genome editing and transgenesis can help in fast development of crop varieties with improved nutraceutical quality components. Gene silencing using RNAi technique also has the potential for use in silencing or blocking pathways of undesired compounds like phytic acid in cereals, toxins, antinutritional factors and catabolic enzymes to improve crop cultivars with respect to nutraceutical qualities (Saxena & Cramer, 2013).

Problems

20.1 Why is it necessary to incorporate speed breeding in modern crop plant improvement programmes?

20.2 How is speed breeding accomplished in backcrossing and single seed descent breeding methods

20.3 What are the problems associated with speed breeding?

20.4 What is responsible for raised temperatures around the globe?

20.5 Give food production problems associated with high global temperatures.

20.6 Describe any two modern breeding methods that can be used to fast-track plant breeding processes to address climate change challenges

20.7 How are QTLs employed to improve nutraceutical qualities of major cereal crops, namely, maize, rice, wheat and sorghum?

References

Alahmad, S., Dinglasan, E., Leung, A., Derbal, N., Voss-Fels, K., Able, J. A., Bassi, F. M., Christopher, J., & Hickey, L. T. (2018). Speed breeding for multiple quantitative traits in durum wheat. *Plant Methods, 14*, 36. https://doi.org/10.1186/s13007-018-0302-y. Creative Commons Attribution International Licence (CC BY).

Anitha, S., Kane-Potaka, J., Tsusaka, T. W., Botha, R., Rajendran, A., et al. (2021). A systematic review and meta-analysis of the potential millets and sorghum for managing and preventing Diabetes mellitus. *Frontiers in Nutrition, 386*. https://doi.org/10.3389/fnut.2021.687428. Creative Commons Attribution International Licence (CC BY).

Anuradha, K., Agarwal, S., Rao, Y. V., Rao, K. V., Viraktamath, B. C., & Sarla, N. (2012). Mapping QTLs and candidate genes for iron and zinc concentrations in unpolished rice of Madhukar x Swarna RILs. *Gene, 508*, 233–240. https://doi.org/10.1016/j.gene.2012.07.054. Elsevier, No. 5631230036466.

Bakala, H. S., Singh, G., & Srivatava (2020). Smart breeding for climate resilient agriculture. https://doi.org/10.5772/intechopen.94847. Creative Commons Attribution International License (CC BY).

Bouis, H. E., & Saltzman, A. (2017). Improving nutrition through biofortification: A review of evidence from HarvestPlus 2003 through 2016. *Global Food Science, 12*, 49–58. https://doi.org/10.1016/j.gfs.2017.01.009. Creative Commons Attribution International License (CC BY).

Cazzola, F., Bermejo, C. J., Guindon. M. F., & Cointry, E. (2020). Speed breeding in pea (*Pisum sativum* L.), an efficient and simple system to accelerate breeding programs. *Euphytica, 216*, 178. https://doi.org/10.1007/s0681-020-02715-6. Springer Nature, No. 5626071097582.

Forster, B. P., Till, B. J., Ghanin, A. M. A., Huynh, H. O. A., Burstmayr, H., & Caligari, P. D. S. (2014). Accelerated plant breeding. *CABI Reviews, 9*, 1–16. https://doi.org/10.1079/PAVSNNR20149043. Creative Commons Attribution International License (CC BY).

Fotev, Yu, V., Artemyeya, A. M., & Zvereva, A. O. (2012). Genetic resources of vegetable crops: From breeding non-traditional crops to functional food. *Vavilov Journal of Genetics and Breeding, 25*(4), 442–447. https://doi.org/10.18699/VJ.21.049. Creative Commons Attribution International License (CC BY).

Garcia-Oliveira, A. L., Chander, S., Ortiz, R. Menkir, A., & Gedil, M. (2018). Genetic basis and breeding perspectives of grain iron and zinc enrichment in cereals. *Frontiers in Plant Science, 9*, 937. https://doi.org/10.3389/fpls.2018.00937. Creative Commons Attribution International License (CC BY).

Ghosh, S., Watson, A., Gonzalez-Navarro, O. E., Ramirez-Gonzalez, R. H., Yanes, L., Mendoza-Suarez, M., Simmonds, J., Wells, R., Rayner, T., Green, P., Hafeez, A., Haytas, S., Melto, R. E., Steed, A., Sarkar, A., Carter, J., Perkins, L., Lord, J., Tester, M., Osbour, A., Moscou, M. U., Nicholson, P., Harwood, W., Martin, C., Domoney, C., Uauy, C., Hazard, B., Wulff, B. H. H., & Hickey, L. T. (2018). Speed breeding in growth chambers and glasshouses for crop breeding and model plant research. *Nature Protocols, 13*, 2944–2963. https://doi.org/10.1038/s41596-018-0072-z. PMID:30446746. Springer Nature, No. 5626090210720.

Holme, I. B., Gregersen, P. L., & Brinch-Pedersen, H. (2019). Induced genetic variation in crop plants by random or targeted mutagenesis: Convergence and differences. *Frontiers in Plant Science, 10*. Article1468. https://doi.org/10.3389/fpls.2019.01468. Creative Commons Attribution International Licence (CC BY).

Jin, L. Xiao, P., Lu, Y., Shao, Y., Shen, Y., & Jinsong, B. (2009). Quantitative trait loci for brown rice colour, phenolics, flavonoid contents, and antioxidant capacity in rice grain. *Cereal Chemistry Journal, 86*, 609–615. https://doi.org/10.1094/CCHEM-86-6-0609. John Wiley and Sons, No. 5626110372430.

Kasha, K. J., & Kao, K. N. (1970). High frequency haploid production in barley (*Hordeum vulgare* L). *Nature, 225*, 874–876. https://doi.org/10.1038/225874a0. Springer Nature, No. 5700900372196.

Meuwissen, T. H. E., Hayes, B. J., & Goddard, M. E. (2001). Prediction of total genetic value using genome-wide dense marker maps. *Genetics, 157*, 1819–1829. https://doi.org/10.1093/genetics/157.4.1819. Oxford University Press, No. 5626070526316.

Mobini, S. H., Lulsdorf, M., Warketin, T. D., & Vandenberb, A. (2015). Plant growth regulators improve in vitro flowering and rapid generation advancement in lentil and faba bean. *Vitro Cell Developmental Biology—Plant, 51*, 71–79. https://doi.org/10.1007/s11627-014-9647-8. Springer Nature, No. 5775321287749.

Nayak, S. N., Aravanind, B., Malavalli, S. S., Sukanth, B. S., Pornima, R., Bharati, P., Hefferon, K., Kole, C., & Puppala, N. (2021). Omics technologies to enhanced plant based functional foods: An overview. *Frontiers in Genetics, 12*, 742095. https://doi.org/10.3389/fgene.2021.742095. Creative Commons Attribution International License (CC BY).

Potts, J., Jangra, S., Michael, V. N., & Wu, X. (2023). Speed breeding for crop improvement and food security. *Crops, 2023*(3), 276–291. https://doi.org/10.3390/crops3040025. Creative Commons Attribution International License (CC BY).

Rawat, N., Tiwari, V. K., Singh, N., Randhawa, G. S., & Singh, K. (2009). Evaluation and utilisation of *Aegilops* and wild *Triticum* species for enhancing iron and zinc content in wheat. *Genetic Resources and Crop Evolution, 56*, 53–64. https://doi.org/10.1007/s10722-008-9344-8. Springer Nature, No. 5626101293107.

Razzaq, A., Kaur, P., Akhter, N., Wani, S. S., & Saleem, F. (2021). Next-generation breeding strategies for climate-ready crops. *Frontiers in Plant Science, 12*, 620420. https://doi.org/10.3389/fpls.2021.620420. Creative Commons Attribution International License (CC BY).

Samantara, K., Bohra, A., Mohapatra, S. R., Prihatini, R., Asibe, F., et al. (2022). Breeding more crops in less time: A perspective on speed breeding. *Biology, 11*, 275. https://doi.org/10.3390/biology11020275. Creative Commons Attribution International License (CC BY).

Saxena, A., & Cramer, C. S. (2013). Metabolomics: A potential tool for breeding nutraceutical vegetables. *Advances in Crop Science and Technology, 1*, 106. https://doi.org/10.4172/2329-8863-1000106. Creative Commons Attribution International License (CC BY).

Shanmugavel, P., Ramasamy, G., Vellingiri, G., Marimuthu, R., & Thiyagarajan, K. (2022). Speed breeding: A propitious technique for accelerated crop improvement. In H. Wang (Ed) *Case Studies of Breeding Strategies in Major Plant Species, IntechOpen 2023*. https://doi.org/10.5772/intechopen.105533. Creative Commons Attribution International License (CC BY).

Shetti, P., Sagare, D. B., Surender, M., & Reddy, S. S. (2020). Development of lysine and tryptophan rich maize (*Zea mays*) inbreds employing marker assisted backcross breeding. *Plant Gene, 23*, 100236. https://doi.org/10.1016/j.plgene.2020.100236. Elsevier, No. 5626100765309.

Shvachko, N. A., Loskutov, I. G., Semilet, T. V., Popov, V. S., Kovaleva, O., & Konarev, A. V. (2021). Bioactive components in oat and barley grain as a promising breeding trend for functional food product. *Molecules, 26*, 2260. https://doi.org/10.3390/molecules26082260. Creative Commons Attribution International License (CC BY).

Singh, B., Singh, J. P., Singh, N., & Kaur, A. (2017). Saponins in pulses and their health promoting activities: A review. *Food Chemistry, 233*, 540–549. https://doi.org/10.1016/j.foodchem.2017.04.161. Elsevier, No. 5776081169813.

Varshney, R. K., Singh, V. K., Kumar, A. Powell, W., & Sorrells, M. E. (2018). Can genomics deliver climate change ready crops? *Current Opinion in Plant Biology, 45*, 205–211. https://doi.org/10.1016/j.pbi.2018.03.007. Creative Commons Attribution International License (CC BY).

Watson, A., Ghosh, S., Williams, M. J., Cluddy, W. S., Simmonds, J. R., et al. (2017). Speed breeding: A powerful tool to accelerate crop research and breeding. *Nature Plants, 4*, 23–29. https://doi.org/10.1101/161182. Springer Nature, No. 5626080662385.

Chapter 21
Varietal Evaluation and Adaptability

Plant breeding programmes target production of new varieties which excel in some defined characters as compared to the existing varieties and that each new variety must be distinct from the already existing cultivars and other new varieties in at least one character. Before decision is made for release of the new variety and its seed multiplied for distribution the breeder conducts tests to compare it with other established varieties and to verify the superiority, worthiness in terms of value, uniqueness and distinctiveness of the new variety. Evaluations include tests for yield performance, adaptability, resistance to the prevailing biotic and abiotic stress factors and quality attributes. The tests are done using the recommended agronomic practices in at least one season by comparing the new cultivars and other cultivars under appropriate environments. There is need for good knowledge and special attention for distinctive characters of the established varieties and the new one and lists of such characters should be prepared to facilitate the comparisons. Data may be taken for stem height, days to flowering, days to maturity, ear height for maize, yield, lodging resistance, disease and pest resistance, resistance to abiotic stress factors, quality measurements, physiological characters, etc. The new variety should be distinguished distinctly by defining its clearly visible unique characteristics which are easy to score. If the new variety is similar in all respects to the already established varieties there is no need releasing it to the farmers. The variety must have a stable genotype and maintain its genetic integrity under multiplication, seed certification and in farmers' fields and it should be reproducible over several generations without losing its uniqueness. Pure line varieties in self-pollinated crops are not only uniform but very stable. The new variety should have a high level of homogeneity, especially for quality and key agronomic traits. Uniformity is easier to achieve for pure line varieties in self-pollinated crops and F_1 hybrids between inbred lines than for multiple crosses like three-way, double crosses, synthetic and composite varieties and other open-pollinated varieties in cross pollinated crops. The new variety will be preferred by farmers if it is superior with respect to agronomic traits like yield, maturity, response to soil amendments, resistance and tolerance to the prevailing pests, diseases and abiotic stress factors.

P. O. Ayiecho and J. O. Nyabundi, *Conventional and Contemporary Practices of Plant Breeding*, https://doi.org/10.1007/978-3-031-74998-8_21

The new variety must be non-shattering and non-sprouting at maturity, lodging resistant, and have required value of produce. The harvested produce from the new variety must have the right quality for its purpose, e.g., wheat baking quality, barley malting quality, sugarcane sucrose content, cotton lint quality and meet market requirements. Good quality makes the produce easy to sell and makes it fetch high market price because of favourable value, assuring farmers of good returns. When superiority, distinctiveness, yield and other agronomic and market value, adaptability and homogeneity of the new variety are established as acceptable it is released formally as a commercial variety. Though evaluations for traits with simple inheritance and high heritability do not require replication yield evaluations often entail replicated tests.

21.1 Varietal Comparative Yield Trials

In varietal development programs the breeder is usually faced with the problem of a range of environmental conditions for which to develop crop genotypes that are appropriately adapted and stable but flexible enough for good performance in the changing environments. Environmental conditions change from one place to another and from year to year, posing a major challenge for development of the most appropriately adapted and high yielding varieties. Once a set of promising varieties for a given crop have been identified the breeder has to identify the most superior genotype to be released for commercial utilisation in a particular environment or a given set of environments by conducting appropriately designed comparative adaptability and yield performance trials. In such trials the new variety/varieties is/are assessed and compared with the established standard and other locally adapted varieties. The breeder conducts the yield performance and adaptability tests in collaboration with other relevant agricultural scientists like entomologists, pathologists, nutritionists, agronomists, plant physiologists, etc. Though the main focus of the comparative trials is yield performance and adaptability, other characters like maturity period, lodging resistance, quality attributes, disease and pest resistance are assessed as well. Trials to be conducted include on-station (preliminary) yield trials, multi-location trials, national trials and on farm trials.

On-station trials, also referred to as preliminary trials, are conducted by the plant breeder at the station replicated for a few years to compare the newly developed genotypes with the established varieties or other potential varieties. These yield trials also help to reduce the number of genotypes to be evaluated in multi-locational tests. In multilocation trials a sample of sites in regions for which the new genotypes are developed are used for the tests and only the most promising genotypes are subjected to the trials to save on resources and space. The replications per site usually range between three to four with application of recommended agronomic practices. The region or country for which the crop is meant may be divided into zones based on agroecological conditions and the new cultivars are evaluated at a number of sites within the zone for which they have been developed. For example, in Kenya the country is grossly subdivided into high altitude high rainfall regions, medium altitude medium rainfall regions and low altitude medium low rainfall regions for purpose of rainfed agriculture. Maize varieties are often developed in line with such conditions for optimum production. The multi-location trials are used to assess the adaptability and stability of the new varieties in the environments for which they are meant before any of them is recommended for release.

National trials are conducted in the various regions where the crop is grown across the country. One or a few outstanding variety/varieties from each region are submitted for national trials for comparison of their performances and adaptability. Appropriate recommended agronomic practices and replications are used. National variety performance trials test for wide adaptation at a number of representative locations with agroecological conditions variation. Cultivars of a crop can be grouped into clusters where different clusters reflect different agroecological adaptation groups for the cultivars to be tested.

On-farm trials are adoptive trials which are used to demonstrate to the farmers the superiority of the new variety/varieties on selected farms to popularise the new variety/varieties prior to release. They are conducted on selected farms using prevailing or recommended agronomic practices with full participation of the farmers from land preparation to harvesting. Since the farm for the trials should be accessible to other farmers and the scientists in charge of the trials the farmer whose farm is selected for the trials should be an individual in good relationship with the rest of the farming community so that other farmers can freely visit the farm to learn from the demonstrations. It is essential that the farmer understands the trial well so that he/she can explain to other farmers the trial that is undertaken. The trial at the selected farm is usually not replicated. On-farm trials improve cost-effectiveness of research and is an important component of plant breeding programs. These trials help in accomplishment of rigorous testing for the new varieties, understanding of the farmers' situation in regard to adoption of the new varieties, pass the technology to the farmers in a way they understand and help the breeder get fast feedback from the farmers. On-farm trials also test the suitability of the new varieties in the farming system and should illustrate that the new varieties do not require extra labour and additional resources as compared to other varieties of the same crop already under cultivation.

21.2 Varietal Adaptability

The phenotypic expression of a cultivar, hence its performance, is a product of its genotype (G) and environment (E) in which it is grown and relative performances of different genotypes will vary from one environment to another. The interplay between the cultivars' genotypes and the environment ($G \times E$) the cultivars are exposed to is responsible for the variation in the performances of the cultivars if they are grown in different environments. The phenotypic performance of a cultivar, P, is the sum of three components, G, E and $G \times E$, that is,

$$P = G + E + G \times E.$$

The magnitude of G × E depends on the range of environments in which the trials for the cultivars are conducted, which includes locations, years, seasons, agronomic practices and soil amendments, use of pesticides and other non-genetic factors. In principle, the performance of any cultivar relative to other cultivars grown with it in the same environment should be consistent. However, genotype-environment interaction leads to varied ranking of the performances of different cultivars from one environment to another. Developing a stable variety for a large region with high heterogeneity of environmental conditions is challenging and sometimes self-defeating for some breeding objectives since varieties with general adaptability across a arrange of environments are often not the highest yielding. Varieties developed should be flexible enough and have some plasticity to withstand the environmental fluctuations like the often-unpredictable weather changes which include daily and seasonal temperature variations, rainfall, humidity, drought incidences and cloudiness. While conditions like topography and soil type for a given place do not change, they will vary from place to place even within a region or agro-ecological conditions and affect cultivar performances across the test environments.

Breeders have to contend with such problems when developing improved varieties and also address such questions as to whether is it advisable to develop varieties with specific adaptability to certain specific environments or varieties with general and wide adaptation over a wide range of environments despite G × E interactions. Wide or general adaptation refers to consistently and low but stable performance of genotypes across a set of environmental conditions. Specific adaptation refers to a good genotypic performance in a specific environment because the genotype has restricted adaptation to a limited range of environments and performs extremely well in certain environments but very poorly in other environments. With such challenges the breeder is to make the choice of varieties having the appropriate adaptation through adaptability tests by growing the genotypes in a range of potential target environments to identify the best adapted and most superior varieties.

Adaptability for a set of genotypes in a range of environments is quantified by G × E interaction effects whereby large G × E effects are due to specific adaptability while low G × E interactions are due to general adaptability. Suitability of the variety will also be determined by absolute performance and a stable and poor performing variety is not desirable even if it has good adaptability in a particular environment.

In adaptability tests for a number of varieties the evaluations are replicated in years and locations and the analyses are done using simple operational linear random, mixed or fixed models. In most varietal adaptability tests the fixed or mixed model is used where the varieties are considered as a fixed set. In the tests the genotype × year interaction, genotype × location interaction and genotype × year × location interaction comprise G × E interaction. The significance of G × E interaction components in the analysis of variance implies that genotypes vary in their relative rankings in different environments. The variation of relative performances across the environments necessitates clustering of the environments, especially clustering of the locations into regions, and assessment of the genotypes for specific adaptations in different environment clusters. Such clustering is based on the relative performances of the varieties in the locations and the varieties can be selected for specific adaptation to the various environment clusters, especially for large regions in which the different clusters have distinct and sharp differences in terms of climate and altitude. Adaptability of the genotypes is assessed in terms of stability of performance by regressing the individual varietal means to the environmental means (overall mean of the genotypes in each environment) in which the regression coefficient and deviation of each variety performance are used as measures of adaptability. Stable varieties have regression coefficients close to 1.00 and deviations of actual varietal performances from the expected performances based on the regression analysis as small as possible. Such parameters are also employed to place the varieties in clusters for adaptation and cultivation in different environments.

Problems

21.1 What are the different trials a new variety is subjected to before being considered for release?

21.2 What do plant breeders achieve by conducting on farm tests?

21.3 Distinguish between general and specific adaptability as regards varietal performance

Chapter 22
Varietal Release, Seed Multiplication, Distribution and Quality Control

The impact of improved crop varieties on agricultural production is conditioned on efficient and reliable seed production and distribution to farmers. There has to be a well-established and well-coordinated seed multiplication and distribution system to supply farmers with good quality seed for planting. The seed is usually multiplied and distributed to the farmers by a commercial or public institution. In addition to a reliable seed multiplication and distribution system, the variety must be maintained in genetically stable and pure form.

22.1 Releasing of New Variety

Once the various necessary tests to ascertain the superiority of a new variety are accomplished the outstanding variety is identified as a candidate for release as a new commercial variety based on the criteria established by the variety release agency for each crop and type of variety. The respective breeder proposes to the board or the authority mandated to release new crop varieties that the new variety be released as a new commercial variety, giving the necessary data and specifying the area or region and agro-ecological conditions for which the variety is meant and key attributes like yield potential, adaptability and resistance to specific diseases and pests. The breeder also provides information about the parentage or pedigree, the kind of variety, that, is whether it is pure line, hybrid or synthetic variety. The proposal is reviewed by the board or the authorised agency which makes a decision as to whether the variety may or may not be recommended for release and registered. The review board may also recommend an evaluation test before the decision to release the proposed variety. Once the decision to release the new variety has been made the breeder suggests a simple name for the variety which may comprise of one or two or three words or letters and numbers and thereafter seed multiplication process is initiated.

P. O. Ayiecho and J. O. Nyabundi, *Conventional and Contemporary Practices of Plant Breeding*, https://doi.org/10.1007/978-3-031-74998-8_22

22.2 Seed Categories and Multiplication

There are usually four categories of seed in the seed multiplication chain, namely, breeder seed, foundation (prebasic) seed, registered (basic) seed and certified seed. In addition to these, there may be the original breeder's parental material, often referred to as the nucleus seed held by the breeder, from which the breeder seed is obtained. It is the responsibility of the breeder or the breeding institution to ensure that the genetic and physical purity of the variety is maintained by ensuring that there is no genetic change in the variety through cross-pollination or physical mixing with other cultivars. Before any of the seed categories is used to produce the next category of seed its genetic purity must be verified in grow-out tests and in seed multiplication or rejuvenation plots. This is particularly important for breeder seed. The original parental nucleus seed or breeder seed is planted in isolated plots in which other varieties of the same crop had not been grown for a specified period of time. Vegetatively propagated crop varieties can easily be maintained genetically pure through continued clonal propagation unless there is accidental physical mixing of planting materials or occasional mutations. Variants arising from such incidences are easily eliminated through clonal selection. The maintenance of varietal genetic purity and integrity is easier for self-pollinated crops than for cross-pollinated crops because of pollen flow over long distances for the latter. Hybrids in cross-pollinated crops are maintained pure by maintaining the genetic and physical purity of the component parent nucleus seed and breeder seed. Similarly, integrity of a self-pollinated crop variety is ensured by strict maintenance of genetic and physical purity of its parental nucleus seed or breeder seed (Fig. 22.1). For self-pollinated or cross-pollinated crops only those plants showing characteristics of the pure line variety, parental genotypes or open-pollinated variety to be used in hybrid or synthetic variety are harvested when the material is grown to ensure complete genetic and physical purity. A large number of plants that are typical of the genotype under maintenance and purification process are selected and any doubtful plants or off-types and those showing disease and pest attack are destroyed and discarded. The seed of the selected plants are grown in progeny rows or plots for further selection and varietal cleaning and purification by eliminating plants or lines that do not have the characteristics and features of the genotype to be multiplied. Pure lines or inbred lines showing evidence of segregation are discarded. After adequate cleaning and purification has been accomplished seed is harvested and bulked for seed multiplication to derive breeder seed and other subsequent categories of seed. Inbred lines are maintained through sib-mating to avoid excessive inbreeding.

22.2.1 Breeder Seed

Before a new variety is released the breeder makes a limited increase of the seed of the new variety, referred to as the breeder seed. The breeder seed is the botanical

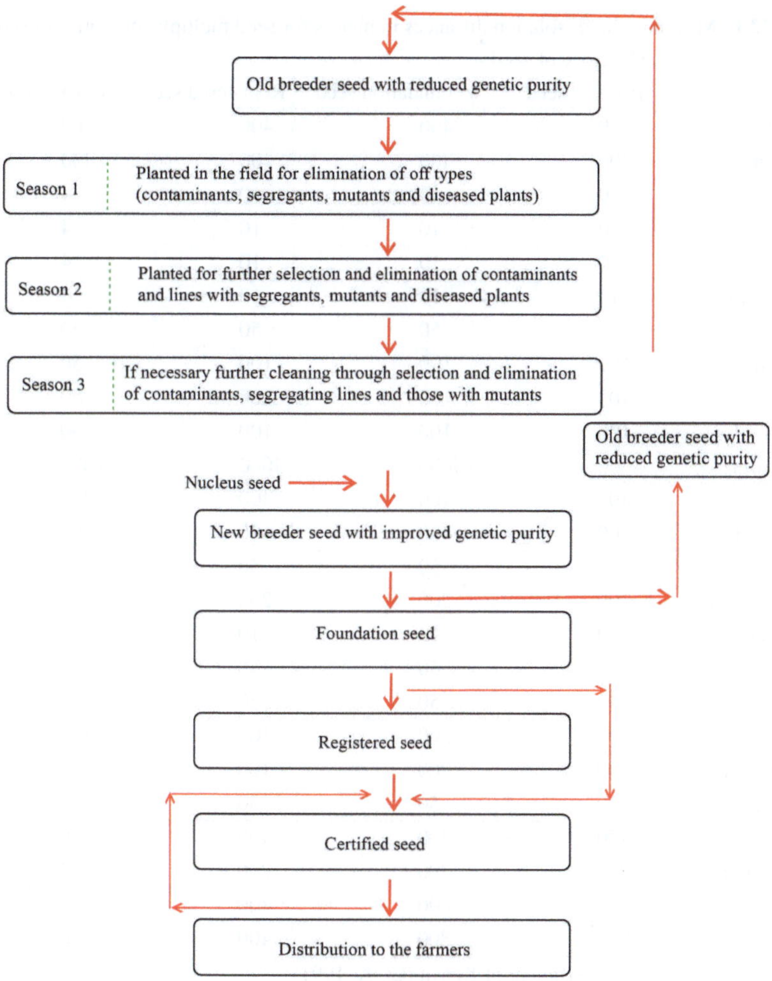

Fig. 22.1 The scheme for production and maintenance of various categories of seed

seed or the vegetative propagating material produced by the breeder or the breeding institution. It is produced from the original breeder's parental material, the nucleus seed, held by the breeder. In case of autogamous crops, it is seed of a pure true breeding stock for the pure line variety in question while for hybrid varieties it is seed of the parental genotypes which are highly homozygous inbred lines or open pollinated parents maintained in small and sufficient amounts. The original parental material seed or breeder seed is planted in isolated plots in which other varieties of the same crop had not been grown for a specified period of time. Furthermore, any seed multiplication plots for the parents or any category of seed must be appropriately isolated. Seed maintenance and multiplication fields for various categories of seed

Table 22.1 Minimum field isolation distances in meters for seed multiplication in some crops

Crop	Category of seed			
	Breeder seed	Foundation seed	Registered seed	Certified seed
Maize	400	400	400	200
Sorghum	400	400	400	200
Wheat	10	10	10	4
Barley	10	10	10	4
Oats	10	10	10	4
Finger millet	10	10	10	4
Triticale	50	50	50	20
Tomato	100	100	100	50
Lettuce	100	100	100	50
Potato	100	100	100	50
Sunflower	4000	4000	4000	1000
Safflower	400	400	400	200
Rapeseed	400	400	400	200
Field beans	50	50	50	25
Broad beans	200	200	200	100
Soybean	10	10	10	4
Cowpea	50	50	50	25
Pea	50	50	50	25
Rhodes grass	200	200	100	100
Setaria grass	200	200	100	100
Guinea grass	50	50	50	25
Sudan grass	400	400	400	200
Desmodium	200	200	200	100
Alfalfa	800	800	400	400
Clover	800	800	400	400

With permission from Government of Kenya (GOK, 1991)

for cross-pollinated or self-pollinated crops must be strictly isolated from any other variety of the same crop as recommended by the local seed certifying agency or as required by the law and internationally accepted (Table 22.1). For each variety developed and released by a particular breeder or breeding institution the breeder seed is maintained and grown by the respective breeder/institution in small populations from year to year under strict pollination control conditions in the greenhouse or in isolated plots to help maintain the genetic purity of the variety.

In multiplication of breeder seed in cross-pollinated crops the inbred lines are subjected to inbreeding through self-pollination or sib-mating and the seed harvested may be grown further in another isolated plot for further multiplication under similar pollination control. There is thorough field inspection of breeder seed multiplication plots and any suspect plants and off-types and those showing disease and pest attack are discarded. At harvest the seeds of all the plants which look alike in all respects and are typical of the variety or line are composited together to give fresh clean pure breeder seed which is turn used for production of foundation seed. While the breeder seed is used to produce foundation seed (Fig. 22.1) a portion of it is kept in cold storage to serve as the nucleus seed for further multiplication or any eventuality and prevention of any incidences of mutation.

22.2.2 Foundation (Prebasic) Seed

Foundation seed, derived by multiplying breeder seed (Fig. 22.1), is also produced under the control of the breeder or the breeding institution. It is obtained by multiplication of a pure line variety breeder seed or inbred line breeder seed for use in the hybrid seed production (Figs. 22.2 and 22.3). Since the breeder or breeding institution may not have adequate land and personnel the production of foundation seed is usually done by certified seed growers contracted by the breeder or the breeding institution. They are supervised by the contracting breeder/institution who provides the technical support during the production to ensure total genetic and physical purity without presence of diseases and pests. The growers are approved and certified by the government seed certifying agency and must be farmers of integrity. Foundation seed production follows the standards set by the certifying agency which visits and inspects the seed production fields and also approves the seed if it meets the legally acceptable quality standards. Foundation seed is used for production of registered (basic) seed and in some cases, it is used to produce certified seed.

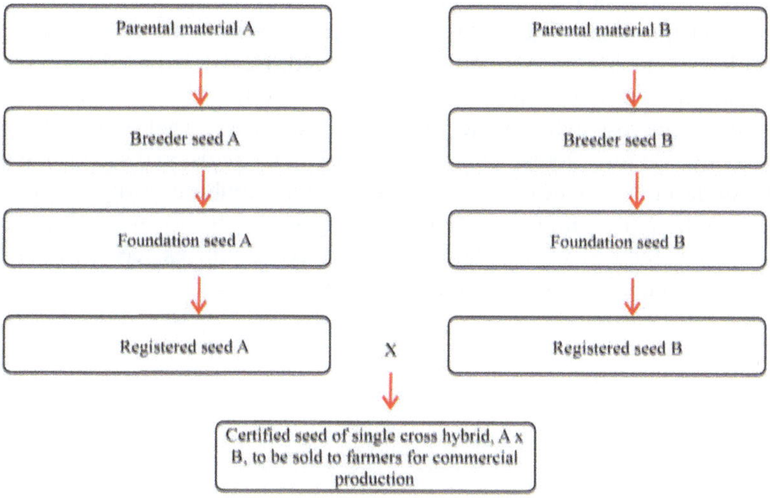

Fig. 22.2 Outline for single cross hybrid seed production with parents A and B

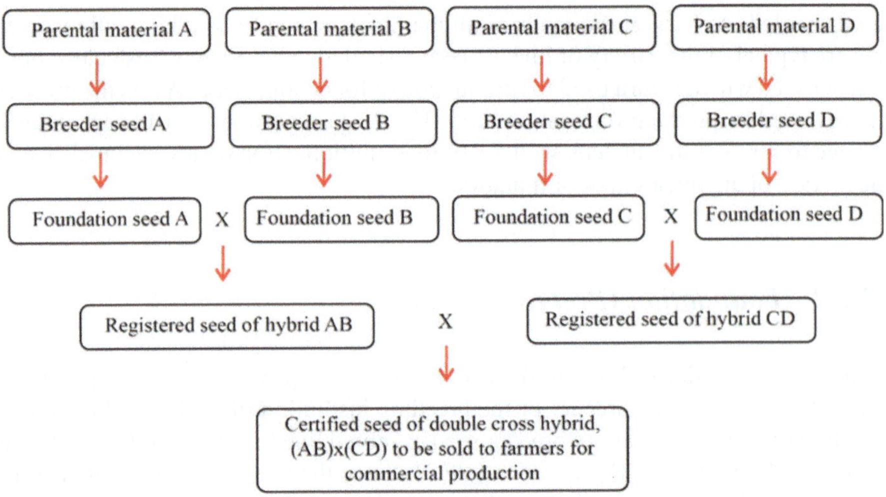

Fig. 22.3 Outline for double cross hybrid seed production with parents A, B, C and D

22.2.3 Registered (Basic) Seed

Registered (basic) seed is derived from foundation seed and is also produced by contracted certified seed growers under the direct control of the breeder or breeding institution or the commercial seed company in charge of seed production and distribution to farmers. It is produced according to the standards set by the seed certifying agency who also supervises the production to ensure that the rules are followed and that genetic and physical purities are maintained without signs of disease and pest attack. The amount of registered seed produced must be sufficient for use in production of adequate certified seed for sale to the farmers. It must meet the qualities required for producing good quality certified seed. Registered seed for a pure line variety takes the form of the pure line seed multiplied from the pure line foundation seed. In a single cross hybrid variety registered seed is the seed of the two parental inbred lines while for a double cross hybrid variety the registered seed is the seed of each the single crosses to be used as the parents for the double cross hybrid (Figs. 22.2 and 22.3). Where the commercial hybrid variety is a hybrid between a single cross and an open-pollinated variety the registered seed comprises seed of the parental single cross and the seed of open pollinated variety.

22.2.4 *Certified Seed*

Certified seed is produced from foundation or registered seed and is multiplied by certified seed growers contracted by the commercial seed company. In some situations, approved clean certified seed can be used to multiply certified seed for pure line varieties. Production of this seed is under the supervision of the seed company and the certifying agency to ensure its genetic and physical purity and absence of disease and pest incidences. Like other categories of seed, certified seed is produced and handled according to the rules and quality standards set by the government seed certifying agency. It is the seed of a pure line variety, a single cross hybrid variety, a three-way or a double cross hybrid produced on a large scale to provide sufficient seed to be sold to farmers for commercial production (Figs. 22.2 and 22.3).

22.3 Seed Certification Process

In Kenya the government seed certifying agency is Kenya Plant Health Inspectorate Services (KEPHIS) which is a government regulatory body with its mandate spelled out in Kenya Plant Health Inspectorate Service Act (2012) and the Seeds and Plant Varieties Act, Cap 326 of the Laws of Kenya (KEPHIS** Information from KEPHIS Managing Director). The agency is established under the Kenya Plant Health Inspectorate Service Act (2012) and is responsible for implementing the Seeds and Plant Varieties Act which provides for quality assurance of propagating material (seeds) in Kenya, among other functions (KEPHIS** Information from KEPHIS Managing Director). This agency sets procedure for the production and quality standards for each category of seed for any given crop according to Laws of Kenya. KEPHIS sets qualification procedure and rules for selecting seed growers to be contracted. The agency also verifies suitability of the farm equipment and stores, ability and integrity of the grower, the general environment and relationship with the neighbouring farming community. KEPHIS enforces the rules through strict supervision and inspection in collaboration with the breeder or the breeding institution and the seed company at various stages of seed crop growth, harvesting, seed processing and packaging to ensure that appropriate measures are taken for good quality seed production to be sold to the farmers.

Any certified selected seed grower is normally contracted to produce seed for only one variety on his/her farm. The grower and his/her farm must be approved by the certifying agency and the field to be used must have a history of not being planted to other varieties of the same crop for a period of time specified by certifying agency or the relevant laws. The freedom of the seed multiplication field from volunteer plants and noxious weeds, diseases and pests, is verified through regular field inspection visits by the personnel of the certifying agency and the contracting seed company/institution. Minimum isolation distance from fields of different varieties of the same crop must be observed as specified for different categories of seed for various crops (Table 22.1). Isolation distances are also influenced by presence of pollination agents and direction and speed of prevailing wind during the growing season. Field inspection visits by the personnel of the certifying agency are made to verify adequate isolation, completeness of rouging, thoroughness of detasseling of female parents in case of maize, complete lack of pollen in the male sterile lines and absence of diseases, pests and noxious weeds during seed crop growth. Inspection is also done during harvesting, cleaning, packaging, storage and other aspects of seed processing. Seed inspection is also done in seed company's facilities where the seed is delivered by the grower for processing and packaging. Samples of each seed lot are tested for impurities, seed borne diseases, presence of broken or damaged seeds, germination and viability, moisture content and other aspects of seed quality before being packaged for sale. Any packaging container/bag should bear an official tag or certificate of the certifying agency and be labelled to indicate crop species, variety, weight, last date of planting and recommended environment and altitude.

Problems

22.1 What are the categories of seed apart from the breeder nucleus seed?

22.2 What criteria must a farmer meet to be contracted to grow certified seed?

22.3 What are the roles of the breeder and certifying agency personnel in the production of the different categories of seed?

References

GOK. (1991). *The Seeds and Plant Varieties Act, CAP 326, 1977 (Revised 1991).* The Laws of Kenya: Government Printer, Nairobi, Kenya. Kenya Government Attorney General permission.

KEPHIS** Information from KEPHIS Managing Director.

MIX
Papier aus verantwortungsvollen Quellen
Paper from responsible sources
FSC® C105338

If you have any concerns about our products,
you can contact us on
ProductSafety@springernature.com

In case Publisher is established outside the EU,
the EU authorized representative is:
Springer Nature Customer Service Center GmbH
Europaplatz 3, 69115 Heidelberg, Germany

Printed by Libri Plureos GmbH
in Hamburg, Germany